BIOLOGY OF THE SNAPPING TURTLE

(Chelydra serpentina)

BIOLOGY *of the* SNAPPING TURTLE *(Chelydra serpentina)*

Edited by ANTHONY C. STEYERMARK
MICHAEL S. FINKLER
RONALD J. BROOKS
Foreword by J. WHITFIELD GIBBONS

THE JOHNS HOPKINS UNIVERSITY PRESS | *Baltimore*

© 2008 The Johns Hopkins University Press
All rights reserved. Published 2008
Printed in the United States of America on acid-free paper
9 8 7 6 5 4 3 2

The Johns Hopkins University Press
2715 North Charles Street
Baltimore, Maryland 21218-4363
www.press.jhu.edu

Library of Congress Cataloging-in-Publication Data

Biology of the snapping turtle (Chelydra serpentina) / edited by
Anthony C. Steyermark, Michael S. Finkler, and Ronald J. Brooks.
    p. cm.
Includes bibliographical references.
ISBN-13: 978-0-8018-8724-6 (hardcover : alk. paper)
ISBN-10: 0-8018-8724-0 (hardcover : alk. paper)
1. Chelydra serpentina.   I. Steyermark, Anthony C., 1968–
II. Finkler, Michael S., 1969–   III. Brooks, Ronald (Ronald J.)
QL666.C539B56   2007
597.92'2—dc22   2007018746

A catalog record for this book is available from the British Library.

Frontispiece: A female snapping turtle nesting on the Edwin S. George
Reserve in southeastern Michigan. Photo by O. M. Kinney.

Special discounts are available for bulk purchases of this book.
For more information, please contact Special Sales at 410-516-6936
or specialsales@press.jhu.edu.

CONTENTS

FOREWORD

A BOOK ABOUT SNAPPING TURTLES must be a book of contrasts—the old and the new—a mix of an ancient lineage and twenty-first-century technology.

As a taxonomic group, snapping turtles are old, reminiscent in appearance of prehistoric creatures that most people know from drawings in books about dinosaurs. The fossil record confirms that animals morphologically similar in appearance to, and no doubt the ancestors of, *Chelydra serpentina* roamed the lands and swam in the fresh waters of North America almost 90 million years ago during the Late Cretaceous. In life, most adult snappers simply look like they have lived a long time, and to be sure, many of them have.

Nonetheless, as with other species of freshwater turtles, the scientific study of snapping turtles now involves a suite of new approaches, whereby much of our scientific knowledge has been acquired through modern technology. In fact, individual snapping turtles alive today hatched from their eggs and made their way to the water before many of today's sophisticated techniques had ever been used in the field of ecology. Radiotelemetry, x-ray photography, and DNA analyses are but a few of the modern procedures that have contributed to advancing our understanding of the distribution and abundance patterns of the species, as revealed through information on snapping turtle ecology and population dynamics.

All living members of the snapping turtle family, Chelydridae (and indeed most other turtles and tortoises), are characterized by traits that are vital considerations for conservation biologists. Snapping turtles belong in a special class because of delayed maturity in juveniles, near invulnerability to natural predators among adults, and their potential for surviving and reproducing for decades beyond maturity.

Clearly, snappers have evolved to persist naturally with a dependence on extended longevity.

Snapping turtle populations are further limited by each female typically producing no more than a single clutch of eggs annually, a high threat of nest predation by terrestrial predators, and high hatchling vulnerability to aquatic predators. Mortality within populations is frequently increased above natural rates by removal of adults through commercial trapping and as a result of mounting numbers of highway deaths. The situation is exacerbated further by an increase in numbers of raccoons, the quintessential turtle nest robber, as a consequence of being predators unintentionally subsidized by human activities throughout most of the snapping turtle's geographic range. Confirmation of abnormally high snapping turtle mortalities on a broad scale calls to question whether sustainability of local or regional populations of such a long-lived species is possible without active management of populations and modification of highway impacts. Conservation biologists need to keep a protective eye on what traditionally has been one of the most ubiquitous and wide-ranging turtle species in North America. The scientific foundation of ecologically significant information pertaining to reproduction, survivorship, and other life-history traits that are provided in this book will serve as a major contribution to science-based management decisions.

Clearly, this work by Steyermark, Finkler, and Brooks, with contributions from their knowledgeable colleagues, will become a classic among books on the life history, general biology, and ecology of turtles. Among turtle biologists, *Biology of the Snapping Turtle* is certain to achieve the walk-of-fame status enjoyed by the books on alligator snappers, ridley sea turtles, and sliders that have synthesized a vast and diverse body of information about a single species of turtle.

J. Whitfield Gibbons
University of Georgia
Savannah River Ecology Laboratory
Aiken, South Carolina

CONTRIBUTORS

Ralph A. Ackerman
Iowa State University

Abdulaziz Y. A. AlKindi
Sultan Qaboos University

Barbara A. Bell
Drexel University

Ronald J. Brooks
University of Guelph

Justin D. Congdon
Savannah River Ecology Lab

Carl H. Ernst
George Mason University

Michael A. Ewert
Indiana University Bloomington

Michael S. Finkler
Indiana University Kokomo

Matthew K. Fujita
University of California, Davis

Eugene S. Gaffney
American Museum of Natural History

David A. Galbraith
Royal Botanical Gardens

Robert E. Gatten, Jr.
University of North Carolina at Greensboro

Judith L. Greene
Savannah River Ecology Lab

J. Howard Hutchison
University of California, Berkeley

John B. Iverson
Earlham College

Fredric J. Janzen
Iowa State University

Jason J. Kolbe
Washington University

David B. Lott
Clarion University of Pennsylvania

Ibrahim Y. Mahmoud
Sultan Qaboos University

Don Moll
Southwest Missouri State University

Scott A. Reese
Kennesaw State University

Todd A. Rimkus
Marymount University

H. Bradley Shaffer
University of California, Davis

James R. Spotila
Drexel University

David E. Starkey
University of Central Arkansas

Anthony C. Steyermark
University of St. Thomas

Gordon R. Ultsch
University of Alabama

Nigel H. West
University of Saskatchewan

BIOLOGY OF THE SNAPPING TURTLE

(Chelydra serpentina)

PART 1

TAXONOMY AND SYSTEMATICS

Systematics, Taxonomy, and Geographic Distribution of the Snapping Turtles, Family Chelydridae

CARL H. ERNST

SNAPPING TURTLES OF THE family Chelydridae are among the oldest North American chelonians. Today members of this family occur only in the Americas, although fossil chelydrid species reveal they were once present in Europe and Asia. The extant species are found in a variety of aquatic habitats ranging from bogs and marshes to deep rivers and lakes. The best-known species, *Chelydra serpentina*, has been the subject of numerous anatomical, physiological, behavioral, and ecological studies, but the biology of the other living species is less well known. This chapter reviews the systematics of the family and examines the living taxa.

SYSTEMATICS OF THE FAMILY CHELYDRIDAE (GRAY 1870)
Nomenclatural History

The first group name attributed to snapping turtles was Chelydrae Gray 1831b:4, but it was used as a plural form for turtles of the genus *Chelydra* Schweigger 1812:292 and as such is not available for the family name. Swainson (1839:116) used the name Chelidridae, but it is based on *Chelidra* Bonaparte 1831:68, an emendated, objective, junior synonym of *Chelydra* Schweigger 1812 and is also not available. The name Chelydroidae Agassiz 1857:341 is the early acceptable name for the family of New World snapping turtles, but it is not orthographically proper. Therefore, Chelydridae Gray 1870:17 is the earliest name with proper orthography attributed to the family, and it is universally accepted as such today.

Characteristics

Turtles of this family have large bodies, large heads, powerful jaws, and slightly emarginate skulls. An epipterygoid bone is present in the cranium, and the frontal bone does not enter the orbit. No contact exists between the parietal and squamosal bones. The carotid artery passes through the pterygoid. The maxilla is not connected to the quadratojugal, and only rarely is its crushing surface ridged. The dentary closing mechanism articulates on a trochlear surface of the otic capsule and the stapes are both surrounded by the quadrate. A secondary palate is absent, and the tip of the upper jaw is hooked. Only one vertebra is amphicoelous (biconvex) in the cervical vertebrae series. The neck withdraws the head in the cryptodiran manner. The tenth dorsal vertebra lacks ribs and is excluded from the sacral complex. Most caudal vertebrae are opisthocoelous. The well-developed, rough carapace is weakly keeled and strongly serrate posteriorly; it does not become fully ossified until late in life. Eleven peripheral bones on each side and a nuchal with costiform processes form its rim. The midline of the carapace is composed of seven to eight neural bones and one to two suprapygals. The carapace is connected to a reduced, cross-shaped (cruciform), hingeless plastron by a narrow bridge. The plastron buttresses articulate loosely with the costal bones of the carapace. The plastron has a T-shaped entoplastron, a median plastron fontanelle, and narrow epiplastra, but no mesoplastron. The hypoplastron bone contributes more to the bridge than does the hyoplastron. Inframarginal scutes are present, and *Macrochelys* has several supramarginals. The abdominal scutes do not meet at the plastral midline but instead are shifted on to the bridge. They are separated from the marginal scutes by a series of submarginal scales. The pelvic girdle is loosely articulated with the plastron. Its ilium lacks a thecal process, and the pubis has parallel pectineal processes. The limbs are well developed, and the toes are webbed and heavily clawed. The saw-toothed tail is as long or longer than the carapace; it is composed of amphicoelous and opisthocoelous vertebrae with well-developed hemal spines.

Distribution

This semiaquatic, New World family ranges east of the Rocky Mountains from southern Canada south to Texas (and possibly northern Mexican tributaries of the Rio Grande), with disjunct populations in the Atlantic drainage of lowland Mexico from southern Vera Cruz, Yucatan, and southern Campeche to Belize, Guatemala, and western Honduras, and from Nicaragua and Panama to Colombia and Ecuador.

Generic Composition

The family Chelydridae consists of two living genera, *Chelydra* Schweigger 1812:292 and *Macrochelys* Gray 1856a:200,

and two fossil genera: *Protochelydra* Erickson 1973:1, from the North American Paleocene, and *Chelydropsis* Peters 1868:73, from the Oligocene to Miocene of Europe, the Middle East, and possibly Asia. The extant genera *Macrochelys* and *Chelydra* date from the Miocene and Pleistocene, respectively.

Relationships

Biochemical and morphological evidence shows that Chelydridae is the oldest of the existing families of cryptodiran turtles, and all other extant cryptodirans are their sister group. Chelydrids' closest living relative appears to be the Asian big-headed turtle, *Platysternon megacephalum*, family Platysternidae. Some combined DNA and morphological data support inclusion of *P. megacephalum* in the Chelydridae (Gaffney 1975; Shaffer et al. 1997), but other data do not (see Schaffer et al., Chapter 4). Karyotypical evidence suggests that *P. megacephalum* is more closely related to the Emydidae (Haiduk & Bickham 1982; Bickham & Carr 1983). In addition, *Platysternon* has two amphicoelous cervical vertebrae, as do only the Emydidae, Geoemydidae, and Testudinidae, and lacks the cruciform plastron, narrow epiplastra, T-shaped entoplastron, serrate posterior carapace rim, a long costiform process on the nuchal bone, separated abdominal scutes, and parallel pectinal processes of the pubis found in the Chelydridae (Whetstone 1978b).

Key to the Living Genera of Chelydridae

1a. A row of supramarginal scutes above the marginals on each side; carapace with three prominent keels extending the entire length; upper jaw strongly hooked; tail with three prominent longitudinal rows of low tubercles . . . *Macrochelys*

1b. No supramarginal scutes above the marginals on each side of the carapace; carapace with keels not extending the entire length; upper jaw only slightly hooked; tail with a prominent medial row of tubercles, and two much lower lateral rows . . . *Chelydra*

CHELYDRA (SCHWEIGGER 1812) SNAPPING TURTLES
Taxonomic History

Snapping turtles were first assigned to the genus *Testudo*, in an early designation for all turtles, by Linnaeus (1758:199). Schweigger (1812:292) reassigned them to the genus *Chelydra,* for which Fitzinger (1843:29) designated the type species *Testudo serpentina* Linnaeus 1758:199. Later Merrem (1820:23) placed them with other freshwater turtles in the poorly defined genus *Emys.* From 1822 to 1835, a series of new generic names, for which the type species was either *Testudo serpentina* Linnaeus 1758:199 or *Chelydra serpentina* (Linnaeus 1758:199), were proposed for snapping turtles: . . . *Rapara* (Gray 1825:210), *Saurochelys* (Berthold 1827:90), *Cheliurus* (Rafinesque 1832:64), *Hydraspis* (Fitzinger 1835:125), and *Emysaurus* (Duméril & Bibron 1835:348). The last generic name applied to snapping turtles was *Devisia,* for which the type species, *Devisia mythodes* Ogilby 1905:11 (= *Chelydra serpentina* [Linnaeus 1758]), had the erroneous type locality "Fly River, British New Guinea." The name *Chelydra* Schweigger 1812:292 has priority over all names proposed after 1812 (see review in Ernst et al. 1988) and has been used regularly since 1857.

Characteristics

Snapping turtles have a massive, slightly rounded carapace with its posterior rim strongly serrate and three low keels composed of knobs located well behind the centers of the scutes. The keels do not extend the entire length of the carapace and become smooth with age. The scutes bear growth annuli in younger individuals and also become smooth with age. The vertebral scutes are broader than long, and the fifth scute is laterally expanded. The cervical scute is short and broad, and 12 marginal scutes occur on each side. No supramarginals are present. Beneath the vertebrals are a series of eight quadrilateral or hexagonal neurals and two suprapygals. The costal bones are reduced, leaving fontanelles between them and the 11 peripheral bones on each side. The tenth dorsal vertebra lacks ribs, and only the eighth cervical vertebra is amphicoelous (biconvex). The plastron is reduced and the bridge small (10% or less than plastron length), creating a cruciform appearance. The abdominal scutes are reduced and confined to the bridge and they do not meet at the plastron midline. The abdominals are separated from the marginals by two to three inframarginals. The entoplastron is T-shaped, and a medial fontanelle is present. The most common plastral scute formula is: anal > <humeral > pectoral > femoral > gular > abdominal. The head is large, with a blunt to acute, slightly projecting snout, a slightly hooked upper jaw, and large dorsolateral orbits. It is posteriorly emarginate, but the temporal region shows little emargination. The frontal bone does not enter the orbit. The squamosal bone touches the large postorbital, but not the parietal. The maxilla does not contact the quadratojugal, and the quadrate completely encloses the stapes. The prootic bone forms a portion of the roof of the internal carotid canal. No secondary palate is present, but the vomer has a ventromedial ridge. The premaxilla and maxilla are ridgeless. No dermal projections are present on the side of the head, but barbels are present on the chin and dermal tubercles on the neck. The tongue lacks a wormlike process. The long tail has a high medial keel of tubercles flanked by two much lower lateral rows of tubercles. The toes are webbed and heavily clawed. Cloacal bursae are present. The karyotype contains 52 diploid chromosomes: 24 macrochromosomes and 28 microchromosomes (Stock 1972; Bickham & Baker 1976; Killebrew 1977, Haiduk & Bickham 1982).

Males grow larger than females (to 49.4 cm; Gerholdt & Oldfield 1987). The male vent is usually situated posterior to the carapace rim, and males have longer preanal tail lengths (commonly over 120% of the length of the posterior plastron lobe) than females (preanal length usually less than 110% of the length of the posterior plastron lobe).

Distribution

Individuals of *Chelydra* are found from Nova Scotia, New Brunswick, and southern Quebec, west to southeastern Alberta, and southward east of the Rocky Mountains to southern Florida and Texas (and possibly northern Mexican tributaries of the Rio Grande), and from southern Vera Cruz, Mexico south through Central America and northwestern South America to Ecuador.

Fossil Record

The geological history of the genus *Chelydra* begins in the Pliocene Hemphillian and Blancan and continues through the Pleistocene Irvingtonian and Rancholabrean to the Recent (see Hutchison, Chapter 2).

Species Composition

Traditionally *C. serpentina* has been considered polytypic on morphological grounds, with four subspecies: *serpentina* (Linnaeus 1758:199), *acutirostris* Peters 1862:627, *rossignonii* (Bocourt 1868:21), and *osceola* Stejneger 1918:89 (Medem 1977; Wermuth & Mertens 1977; Gibbons et al. 1988; Ernst & Barbour 1989; Ernst et al. 1994). A study by Phillips et al. (1996) of restriction endonuclease fragment patterns of mitochondrial DNA (mtDNA) and protein electrophoresis indicated moderate mtDNA sequence differences between North American and tropical *C. serpentina*. The tropical American subspecies *acutirostris* and *rossignonii* were more distinct, both from each other with a minimum of 1.7% sequence divergence and from North American *serpentina* and *osceola* with an average 4.45% sequence divergence. The degree of allozymic variation among the four subspecies was inconclusive. Phillips et al. (1996) thought that the mtDNA data supported species-level distinctness of *acutirostris* and *rossignonii* from each other, and from the two North American subspecies. In contrast, North American *serpentina* and *osceola* were found to be closely related, differing by only a maximum of 0.5% sequence divergence, and this has been substantiated in other molecular studies (Walker & Avise 1998; Walker et al. 1998; Shaffer et al. Chapter 4).

Sites & Crandall (1997), however, questioned the validity of full species recognition of *acutirostris* and *rossignonii* by Phillips et al. (1996). They thought their study was flawed because it failed to present any species concept as a testable hypothesis. Their data were collected in the absence of any conceptual framework for diagnosing species boundaries, so the species boundaries were determined in a nonrigorous, post hoc manner and, therefore, should not be accepted. The absence of specific criteria for species diagnosis led to flaws

Map 1.1. Distribution of *Chelydra serpentina*

in sampling design and the collection and analysis of the data. Sites & Crandall briefly outlined three different lineage-based operational species concepts (phylogenetic, concordance, and cohesion), and presented an alternate interpretation of the data presented by Phillips et al. Sites & Crandall concluded that (1) species status may not be warranted for the tropical taxa, (2) more detailed analysis should be made of the North American taxa because distinct lineage may be obscured by poor laboratory technique or introgression of mtDNA, and (3) the *acutirostris* from Ecuador may deserve species status based on fixed nuclear isozyme loci. They also suggested procedures for implementing rigorous, lineage-based species concepts and methods for the collecting and handling of data.

Despite their objections to the methods used by Phillips et al. (1996), the experimental results of Sites & Crandall (1997) do not contradict those of Phillips et al. (1996). Because of this, the genus *Chelydra* is considered to be polytypic in this chapter, and consisting of the live species *C. serpentina, C. acutirostris,* and *C. rossignonii.*

Key to the Living Species of Chelydra

1a. Length of plastron forelobe (from level of hyo-hypoplastral suture to the anterior tip) less than 40% of carapace length . . . *C. serpentina*

1b. Length of plastron forelobe greater than 40% of carapace length . . . 2

2a. Anterior width of third vertebral scute greater than 25% of

maximum carapace width; neck with long, pointed tubercles . . . *C. rossignonii*

2b. Anterior width of third vertebral scute less than 25% of maximum carapace width; neck with rounded, wartlike tubercles . . . *C. acutirostris*

CHELYDRA SERPENTINA (LINNAEUS 1758) NORTH AMERICAN SNAPPING TURTLE
Nomenclatural History

Chelydra serpentina was described by Linnaeus in 1758:199, and named *Testudo serpentina*. Since then it has been placed in several genera (see above), but the combination *Chelydra serpentina* (Schweigger 1812:292) has priority. Other names considered synonymous are: *Testudo longicauda* Gray 1831a:36, a nomen nudum, inaccurately attributed to Shaw 1802; *Testudo serrata* Gray 1831b:36, a nomen nudum, inaccurately attributed to Pennant 1787; *Chelydra emarginata* Agassiz 1857:413 and *Devisia mythodes* Ogilby 1905:11 (see above).

Characteristics

The brown to olive-brown carapace (to 49.4 cm) is tricarinate, posteriorly serrate, and slightly rounded. In adults of some populations and in juveniles the carapace scutes may be streaked with red and/or yellow radiations. The three keels become progressively lower with age. The anterior width of the third vertebral scute is less than 33% of the maximum carapace width. The small plastron is yellowish to tan in adults, but gray with white mottling in hatchlings and small juveniles. Inframarginal scutes are usually absent; if present, there are only one to two. The bridge is usually less than 8% (6–10%) of the carapace length. The gular scute is normally undivided. The abdominal scute is usually twice as broad as long. The head is large, with a short, only slightly projecting snout and a slightly hooked upper jaw. The jaws are marked with dark bars. Normally only two chin barbels are present. The dorsal surface of the neck may bear long, pointed or short, blunt tubercles. The skin is tan, gray, or black above, yellowish below, with white flecks in some juveniles (in particular, hatchlings and juveniles). The karyotype consists of 52 diploid chromosomes (24 macrochromosomes and 28 microchromosomes); the third pair has a conspicuous short arm, secondary constriction, and is telocentric (Stock 1972; Bickham & Baker 1976; Killebrew 1977; Haiduk & Bickham 1982). Males are larger than females and have the anal opening posterior to the carapace rim (also see genus description).

Distribution

Chelydra serpentina range from Nova Scotia, New Brunswick, and southern Quebec to southwestern Saskachewan, and southward east of the Rocky Mountains to southern Florida and Texas (Map 1.1). It possibly also occurs in Mexican tributaries of the Rio Grande River.

Fossil Record

This species has a long and broadly distributed North American fossil record, dating from the Pliocene Hemphillian and Blancan of Kansas (Galbreath 1948; Hibbard 1963; Schultz 1965), and Blancan of Nebraska (Holman & Schloeder 1991). Pleistocene remains have been recorded from the Irvingtonian of Kansas (Hibbard & Taylor 1960; Holman 1972, 1986a, 1986b; Preston 1979), Maryland (Hay 1923), Nebraska (Preston 1979; Ford 1992), and Oklahoma (Preston 1979); and the Rancholabrean of Florida (Hay 1916; Weigel 1962; Holman 1978), Illinois (Holman 1966), Indiana (Graham et al. 1983; Richards et al. 1987; Holman 1992; Holman & Richards 1993), Kansas (Holman 1972; Preston 1979), Michigan (Wilson 1967; Holman 1988; Holman & Fisher 1993), Mississippi (Holman 1995), Missouri (Parmalee & Oesch 1972; Holman 1995), Nebraska (Preston 1979), Nevada (Van Devender & Tessman 1975), Ohio (Holman 1986b, 1997; Hansen 1992), Pennsylvania (Hay 1923), South Carolina (Bentley & Knight 1998), Tennessee (Corgan 1976), Texas (Holman 1964; Preston 1979), and Virginia (Guilday 1962; Holman & McDonald 1986; Fay 1988).

Geographic Variation

Two subspecies are currently recognized. *Chelydra serpentina serpentina* (Linnaeus 1758:199) (Fig. 1.1), the eastern snapping turtle, to 49.4 cm, ranges from southern Canada south to southern Georgia, and west to the Gulf Coast of Texas. Its plastron forelobe (from the level of the hyo-hypoplastral suture to the anterior tip) is less than 40% of the maximum carapace width, the anterior width of the third vertebral scute is much less than the height of the second pleural scute and less than 33% of the combined length of the five vertebral scutes, its neck has rounded, wartlike tubercles on the dorsal surface, and there are flat juxtaposed scales on its head. *C. s. osceola* (Fig. 1.2) (Stejneger 1918:89), the Florida snapping turtle, to 43.8 cm, is only found in peninsular Florida (it is not found in Yucatan, Mexico, as suggested by Zuurmond & Netten 1983:53). Its plastron forelobe is usually less than 40% of the maximum carapace width, the width of its third vertebral scute is at least as long as the height of the second pleural and about 33% of the combined length of the five vertebral scutes, the neck has long pointed tubercles on its dorsal surface, and granular scales and a few low tubercles are present on the temporal and occipital regions of the head.

Some researchers have considered *C. s. osceola* a full species, rather than merely a subspecies of *C. s. serpentina* (Richmond 1958), but it has more frequently been treated as a subspecies of the latter (Medem 1977; Gibbons et al. 1988; Ernst & Barbour 1989; Ernst et al. 1994) on the basis of reported intergradation between the two races in the Okefenokee region of northern Florida and southern Georgia by Feuer (1971). However, Feuer's (1971) sample size was small, and he only considered seven snapping turtles intergrade. A more thorough morphological study of the *Chelydra* in northern Florida and southern Georgia could solve this problem. Recently, Walker et al. (1998) found that the control-region sequences in the mtDNA of *C. serpentina* from ten southeastern states, including peninsular Florida, have a single, predominant haplotype, and the two rare variants they detected were nearly identical with the common

Fig. 1.1. *Chelydra serpentina serpentina.* Courtesy Roger W. Barbour.

Fig. 1.2. *Chelydra serpentina osceola*. Courtesy Roger W. Barbour.

Fig. 1.3. *Chelydra rossignonii*. Courtesy Roger W. Barbour.

genotype of the species. This suggests only subspecific variation for *C. s. osceola*.

The Florida Pleistocene species *C. laticarinata* Hay 1916:72 and *C. sculpta* Hay 1916:73 were described before *C. s. osceola* Stejneger 1918, and in general are agreed to represent the same turtle (Richmond 1958; Feuer 1971; Mlynarski 1976; Smith & Smith 1980). Because the names *laticarinata* and *sculpta* had been used infrequently, and *osceola* was the name commonly applied to the *Chelydra* of Florida, Smith & Smith (1983) petitioned the International Commission on Zoological Nomenclature (ICNZ) to suppress the names *laticarinata* and *sculpta* and conserve *osceola* for the taxon. In 1986, the ICNZ gave nomenclatural precedence to *osceola* over both *laticarinata* and *sculpta*.

CHELYDRA ROSSIGNONII (BOCOURT 1868:121) MEXICAN SNAPPING TURTLE
Nomenclatural History

Bocourt (1868) originally described this turtle under the name *Emysaurus rossignonii*. Gray (1870:64) assigned it to the genus *Chelydra,* but as *Chelydra serpentina* var. *mexicanae* (Fig. 1.3). Cope (1872:23) returned it to species level, *C. rossignonii,* but Mertens et al. (1934:59) relegated it to a subspecies of *C. serpentina, C. s. rossignonii,* where it remained until Phillips et al. (1996:402) proposed it represented a full species, *C. rossignonii.*

Characteristics

This large turtle reaches a maximum carapace length of 38.9 cm, but most adults are 20–30 cm long. The adult carapace is unicolored brown to olive or olive-black; that of juveniles may contain light radiations or small spots on each scute. It is rounded, with three longitudinal keels (which usually disappear with age), and sharp posterior serrations. The anterior width of the third vertebral scute is more than 25% of the maximum carapace width. Its plastron forelobe is longer than 40% of the maximum carapace width; the an-

terior width of the third vertebral is more than 25% of the maximum carapace width. The plastron is cream to yellow, tan or gray in adults, but that of juveniles may have light and dark mottling. The length of the plastron forelobe is usually more than 40% of the carapace width, and the bridge is 6–8% of the carapace length. The gular scute is subdivided into two in most individuals, and three to four inframarginals may be present. The abdominal scute is usually twice as broad as long. The head is large, with a narrow, pointed snout, and four to six chin barbels. Neck tubercles are long and pointed. Skin is gray.

Males are larger than females and have longer preanal tail lengths and the vent is usually situated beyond the posterior rim of the carapace.

Distribution

This turtle is found in the Atlantic lowland watersheds from central Vera Cruz, Mexico south across the base of the Yucatan Peninsula and southern Campeche to western Belize, Guatemala, and west-central Honduras (Map 1.2).

Fossil History

None.

Geographic Variation

No study of geographic variation has been undertaken.

CHELYDRA ACUTIROSTRIS (PETERS 1862:627) SOUTH AMERICAN SNAPPING TURTLE
Nomenclatural History

Chelydra acutirostris (Fig. 1.4) was originally considered a variety of *Chelydra serpentina, C. serpentina* var. *acutirostris,* by Peters (1862), but Babcock (1932:874) elevated it to a full species, *C. acutirostris.* Müller (1939:98) retained it as a sub-

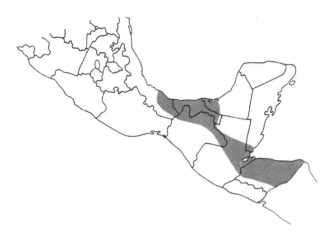

Map 1.2. Distribution of *Chelydra rossignonii*

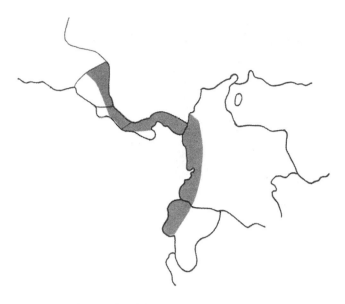

Map 1.3. Distribution of *Chelydra acutirostris*

species of *C. serpentina, C. s. acutirostris,* where it remained until Phillips et al. (1996:402) returned it to full species. The name *Chelydra angustirostris* Dunn 1945:316 is an ex errore.

Characteristics

Record carapace length is 41 cm, but most adults probably have 20- to 30-cm carapaces. The brown to olive, dark brown, olive gray, or black carapace has a few light radiations or small spots in young individuals, but is unicolored in older adults. It is slightly rounded with three low keels (which may disappear with age), and sharp posterior serrations. The anterior width of the third vertebral scute is less than 25% of the maximum carapace width. The adult plastron is yellow, tan, or gray; the plastron of juveniles has a light-dark mottled pattern. The plastron forelobe is usually longer than 40% of the maximum carapace width, and the bridge is 6–8% of the carapace length. The gular scute is subdivided into two, and three to four inframarginals are present. The abdominal scute is usually twice as broad as long. The large head has a narrow pointed snout, and usually four to six chin bar-

bels. Neck tubercles are rounded and wartlike. The skin is gray to olive-black, or dark brown.

Males are larger than females and have longer preanal tail lengths, and the vent is situated beyond the posterior carapace rim.

Distribution

This species ranges from at least Nicaragua, and possibly southern Honduras, southward through the Caribbean watersheds of Central America to the Pacific drainages of Colombia and Ecuador (Map 1.3).

Fossil Record

None.

Geographic Variation

No subspecies are currently recognized, but comparisons should be made between the Central American populations and those from the southern end of the range in Ecuador.

MACROCHELYS (GRAY 1856A) ALLIGATOR SNAPPING TURTLES
Taxonomic History

The oldest available name for the alligator snapping turtle is *Testudo planitia* Meuschen 1778 (often mistakenly attributed to Gmelin 1789; see Bour 1987), now considered a *nomen rejectum,* (ICZN 1963) because its type species, *Testudo planitia* Meuschen 1778, is considered a synonym of *Pelomedusa subrufa* (Lacèpéde 1788). In 1835, Troost (in Harlan 1835) described *Chelonura temminckii* from western Tennessee. Later A. Duméril (in Duméril & Duméril 1851) provided a further description, based on several drawings by

Fig. 1.4. *Chelydra acutirostris.* Courtesy William W. Lamar.

Fig. 1.5. *Macrochelys temminckii.* Courtesy Carl H. Ernst.

Fig. 1.6. *Macrochelys temminckii* tongue lure. Courtesy Roger W. Barbour.

Lesueur in 1834 of a living turtle from New Orleans and a shell sent to Paris by him that same year, and assigned the turtle to the genus *Emysaurus.* Gray (1856a:200) created the new genus *Macrochelys* for the alligator snapping turtles, and in a later publication that year (1856b:48) referred to the genus as *Macroclemys.* The final synonymous name applied to the alligator snapping turtles was *Gypochelys* by Agassiz (1857:413). *Macrochelys* is the generic name by priority (Webb 1995). More detailed taxonomic histories of *Macrochelys* are given by Bour (1987), Pritchard (1989), and Webb (1995).

Characteristics

The *Macrochelys* are some of the largest and heaviest freshwater turtles in the world. Living individuals attain a carapace length up to 80 cm and weigh more than 100 kg. Beneath the vertebrals is a series of six to nine, usually hexagonal neural bones and one suprapygal. The pygal bone may be medially divided, and 11 peripheral bones are present on each side of the carapace. Costal bones 1–8 are reduced laterally, leaving small spaces, usually fontanelles, between them and the peripherals in juveniles; these close in adults. The cruciform plastron is hingeless, and the bridge short (7–9% of plastron length). A T-shaped entoplastron, and a median fontanelle are present. The skull is large with little temporal emargination. The powerful jaws lack ridges on the premaxillae and maxillae. The upper jaw is strongly hooked, and the maxilla does not contact the quadratojugal. The stapes is completely surrounded by the quadrate. Squamosal-postorbital contact exists, but not squamosal-parietal contact. No secondary palate is present. The vomer has a weak ventral keel, and the foramen housing the facial nerve lies entirely within the pterygoid. The prootic bone forms little of the roof of the internal carotid canal, which continues to pass through the pterygoid bone. No secondary palate is present, and only a weak medial ridge is present on the ventral surface of the vomer. The cervical vertebral series contains only one amphicoelous vertebra.

Distribution

The genus is strictly North American, and occurs only in the Mississippi River and Gulf Coastal drainages. It ranges south from Kansas, Iowa, southern Illinois, and southern Indiana to the Gulf coastal plain from eastern Texas to southeastern Georgia and northern Florida.

Species Composition

The single living species is *Macrochelys temminckii* (Troost, in Harlan 1835:158), described below. Two fossil species are also recognized: *M. schmidti* Zangerl 1945:5, from the Miocene (Hemphillian) Hemmingfordian and Marshland deposits in western Nebraska and northwestern Kansas (Zangerl 1945; Whetstone 1978a; Parmley 1992; Lovich 1993), and *M. auffenbergi* Dobie 1968:59, from the Pliocene (Hemphillian) of Alachua County, Florida.

MACROCHELYS TEMMINCKII (TROOST, IN HARLAN 1835) ALLIGATOR SNAPPING TURTLE
Characteristics

Macrochelys temminckii (Fig. 1.5) is the largest freshwater turtle in North America, growing to a carapace length of 80 cm (Pritchard 1980) and weight to 113 kg (Pawley 1987). Its large, dark brown or dark gray carapace has a roughened surface, three knobby longitudinal keels, and posterior serrations. The vertebral scutes are broader than long, and there is a short, broad cervical scute and 23 marginal scutes. Three to eight (usually three) supramarginals form a row between the marginals and first three pleural scutes on each side. The bridge and plastron are mottled gray. The abdominal scutes are reduced, do not usually meet at the plastron midline, and are separated from the marginal scutes by a series of inframarginals. The average plastron scute formula is femoral > <pectoral > anal > humeral > gular > abdominal. The head is large with a pointed snout, powerful

Map 1.4. Distribution of *Macrochelys temminckii*

jaws, a strongly hooked upper jaw, and large, lateral orbits. Numerous dermal projections are present on the side of the head, around the eyes, and on the chin and neck. The tongue has a unique, wormlike process (Fig. 1.6), which is used to lure fish into the turtle's mouth. Both ends are free, as the "lure" is attached at its center to a muscular base on the tongue. This allows it to be contracted and expanded, imitating the movements of a worm. The two ends of the projection are single in adults, but may be branched in juveniles. The skin is dark brown to gray above, but lighter below; dark blotches may be present on the head. The tail is about as long as the carapace, and has three dorsal rows of low tubercles; its ventral side is covered with small scales. The karyotype consists of 52 diploid chromosomes (14 metacentric or submetacentric, and 10 telocentric or subtelocentric; the third pair lacks the conspicuous short arm, secondary constriction, and telocentric condition found in *Chelydra*), 26 microchromosomes, and a unique pair of large metacentric chromosomes (Killebrew 1977; Haiduk & Bickham 1982). Males have longer preanal tail lengths than females, and the vent is posterior to the carapace rim.

Distribution

Macrochelys temminckii is restricted to watersheds flowing into the Gulf of Mexico (Map 1.4) (see genus distribution above).

Fossil Record

Fossils of this species date from the Pliocene (Clarendonian) of South Dakota (Zangerl 1945); Pliocene (Blancan?) of Kansas (Hibbard 1963); Pleistocene (Irvingtonian) of Florida (Hulbert & Morgan 1989); Pleistocene (Rancholabrean) of Florida (Auffenberg 1957), and Mississippi (Daly 1992); and Recent deposits in Texas (Hay 1911).

Geographic Variation

No subspecies are currently recognized. However, Roman et al. (1999) sequenced 420 mtDNA base pairs in individuals from 12 drainages within the geographical range. Their results indicate considerable phylogeographic differences and strong separations among watersheds, with 8 of 11 haplotypes river-specific. Three groups—eastern, central, and western—were indicated in the mtDNA genealogy that corresponds to recognized biogeographic provinces. The *M. temminckii* from the Suwannee River of Florida are particularly divergent.

2

History of Fossil Chelydridae

J. HOWARD HUTCHISON

THE FOSSIL RECORD OF THE CHELYDRIDAE is spotty and, in general, fossils are scarce relative to trionychoids and testudinoids. Nevertheless, the record of the family extends back to at least the Turonian (Late Cretaceous, about 90 million years ago [Ma]) in North America. The family Chelydridae exhibits a striking resemblance to a group of Cretaceous and Paleogene turtles from Asia generally grouped under the family Macrobaenidae (Sukhanov 2000). The Chelydridae is probably derived from a member of the Macrobaenidae (Ckhikvadze 1973; Broin 1977) that Peng & Brinkman (1993), Sukhanov (2000), and Parham & Hutchison (2003) consider as a paraphyletic group.

The fossil record of undescribed chelydrids in North America begins in the early Late Cretaceous and is limited to North America until the Eocene. Eaton et al. (1999a) and Hutchison (2000) list the presence of an undescribed taxon in the early Late Cretaceous (Turonian) of Utah. Hutchison & Archibald (1986), Brinkman (1990), Rodriquez-de la Rosa & Cevallos-Ferriz (1998), Peng et al. (2001), and Holroyd & Hutchison (2002) note the presence of two or more undescribed genera from the Late Cretaceous (Judithian, about 76 Ma and Lancian, about 70 Ma) to early Paleocene (Puercan, about 64 Ma) of the Rocky Mountain region and south into northern Mexico. The record of described species of North American Chelydridae is largely limited to the Rocky Mountain region and eastward. The earliest of these, *Denverus middletoni* Hutchison and Holroyd 2003, is from the earliest Paleocene (Puercan, about 65 Ma) but the first reasonably complete material is *Protochelydra zangerli* Erickson 1973 from the later Paleocene (Tiffanian, about 57 Ma). The next oldest described taxa are species assigned to the extant genera *Chelydra* (Galbreath 1948) and *Macrochelys* (Zangerl 1945). *Macrochelys* is represented by a series of four species beginning in the early Miocene (Hemingfordian, about 19 Ma) and extending into the present. Confirmed records of *Chelydra* are limited to the late Miocene (Hemphillian, about 7 Ma) to the present, but only the extant species is recognized in the fossil record. *Chelydra serpentina* is more widespread in the Pleistocene of the United States than today and occurs in the late Pleistocene (Rancholabrean, >0.5 Ma) of the Snake River (Pinsof 1998) and Colorado River drainages (Van Devender & Tessman 1975). The introduction of the *Chelydra* into northern South America (Ernst, Chapter 1) probably dates to the Pliocene or Pleistocene.

The most northerly record of a chelydrid occurs on Ellesmere Island, Canada (J. H. Hutchison, pers. observ.) in the early Eocene (Wasatchian, 54 Ma). Eaton et al. (1999b) note the presence of a new chelydrid from the latest Eocene (Duchesnean about 38 Ma) of Utah that exhibits similarities to chelydrines, especially *Macrochelys*. In addition to these, there are several Tertiary records of chelydrids on the Pacific drainage of North America that range in age from Late Paleocene (Clarkforkian, about 56 Ma; Hutchison & Pasch 2004) to late Miocene (Hemphillian; Hutchison, pers. observ.). Hanson (1996) reported the presence of a chelydrid from the late Eocene of Oregon. Largely unpublished Oligocene (Arikareean, about 25 Ma) to late Miocene (Hemphillian, about 7 Ma) records of *Chelydra*-like chelydrines from California, Oregon, and Nevada (Hutchison 1994) indicate a long history on the Pacific slope and may indicate that the origin of *Chelydra* is outside of its present range.

The enigmatic *Archerontemys heckmani* Hay 1899 from the Eocene of Washington is based on a carapace. Hay (1908a) placed it in the Chelydridae but it exhibits no diagnostic characters of the family. The specimen is crushed and embedded in shale. The plastron is probably present but unprepared. The areas of costals five and six and costal one are buckled up from below and indicate the presence of strong plastral buttresses, a feature absent in chelydrids. *Archerontemys* is here removed from the Chelydridae and placed in the Testudinoidea, *incertae sedis*.

The fossil record of chelydrids in Eurasia is more limited in time but more densely represented by nominal taxa. The record of Chelydridae begins in the Middle Eocene (Bartonian, about 39 Ma) (Lapparent de Broin 2001). The earliest described taxon is in the form of the contentious genus *Chelydrasia* Ckhikvadze 1999 that appears in the early Oligocene of Europe and Central Asia. A second genus, *Chelydropsis* Peters 1868, appears in the middle Miocene (Orleanian, about 19 Ma) of Europe and is apparently derived from *Chelydrasia*. *Chelydrasia* or its descendants persist into the Pliocene of Central Asia and *Chelydropsis* into the Pliocene of Europe and Asia Minor.

The fossil record, as presently known, supports the origin of the Chelydridae in North America. The presence of a specimen referred to *Protochelydra* from the late Paleocene (Clarkforkian) of Alaska (Hutchison & Pasch 2004) and un-

described material from the early Eocene (Wasatchian) of Ellesmere Island, Canada, suggests that chelydrids may have entered Eurasia as early as late Paleocene or early Eocene during a period of high intercontinental faunal interchange (McKenna 1983). The presence of a chelydrine-like form in the late Eocene of North America coupled with chelydropsines in the middle Eocene of Europe and Asia suggest that these two clades were distinct by the later Eocene and probably enjoyed largely separate histories thereafter. Presence of *Chelydra*-like forms in the late Oligocene (about 25 Ma) of North America, *Macrochelys* in the early Miocene (Hemingfordian, about 18 Ma), and a new genus with similarities to *Macrochelys* in the latest Eocene (Duchesnean, about 39 Ma) of North America indicates that the two extant clades were distinct by the latest Eocene.

Gaffney (1975) and Gaffney & Meylan (1988) provided a controversial cladistic arrangement of the genera based primarily on features of the skull in which they placed *Platysternon* as a terminal member of the Chelydridae. This is supported by Brinkman & Wu (1999) on three skull features (a fourth character, no. 17 cited in figure 3, does not apply according to their appendix). Shaffer et al. (1997) provided admittedly weak support for a *Chelydra-Platysternon* group (*Macrochelys* not sampled) based on the combined molecular data for 12S rDNA and cytochrome *b*. This arrangement is not supported by features of the shell (Gaffney & Meylan 1988; Lapparent de Broin 2000), serology (Frair 1972), molecular cytochrome *b* (Shaffer et al. 1997), molecular 12S rDNA (Shaffer et al. 1977), other molecular data (Schaffer et al., Chapter 4), vertebral articulations (Williams 1950; Whetstone 1978a), or karyotypes (Haiduck & Bickham 1982; Bickham & Carr 1983) and was most recently rejected in detail by Lapparent de Broin (2000) who regarded the similarities of the skull of *Platysternon* to *Macrochelys* and *Chelydropsis* as parallelisms. I follow the latter authors in excluding *Platysternon* from the Chelydridae. Lapparent de Broin (2000) recognized two major monophyletic groups: one consisting of the Eurasian *Chelydropsis* (including *Chelydrasia*) and the other of the North American *Protochelydra*, *Macrochelys*, and *Chelydra*. My reevaluation of the characters and additional material of *Protochelydra* alters the relative position of these cladistic arrangements and results in *Chelydra* and *Macrochelys* being sister taxa, with *Chelydropsis-Chelydrasia* and *Protochelydra* (and *Denverus*) as progressively more distant out-groups. The latter arrangement is also more congruent with the known fossil history of the genera.

In the following diagnoses, *Protochelydra* is assumed to have primitive state for the characters used in the diagnoses. Characters marked with an asterisk (*) are regarded as derived in comparison with *Protochelydra* or local out-groups. Characters marked with a double asterisk (**) are regarded as autapomorphies of the group in question. Terminology of the shell and skull follows Zangerl 1969 and Gaffney 1972, respectively. In addition to major epoch and stage terminology, North American and European Land Mammal Ages and European Mammal numbered zones for Paleogene (MP#) and Neogene (MN#) are included where known.

CHELYDRIDAE (AGASSIZ) (GRAY 1870)
Taxonomic History

Ernst (Chapter 1) provides the nomenclatural history of the family name.

Osteological Diagnosis of the Fossil and Recent Genera

Chelydrids differ from the macrobaenids in the following derived features: development of long costiform processes of the nuchal, abdominal scales not meeting in the midline, loss of the foramen basisphenoidale, the foramen posterius canalis corotici interni entering the posterior part of the skull and floored ventrally by the pterygoid, frontal excluded from the orbit, squamosal and parietal separated by the temporal emargination or postorbital. Chelydrids share a suite of primitive features with the macrobaenids, such as a long tail with both procoelous and ophithocoelous caudal vertebrae separated by an amphicoelous vertebra, anal-femoral sulcus crosses the hypoplastron-xiphiplastron suture, cruciform shape of the plastron, plastral-carapacial suture confined to peripherals, entoplastron distinctly longer than wide, nasals lost, biconvex fourth cervical vertebra, musk ducts exiting ventrally along the plastral-carapacial suture, cervical scale distinctly wider than long, and primitively three or more contiguous inframarginals.

Key to the extinct and living genera based on the shell

1a. Bridger peripherals greatly thickened medially, costals strongly sculptured with longitudinal ridges . . . *Denverus*

2a. Bridger peripherals not thickened medially, costals not strongly sculptured with longitudinal ridges . . . 2

2a. Posterior margin of fifth vertebral scale lies well onto the suprapygal; posterior peripherals only very weakly notched . . . *Protochelydra*

2b. Posterior margin of fifth vertebral scale lies very near or only slightly overlapping the suprapygal-pygal sulcus, posterior peripherals distinctly notched . . . 3

3a. Gular scales of the plastron lying well forward of the anterior tip of the entoplastron, entoplastron diamond-shaped, suprapygal 1 trapezoidal, costiform process of the nuchal terminates anterior to the distal end of costal rib 1 . . . Chelydropsinae 4

3b. Gular scales of the plastron lying near or on the anterior tip of the entoplastron, entoplastron anchor-shaped, suprapygal 1 quadratic, costiform process of the nuchal terminates medial to the distal end of costal rib 1 . . . Chelydrini 4

4a. Anterior lobe of plastron pointed or rounded. . . *Chelydrasia* (abbreviation *Cha.*)

4b. Anterior lobe of plastron truncate . . . *Chelydropsis* (abbreviation *Ch.*)

5a. Caudal notch confined to the pygal, vertebral 1 wider than vertebrals 2–3 . . . *Macrochelys*

5b. Caudal notch extending beyond the lateral margins of the pygal, first vertebral 1 narrower or equal to vertebrals 2–3 . . . *Chelydra* (abbreviation *C.*)

PROTOCHELYDRINAE (GAFFNEY 1975)
Diagnosis

The jugal is large and longer than quadratojugal (Fig. 2.1A). The skull has large posterodorsal and cheek emarginations. The maxilla has a relatively wide triturating surface. The pterygoid "waist" is wide and the ventral surface has posteriorly directed ridges. The maxilla and premaxilla lack a distinct hook. Postorbital-maxilla contact is absent. Lower mandible is undescribed. The plastron lacks fontanelles (Fig. 2.2A). The epiplastron is broad and the gular scales are well anterior to the entoplastron. The entoplastron is diamond-shaped. Three sets of inframarginals are present and there are no supramarginal scales. The plastron is attached to carapace by gomphotic sutures. The carapace lacks a cephalic notch in dorsal view. The costiform process of nuchal terminates anterior to the insertion of costal rib 1. Suprapygal 1 is trapezoidal and the posterior margin of vertebral 5 lies well on suprapygal 2. The posterior carapace margin is only weakly serrated. The hyoplastral buttress extends to peripheral 3 and the hypoplastral buttress to peripheral 8 (peripheral 7* in *Denverus*). Pleural scale 1 does not

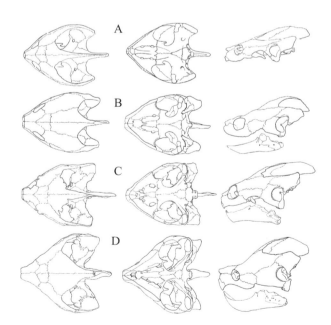

Fig. 2.1. Skull (dorsal, left lateral, and ventral views) and lower mandible (left lateral view) of the Chelydridae known from skulls. (A) *Protochelydra* (*P. zangerli* adapted from Erickson 1973 and Gaffney 1975) (B) *Chelydropsis* (*Ch. murchisoni* adapted from Mlynarski 1980a and Gaffney & Schleich 1994) (C) *Chelydra* (*C. serpentina* adapted from Gaffney 1975) (D) *Macrochelys* (*M. temminckii* adapted from Gaffney 1975). Scales vary.

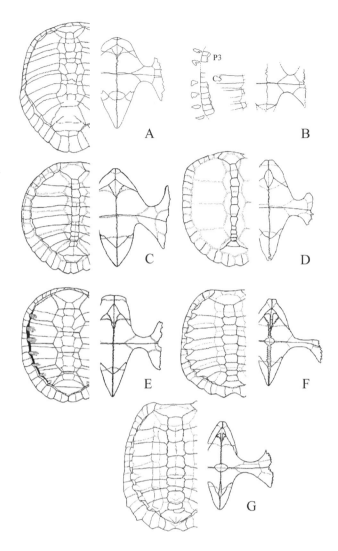

Fig. 2.2. Dorsal view of carapace (dorsal view) and plastron (ventral view) of the five genera of chelydrids. (A) *Protochelydra* (*P. zangerli* modified after Erickson 1982 and Hutchison 1998) (B) *Denverus* (*D. middletoni* adapted from Hutchison & Holroyd 2003) C, costal; P, peripheral. (C) *Chelydrasia* (*Cha. sanctihenrici* adapted from Broin 1977) (D) *"Chelydrasia"* (*Cha. kusnetzovi* adapted from Gaiduchenko and Ckhikvadze 1985 and Ckhikvadze 1985) (E) *Chelydropsis* (*Ch. murchisoni* carapace adapted from Mlynarski 1980a; *Ch. sansaniensis* plastron adapted from Broin 1977). (F) *Chelydra* (*C. serpentina* adapted from Boulenger 1889). (G) *Macrochelys* (*M. temminckii*, carapace from British Museum of Natural History 1860.5.11.5; plastron adapted from Boulenger 1889). Scales vary.

or only slightly overlaps the nuchal. Carapace with or without dorsal carina.

Distribution

Late Cretaceous (Lancian) to early Eocene (Wasatchian) of the eastern slope of the Rocky Mountain region.

Denverus (Hutchison & Holroyd 2003)
Taxonomic History

Hutchison & Holroyd 2003 (figs. 8–9) erected *Denverus* (Fig. 2.2B) for their new species *D. middletoni*.

Distribution

Early Paleocene (Puercan) of Colorado.

Diagnosis

Carapace length reaches a length of about 23 cm. The bridge peripherals are distinctly thickened medially. Hypoplastral buttress terminates on peripheral 7*.

Remarks

Denverus is only known from a partial plastron and carapace. *Denverus* most closely resembles the more poorly known *Protochelydra? caelata* in size and coarse sculpture of the costals but differs in the thickening of the medial side of the bridge peripherals, greater height and roughness of the marginal scale covered areas, and more anterior termination of the hypoplastral buttress.

Protochelydra (Erickson 1973)
Taxonomic History

Erickson (1973) erected *Protochelydra* for his new species *P. zangerli* and placed it in the Chelydrinae. Gaffney (1975) created the subfamily for the genus to distinguish it from the Chelydrinae.

Distribution

Latest Cretaceous (Lancian) of Montana and Wyoming (Holroyd & Hutchison 2002, as *Protochelydra* sp.). Paleocene (Torrejonian, about 62 Ma) of Montana and late Tiffanian (about 57 Ma) of North Dakota. Hutchison 1998 (fig. 18.1) and Holroyd et al. (2001) report *Protochelydra* sp. from the early Eocene (Wasatchian, about 54 Ma) and late Paleocene (Clarkforkian, about 56 Ma) of Wyoming.

Remarks

This is the oldest named genus of chelydrid and is known only from North America. The suggestion of Ckhikvadze (1980) that *Hoplochelys* Hay (1908a) and *Protochelydra* are probably congeneric is untenable in that they are in different families (Hutchison & Bramble 1981) and the additional material of *Protochelydra* confirms this.

Protochelydra zangerli (Erickson 1973)
TAXONOMIC HISTORY

Erickson 1973:1 (figs. 1–11) described *P. zangerli* on the basis of a skull and isolated parts of the plastron, carapace, and appendicular skeleton (Figs. 2.1A and 2.2A). Erickson (1982, figs. 7–8) later figured a carapace and Hutchison (1998, fig. 18.2I) figured a reconstruction of the plastron and carapace. More complete material (Science Museum of Minnesota P76.28.259, fig. 2A) shows that the epiplastron originally assigned to this species is incorrect.

DIAGNOSIS

Carapace length reaches a length of 39 cm and costal sculpture is weak or absent.

DISTRIBUTION

The type locality is from late Paleocene (Tiffanian, about 57 Ma) from North Dakota. Hutchison & Pasch (2003) report *Protochelydra* cf. *P. zangerli* from the late Paleocene (Clarkforkian, about 56 Ma) of Alaska.

Protochelydra? caelata (Hay, 1908b)
TAXONOMIC HISTORY

Hay (1908a:163, plate 27, figs. 1–3) named *Hoplochelys caelata* on the basis of six posterior peripherals and a costal fragment. This species is not a kinosternid but a chelydrid. It is here questionably referred to as *Protochelydra*.

DIAGNOSIS

Carapace length reaches a length of about 12 cm and the costals are distinctly ridged or plicate.

DISTRIBUTION

Middle Paleocene (Torrejonian, about 62 Ma) of Sweetgrass County, Montana.

REMARKS

The species is only known from the type specimen and continued to reside in the genus *Hoplochelys* until now. The peripherals differ from *Hoplochelys* and resemble *Protochelydra* in lack of inflation of the marginal areas that taper evenly to a knife edge, marginal scales cover two-thirds or more of the peripheral depth, and deep pits (groove) for reception of the plastral dentations. The plicate sculpture of the costals is typical of many chelydrids but absent in *Hoplochelys*. The specimen may be either a juvenile of *P. zangerli* or related to the undescribed *Protochelydra* sp. from the late Cretaceous-early Paleocene of Montana. Search of more recent collections from the type locality may yield needed additional material for comparison.

CHELYDROPSINAE (MLYNARSKI 1980)
Taxonomic History

Gaffney (1975) included *Macrocephalochelys* (=*Chelydropsis*) with *Platysternon* in his subtribe Playsternina but, as noted in the introduction, this is abandoned here. Broin (1977) recognized two species groups within *Chelydropsis sensu lato*, the *Chelydropsis decheni-sanctihenrici* and the *Chelydropsis murchisoni* group. Mlynarski (1980a, 1981a) created the Chelydropsinae to include the species referred to as *Chelydropsis* by Ckhikvadze (1971), Broin (1977), and others. Ckhikvadze (1999) created a new genus, *Chelydrasia*, essentially for the *Ch. decheni-sanctihenrici* group. Both genera here retained in the Chelydropsinae.

Diagnosis

The jugal is reduced*, about equally long* as quadratojugal, and is excluded from the orbit by contact** of the postorbital and maxilla (Figs. 2.1B and 2.3G). The skull has moderate* cheek emargination and the temporal passage is roofed over** because of a reduced temporal emargination. The maxillae have relatively wide triturating surfaces. The pterygoid "waist" is wide and with or without* a posteriorly directed ridge on ventral surface. The maxilla and premaxilla lack a hook. The lower mandible has a large and greatly elevated coronoid process and the foramen dentofacialis majus is situated posteriorly and below the coronoid process. Lateral plastron fontanelles may be either present* or absent. The epiplastron is relatively broad with the gular scales well anterior to entoplastron (Fig. 2.2D). The entoplastron is diamond-shaped. Three sets of inframarginals and supramarginal scales may* or may not occur. The plastron is attached to carapace by gomphotic to ligamental* sutures. There is no cephalic notch of the carapace in dorsal view. The costiform process of nuchal terminates anterior to the insertion

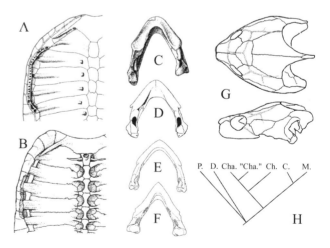

Fig. 2.3. (A) *Cha. sanctihenrici*, right anterior quadrant of carapace, ventral view (adapted from Broin 1977). (B) *M. temminckii*, right anterior quadrant of carapace, ventral view (from British Museum of Natural History 1860.5.11.5). (C) *Cha. sanctihenrici*, lower mandible dorsal view (adapted from Broin 1977). (D) *Ch. murchisoni*, lower mandible dorsal view (modified after Mlynarski 1980a). (E) *Macrochelys temminckii*, lower mandible dorsal view (adapted from Gaffney 1975). (F) *C. serpentina*, lower mandible dorsal view (adapted from Gaffney 1975). (G) *Ch. nopcsai*, skull, dorsal view showing scalation (adapted from Tarashchuk 1971). (H) Cladogram of proposed relationships of chelydrid genera. Scales vary.

of costal rib 1. Suprapygal 1 is trapezoidal and vertebral 5 may* or may not overlap the pygal. The posterior margin of the carapace is distinctly serrated*. The hyoplastron buttress terminates on peripheral 3 or 4* and the posterior buttress terminates on peripheral 7* or eight. Pleural scale one does not overlap the nuchal. The carapace is acarinate to weakly tricarinate*.

Distribution

Eocene to Pliocene of Europe, Neogene of Turkey, and early Oligocene to Pliocene of Central Asia.

Remarks

The shape of the gular area of the epiplastron is given strong weight by most authors (Ckhikvadze 1971, 1999; Broin 1977; Lapparent de Broin 2001). All the species herein included in *Chelydrasia* have, where known, the anterior end of the plastron more or less pointed compared to the distinctly truncate termination in *Chelydropsis* (=*Chelydropsis murchisoni* group of Broin 1977) and, with the exception of "*Chelydrasia*" *kusnetzovi*, possess the other characters detailed by Broin 1977 for her *Ch. decheni-sanctihenrici* group. The presence of supramarginals in this group (Fuchs 1938, fig. 27, 27b; Broin 1977; Ckhikvadze 1982, fig. 1v) is problematic in that they are not always visible (Mlynarski 1976) and frequently appear to be generally coincident with the lateral carina (Broin 1977, figs. 37, 43), unlike those in *Macrochelys*. Their absence in uncrushed or better preserved material (e.g., *Chelydropsis murchisoni staechi* of Mlynarski 1980a) suggests that some may be artifacts of compression or confused with the lateral carina gutter. Detailed descriptions of skull of the genus are unavailable although Broin (1977, plate 37, figs. 7–8, plate 38, fig. 11) provides good descriptions of some elements.

Chelydrasia (Ckhikvadze 1999)
Taxonomic History

The earliest taxon now referred to this genus is *Chelydra decheni* Meyer 1852. Ckhikvadze (1971) named two new species, *Chelydropsis minax* (genotype) and *Ch. poena,* from Asia and placed all the European species previously referred to *Chelydra* in *Chelydropsis*. Broin (1977) divided *Chelydropsis* into two informal groups, the *Ch. decheni-sanctihenrici* and *Ch. murchisoni* groups. Mlynarski (1980a, 1981a) included *Ch. minax* and *Ch. poena* in the *Ch. decheni-sanctihenrici* group of Broin. In a brief diagnosis and discussion, Ckhikvadze 1999 specifically removed *Ch. minax* and *Ch. sanctihenrici* Broin 1977 to *Chelydrasia* (Figs. 2.2C–D, 2.3A, C). He also indicated that *Ch. manuascensis* Bergounioux 1936 most probably belongs and *Ch. decheni* possibly belongs to the genus. He did not discuss the fate of the other two Asian species *Ch. poena* and *Ch. kusnetzovi*. Lapparent de Broin (2001) rejects inclusion of the European species of the *Ch. decheni-sanctihenrici*

group in *Chelydrasia* "because of the longer gular part of the epiplastron symphysis in this new genus." Whatever the merits of this character, by itself, it is not sufficient reason to exclude the European species from *Chelydrasia*, especially considering the more striking differences between the *Ch. decheni-sanctihenrici* group and *Ch. muchisoni* groups in the shape of the gular and other differences noted by Broin (1977). Because *Ch. decheni, Ch minax, Ch. sanctihenrici,* and *Ch. manuascensis* (Bergounioux 1936) form core species of the *Ch. decheni-sanctihenrici* group of Broin (1977), I conclude that *Chelydrasia* is the taxonomic equivalent of this group.

Diagnosis

The free perimeter dimension of anterior peripherals is about as long as deep. The serrations of posterior peripherals are relatively short and weak on peripherals 8–9. The caudal notch of pygal (Fig. 2.2C) is small and one-half or less of the maximum width of the pygal (except in "*Cha.*" *kusnetzoi,* Fig. 2.2D). The hyoplastral buttress is shorter than the hypoplastral buttress and extends to peripheral 3 (except in "*Cha.*" *kusnetzovi*). The epiplastra are relatively narrow and without a distinct convex angle near gular-humeral sulcus. The gular is distinctly narrower than humeral width on epiplastron. The xiphiplastron is relatively broad with convex lateral margin (except in "*Cha.*" *kusnetzovi*). The bridge region of hyo- and hypoplastron is relatively broad. The carapace ranges in length from 22 to 55 cm.

Distribution

Early Oligocene to Pliocene of Kazakhstan and middle Eocene to early Miocene of Europe.

Remarks

I tentatively consider *Ch. kusnetzovi* as a derived member of *Chelydrasia* because of the shape of the epiplastron and anterior lobe. The genotypic species *Cha. minax* has not been adequately figured or described, and the concept of the genus really relies on the much more complete, much better illustrated, and more thorough descriptions by Broin (1977). The entire group needs revision. Some taxa can not be readily compared feature by feature either because of the lack of comparable elements, poor descriptions, or inadequate figures. As presently constituted *Chelydrasia* is a wastebasket group for those species of the Chelydropsinae that lack the derived features of *Chelydropsis* (*Ch. murchisoni* group of Broin 1977).

Key to the species of *Chelydrasia*

1a. Hyo- and hypoplastron short with hypoplastral buttress shorter than hyoplastral buttress . . . *Cha. kusnetzovi*

1b. Hyo- and hypoplastron relatively long with hyoplastral buttress shorter than hypoplastral buttress . . . 2

2a. Costal-peripheral fontanelles present and radial plications absent on carapace . . . *Cha. decheni*

2b. Costal-peripheral fontanelles absent or unknown and radial plications present on carapace . . . 3

3a. Gular scale about equally as long or longer than broad . . . 4

3a. Gular scale shorter than broad or unknown . . . 5

4a. Gular scale about equally as long as broad and anterior margin rounded. . . *Cha. minax*

4b. Gular scale longer than broad and anterior margin truncate . . . *Cha. poena*

5a. Length of pectoral-humeral sulcus 1.5 times the femoral-abdominal sulcus . . . *Cha. apellanizi*

5b. Length of pectoral-humeral sulcus about two times the femoral-abdominal sulcus length . . . *Cha. sanctihenrici*

Chelydrasia minax (Ckhikvadze 1971)

TAXONOMIC HISTORY

Ckhikvadze (1971: 238, fig. 1b) named *Ch. minax* on the basis of an epiplastron and isolated parts of the shell (only the epiplastron has been figured). Mlynarski (1976) placed the genus in quotations but later (Mlynarski 1980a, 1981a) included it the *Ch. decheni-sanctihenrici* group of Broin (1977). Ckhikvadze (1999) made *Cha. minax* the genotypic species of *Chelydrasia*.

CHARACTERS

Carapace reaches a length of about 50 cm. The anterior margin the gular scale is evenly curved. The carapace has a median carina and marked plications of the scales. The other elements of the plastron are unknown. Details of the carapace are largely unknown but it probably lacks costal-peripheral fontanelles.

DISTRIBUTION

The species is only known from the early Oligocene of the Kustov series of the Zaisan Basin, Eastern Kazakhstan (Ckhikvadze 1971).

REMARKS

As noted by previous authors, this taxon rests only on the cursory description and figure of Ckhikvadze (1971, 1999) and comparison with other species is difficult.

Chelydrasia poena (Ckhikvadze 1971)

TAXONOMIC HISTORY

Ckhikvadze (1971: 238, fig. 1b) named *Ch. poena* on the basis of an epiplastron. The species was included in the *Ch. murchisoni* group by Broin (1977). Mlynarski (1976) synonymized *Ch. poena* with *Ch. minax,* but later Mlynarski (1980a, 1981a) recognized them as distinct species of the *Ch.*

decheni-sanctihenrici group of Broin (1977). Ckhikvadze (1999) makes no mention of the species when he created *Chelydrasia* but it is here included because of the strong similarity of the epiplastron to *Cha. minax.*

DIAGNOSIS

The carapace reaches a length of about 50 cm. The gular scale is truncated anteriorly and relatively larger than *Cha. minax.*

DISTRIBUTION

Early Miocene (in Ckhikvadze 1971; Broin 1977) but listed as early-middle Eocene (in Mlynarski 1980a, 1981a), Zaisan Basin, eastern Kazakhstan.

REMARKS

Chelydrasia poena is apparently known only from a single epiplastron and the "diagnostic" character used by Ckhikvadze (1971) appears to be trivial. Although its very strong similarity to *Cha. minax* supports its inclusion in *Chelydrasia,* little more can be said.

Chelydrasia apellanizi (Murelaga et al. 1999)

TAXONOMIC HISTORY

Murelaga et al. (1999:424, figs. 1–2, 3c) named *Ch. apellanizi* on the basis of a partial hyoplastron and referred isolated parts of the carapace and plastron. They assigned the species to the *Ch. decheni-sanctihenrici* group of Broin (1977). Because this group is here regarded as equivalent to *Chelydrasia* of Ckhikvadze (1999), it is transferred to this genus.

DIAGNOSIS

Carapace reaches a length of about 28 cm. The carapace has simple sulci, radial plications, and costal-peripheral fontanelles; it is thin. The pygal notch is one-third the width of the pygal. The anterior part of the epiplastron is unknown but the posterior terminus is well anterior to the base of the axillary notch. The pectoral-abdominal sulcus is 1.5 times longer than the femoral-abdominal sulcus.

DISTRIBUTION

The species is known only from the early Miocene (MN3), of the Navarra Province of Spain (Murelaga et al. 1999, 2002).

Chelydrasia decheni (Meyer 1852)

TAXONOMIC HISTORY

Meyer (1852:242, plates 28–30, figs. 5–6) described *C. decheni* on the basis of two partial but articulated skeletons and added a nearly complete third specimen later (Meyer 1868, plate 9). Mlynarski (1976) retained the species in *Chelydra* but put the genus in quotations. Ckhikvadze (1971) placed *C. decheni* in the genus *Chelydropsis* and Broin (1977) included it as a core species of her *Ch. decheni-sanctihenrici* group. She also noted that *Emysaurus* (=*Chelydra s. s.*) *meilheuratiae* Pomel 1846 (Maack 1869 placed this species in *Chelydra*) has two characters of *Chelydropsis* (*sensu lato*) and, while insufficient for specific identification, may be questionably referable to *Ch. decheni*. Ckhikvadze (1999) indicated that *Ch. decheni* is possibly a member of *Chelydrasia*, where it is here included.

DIAGNOSIS

Carapace reaches a length of 23 cm. The carapace is thin, has doubled sulci, and costal-peripheral fontanelles, but it lacks radial plications. The pygal notch is one-half the width of the pygal and suprapygal 1 is wider than the second. The epiplastra are long and extend posteriorly to near the base of the axillary notch. The plastral scales are unknown.

DISTRIBUTION

Early Oligocene (Rupelian, about 31 Ma) of Leipzig State, Germany (*Ch. cf. decheni* of Karl (1990, figs. 1–2) and Allier Department, France (*Chelydropsis* sp. = "*meilheuratiae*" = "*decheni*" of Broin 1977 [plate 36, figs. 3–5]) to early Miocene (Late Orleanian, MN5) of Bayern State, Germany (*Ch. cf. decheni* of Groessens van Dyck and Schleich 1985 [figs. 2–3, plate 1, figs. 5–11, plate 2, figs. 2, 5, 13]) and early Miocene (Aquitanian) of Nordrhein Westfalen, Germany (Meyer 1852, 1856).

REMARKS

Meyer (1868, plate 9; 156: plate 9, figs. 4–5) and Broin (1977) present lengthy descriptions of the original material consisting of an articulated specimen in a shale. One of the original specimens is photographed in ventral view by Böhme & Lang (1991, fig. 1), but all the type material needs to be reexamined and redescribed, perhaps with new tools, with attention to details useful for comparison with other nominal species.

Chelydrasia sanctihenrici (Broin 1977)
TAXONOMIC HISTORY

Bergounioux (1936) described three specimens referable to this species and referred them to his new taxon *Broilia massiliensis*. Broin (1977, figs. 37–40; plate 35, figs. 3–4; plate 36, figs. 6–7; plate 37, figs. 4–6) noted that the type of *B. massiliensis* should be included in *Mauremys* but the three referred specimens belong to *Chelydropsis*, for which she created *Ch. sanctihenrici* on the basis of a carapace, parts of the plastron,

mandible, hyoid fragment, scapula, humerus, radius, and various podials (Figs. 2.2C, 2.3A and C). This species is the core taxon of the *Ch. decheni-sanctihenrici* group of Broin (1977). Ckhikvadze (1999) specifically included *Ch. sanctihenrici* as a core species of his *Chelydrasia*. Lapparent de Broin (2001) rejected the inclusion of this species in *Chelydrasia*, but it is here retained in *Chelydrasia* pending a thorough review of the pertinent species.

DIAGNOSIS

Carapace reaches a length of 30 to 40 cm. The carapace is thick and has simple sulci and radial plications but lacks costal-peripheral fontanelles. The pygal notch is less than one-half the width of the pygal. Suprapygal 1 is wider than suprapygal 2. The epiplastra do not extend to near the base of the axillary notch. The gulars are short and truncate anteriorly. The length of the pectoral-abdominal sulcus is about twice that of the femoral-abdominal sulcus.

DISTRIBUTION

Early Oligocene (Rupelian, about 31 Ma) of Lot-et-Garonne, Haute-Garonne, and Tarn-et-Garonne Departments, France (*Chelydropsis* sp. in Broin 1977, but *Ch. sanctihenrici* in Lapparent de Broin 2000) to late Oligocene (Upper Stampian, about 23 Ma) of Bouches du Rhône and Basses Alpes Departments, France (*Chelydropsis* sp. in Broin 1977).

REMARKS

The carapace of the type of *Cha. sanctihenrici* (Broin 1977, fig. 37) has an anomalous scale pattern of the left side with a supernumerary pleural scale between pleurals three and four. The posterior margin of vertebral five usually lies on the second suprapygal but may coincide with the pygal-suprapygal suture or just overlap the pygal. Broin (1977) notes the presence of supramarginals. These appear to be coincident with the lateral carina and are only indicated by dashed lines in her figures.

"Chelydrasia" kusnetzovi (Ckhikvadze in Gaiduchenko & Ckhikvadze 1985)
TAXONOMIC HISTORY

Ckhikvadze named and figured the dorsal view of the carapace *Ch. kusnetzovi* in Gaiduchenko and Ckhikvadze (1985:117, see their fig. 1) (Fig. 2.2D) and figured the plastron in the same year elsewhere (Ckhikvadze 1985, plate 1). Karl (1990) placed this species in the *Ch. pontica* group of Mlynarski (1980a). Ckhikvadze (1999) made no mention of this species when he erected *Chelydrasia*, but it is here questionably referred to as *Chelydrasia* pending clarification of its relationships.

DIAGNOSIS

The carapace reaches a length of 55 cm. The carapace is thick and has simple sulci and lacks radial plications and costal-peripheral fontanelles. The pygal notch is greater* than one-half the width of the pygal. Suprapygal 1 is narrower* than suprapygal 2. The epiplastra do not extend to near the base of the axillary notch. The gular scale is rounded anteriorly and is relatively wide*. The length of the pectoral-abdominal sulcus is about equal that of the femoral-abdominal sulcus.

DISTRIBUTION

The species is known only from the type locality in the middle Pliocene of Pavlodar, Eastern Kazakhstan (Ckhikvadze 1985).

REMARKS

The systematic placement of "Chelydrasia" kusnetzovi is questionable. In his description of Chelydrasia Ckhikvadze (1999) does not mention it and presumably leaves it within Chelydropsis. This association is supported by the reduction in length of the hyo- and hypoplastral buttresses, narrow ziphiplastra, and relatively wide caudal notch of the pygal. Conversely, other features resemble Chelydrasia, such as the pointed anterior lobe, the greater width of the anterior lobe compared with the posterior lobe, narrow neurals, greater width of the plastral bridge, greater reduction of the hypoplastral buttress compared with the hyoplastral buttress, and lack of carapacial and plastral fontanelles. Judging from the sparse figures by Ckhikvadze in Gaiduchenko and Ckhikvadze (1985, fig. 1) and Ckhikvadze (1985, plate 1), "Chelydrasia" kusnetzovi also has unique features for a chelydropsine such as the relatively narrow first vertebral scale and first suprapygal. The derived features that resemble Chelydropsis appear independently in other chelydrids (e. g., Chelydra and Macrochelys) and appear to be general trends in all later chelydrids. In the absence of the fenestration of the shell and uniquely truncate anterior lobe of Chelydropsis, the other features are not convincing evidence for inclusion in Chelydropsis. Geography and the sparse morphological evidence suggest that Chelydrasia is probably independently derived from Oligocene Chelydrasia and may be worthy of a genus of its own when better described.

Chelydrasia sp.
DISTRIBUTION

Middle Eocene (Bartonian, MP16, about 39 Ma) of Oise Department (Lapparent de Broin 2001), late Oligocene (Upper Stampian, about 25 Ma) of Puy-de-Dôme (Broin 1977) and Tarn-et-Garonne Department (Broin 1977, plate 35, figs. 5–6, plate 38, fig. 13), early Miocene (Aquitanian, about 22 Ma), Aud, Allier, and Gers Departments (Broin 1977, plate

36, figs. 1–2; 293), early Miocene (Agenium, MN$_2$, Orleanian, MN4, about 19 Ma) (Broin 1977; Ginsburg et al. 1991; Lapparent de Broin 2000), Loir-et-Cher Department (Broin 1977), Loiret Department (Broin 1977, figs. 41–42, plate 37, figs. 7–10, plate 38, figs. 11–12), and Maine-et-Loire and Indre-et-Loire Departments (Broin 1977, plate 35, figs. 1–2) of France and late Oligocene (Lower Stampian, about 27 Ma) in Rhein-Pfalz in Germany

REMARKS

The material included here is mostly for specimens previously listed as Chelydropsis sp. in the Ch. decheni-sanctihenrici group of Broin (1977).

Chelydropsis (Peters 1868)
Taxonomic History

Peters (1868:73) erected the genus Chelydropsis for his Ch. carinata on the basis of a partial shell (Figs. 2.1B, 2.2E, 2.3D, G). Ckhikvadze (1971) placed all the European species previously referred to Chelydra in Chelydropsis. He also included Macrocephalochelys (Pidoplitschko & Tarashchuk 1960), Leptochelys (Bergounioux & Crouzel 1965), and Broilia (Bergounioux 1936). Broin (1977) erected the Ch. murchisoni group to distinguish the later Miocene and Pliocene species from the Oligocene and early Miocene species then referred to the Ch. decheni-sanctihenrici group. Mlynarski (1980a, 1981a) recognized the two groups of Broin (1977), calling her Ch. murchisoni group the Ch. murchisoni-sansaniensis group, and added a third, the Ch. pontica group, for Ch. pontica. Ckhikvadze (1999) erected Chelydrasia for members of the Ch. decheni-sanctihenrici group, resulting in the Ch. murchisoni group being equal to Chelydropsis sensu strictu. Lapparent de Broin (2000) considered Ch. pontica as a member of the Ch. murchisoni group.

Diagnosis

The free perimeter dimension of anterior peripherals are longer* than deep. The serrations of posterior peripherals are relatively long and strong* on peripherals 8 and 9. The caudal notch of pygal is large and more than one-half* of maximum width of pygal (Fig. 2.2E). The hyoplastral buttress is shorter than the hypoplastral buttress short extends only to peripheral 4*. The epiplastra are relatively wide* and with a distinct angle* at the gular-humeral sulcus. The gular scale is about as wide* as the humeral width on epiplastron. The xiphiplastron is relatively narrow* with a straight lateral margin. The bridge region of hyo- and hypoplastron is relatively narrow. The carapace reaches lengths of 70 cm.

Distribution

Middle Miocene to Pliocene of Europe and Neogene of Asia Minor.

Remarks

Chelydropsis exhibits some convergence on *Chelydra* and *Macrochelys* in reduction of the plastron and development of the posterior serrations and elongation of the peripherals. The possible presence of inframarginals is also similar to *Macrochelys*.

Key to the species of Chelydropsis

1a. Carapace greater than 40 cm . . . *Ch. nopcsai*

1b. Carapace less than 40 cm . . . 4

2a. Plastron without lateral fontanelles and with radial plications on carapace . . . *Ch. carinata*

2b. Plastron with lateral fontanelles and with weak placations . . . *Ch. sansaniensis*

2c. Plastron with lateral fontanelles and no radial carapacial placations . . . *Ch. murchisoni*

Chelydropsis carinata (Peters 1868)
TAXONOMIC HISTORY

Peters (1855, plate 5) described the first material to be referred to this species as *Chelydra* sp. on the basis of a damaged shell. Peters (1868) later erected the new genus and species for this material as *Ch. carinata*. This was more fully described and figured and named again in Peters (1969, fig. 1, plate 1). Mlynarski (1976, fig. 66.2) referred to it as *Ch. carinatus*. Broin (1977) included *Ch. carinata* as a member of her *Ch. murchisoni* group (=*Ch. murchisoni-sansaniensis* group of Mlynarski 1980a (=*Chelydropsis sensu strictu*).

DIAGNOSIS

The carapace is massive and reaches lengths of 30–38 cm. The carapace has simple sulci (not doubled) and apparently radial plications but lacks costal-peripheral fontanelles in adults. The serration of peripherals 8 and 9 is relatively strong. The posterior margin of vertebral 5 lies on suprapygal 2. The plastron is firmly attached to the carapace and lacks lateral fontanelles.

DISTRIBUTION

Middle Miocene (Early Orleanian, MN5, about 19 Ma) of Steirmark Province, Austria.

REMARKS

The nuchal of the type (Peters 1869, plate 1; Mlynarski 1976, fig. 66.1) has two ossifications and a trace of supramarginals. The two ossifications are probably an anomaly or misinterpretation of a crack as a suture.

Chelydropsis murchisoni (Bell 1832)
TAXONOMIC HISTORY

Bell (1832) named *C. murchisoni* on the basis of shell material (Figs. 2.1B, 2.2E in part, 2.3D). Meyer (1845, plates 11–12), 1852) provided lengthy descriptions based on partial skeletons, clarified previous identifications from the type locality, and recognized *C. oeningensis* and *C. murchisoni*. Maack (1869) synonymized *C. oeningensis* with *C. murchisoni*. Fuchs (1938:91, fig. 26) erected *C. allingensis* on partial shell and limb material and described other shell material (fig. 27) as *Macroclemys* sp. Williams (1952) questioned assignment of *C. murchisoni* to *Chelydra* and placed the genus in quotations. Ckhikvadze (1971) placed the *C. murchisoni* in *Chelydropsis* but Mlynarski (1976) retained it in *Chelydra* in quotations. Broin (1977) placed *Ch. murchisoni* as the core species of the *Ch. murchisoni* group. Broin (1977) also noted that the *C. allingensis* and *Macroclemys* sp. reported by Fuchs (1938) apparently represent a single species of *Chelydropsis* but that the material is not sufficiently described for definitive comparison or synonymy with other Miocene species of the genus and placed the species name in quotations. Mlynarski (1980a) unequivocally regarded *C. allingensis* as a synonym of *Ch. murchisoni* and named a new subspecies, *Ch. murchisoni staeschi* (Mlynarski 1980a, figs. 1–18; plates 1–3). Kosatsky & Redkozubov (1989) synonymized *Ch. decheni* and *Ch. allingensis* with *Ch. murchisoni*, but the synonymy of *Ch. decheni* is not supported on morphology or temporal grounds. Lapparent de Broin (2000) notes that the cranial and lower mandible characters used by Mlynarski (1980a) to define *Ch. m. staeschi* are not known in *Ch. m. murchisoni* and thus inadequate for diagnosis of a subspecies.

DIAGNOSIS

Carapace reaches a length of 50 cm. Carapace has peripheral-costal fontanels, simple sulci, but lacks radial plications. The fifth vertebral overlaps the pygal and supramarginals are absent. The serrations of peripherals 8 and 9 are weak to strong. The plastron and carapace are connected by ligament or open gomphotic sutures. The plastron has lateral fontanelles. The hypoplastron is wider than the hyoplastron, and the hyoplastral buttress terminates in peripheral 3.

DISTRIBUTION

Middle to late Miocene (Astaracian, MN7, about 14 Ma) of Württenberg and Bayern States, Germany (Bell 1832; Meyer 1852, plates 26, 30, fig. 7; Fuchs 1938; Broin 1977; Mlynarski 1980a, fig. 1, plate 1; Schleich 1981, 1985; Schleich & Groessens van Dyck 1988; Gaffney & Schleich 1994) and late Miocene of Selesia, Poland (Mlynarski 1981a, plate 2, fig. 4; 1981b, fig. 1, plate 9).

REMARKS

The strength of the peripheral serrations varies from strong (Meyer 1852: plates 26, 30, fig. 7) to weak (Mlynarski 1980a, fig. 1, plate 1). Mlynarski (1980a, figs. 1–18: plates 1–3; 1980b, fig. 8b; 1981a, plate 2, fig. 4; 1981b, fig. 1, plate 9) and Gaffney & Schleich (1994, figs. 1–2, plates 1–4) describe and illustrate the skulls, shell, and postcranial elements. Lapparent de Broin (2001) lists the temporal range of this species into the Pliocene but this may include *Ch. nopscai*.

Chelydropsis sansaniensis (Bergounioux 1935)
TAXONOMIC HISTORY

Broin (1977) notes that the first material probably assignable to this taxon is *Emys sansaniensis* Lartet 1851 but the material is insufficient for species recognition. Bergounioux 1932 described *Broilia denticulata* on the basis of some specifically undiagnostic peripherals and plastral fragments that were later destroyed during World War II. The first material presently referable to *Ch. sansaniensis* was described by Bergounioux (1935) as *Trionyx (Axestus) sansaniensis* based on a fragment of the plastron (Fig. 2.2E in part). Bergounioux & Crouzel (1965, fig. 5) later named a new genus and species, *Leptochelys braneti*, from the same fauna on the basis of a hyo-hypoplastron. Bergounioux & Crouzel (1965, figs. 6–14) also named *Broilia robusta* on the basis of several specimens including a tail, fragments of the carapace, humerus, proximal tibia, digit of the manus, and hyoplastron fragment. They did not designate a holotype, and more than one taxon is present (including a testudinoid, fig. 15); however, they note that the most important (p. 18, fig. 6) of these specimens is the caudal series and this can be regarded as the lectotype. Ckhikvadze (1971) synonymized *Leptochelys* with *Chelydropsis*. Broin (1977, figs. 43–46; plate 37, figs. 1–3; plate 38, figs. 1–10) and Lapparent de Broin (2000, figs. 1–2) noted that *Tr. sansaniensis* Bergounioux 1935 belongs in *Chelydropsis* and also noted that some of the specimens referred to *B. denticulata* Bergounioux 1932 and *B. robusta* Bergounioux & Crouzel (1965, fig. 16) are referable to a single species of *Chelydropsis*, *Ch. sansaniensis* (Bergounioux 1935). She also placed this species in her *Ch. murchisoni* group (*Ch. murchisoni-sansaniensis* group of Mlynarski 1980a).

DIAGNOSIS

Carapace reaches a length of 37 cm. The carapace has costal-peripheral, lateral plastral fontanelles, and weak radial plications; the dorsal sulci are doubled or multiple. The plastral-carapacial articulation is ligamental and the hyoplastron is wider than the hypoplastron. Peripherals 8 and 9 are strongly serrate and supramarginals are present. The posterior margin of vertebral 5 coincides with the pygal-suprapygal suture. The plastron has lateral fontanelles. The hyoplastron buttress terminates on peripheral 4.

DISTRIBUTION

Middle Miocene (Astaracian, MN6, about 14 Ma) of Gers Department, France (Broin 1977, fig. 43–46; plate 37, figs. 1–3; plate 38, figs. 1–10)

REMARKS

Broin (1977) and Lapparent de Broin (2000) clarified the confused taxonomic history and presented the best descriptions of this species.

Chelydropsis nopcsai (Szalai 1934)
TAXONOMIC HISTORY

Szalai (1934:134, plate 4, fig. 22) described the first material that referred to this species as *Trionyx nopcsai* on the basis of a right and left dentary (Fig. 2.3G). Macarovici & Vancea (1960, plate 1, figs. 1–2, plate 2, figs. 7–9) described *Testudo grandis* on the basis of fragments of plastron and carapace. Schmidt (1966:25, figs. 1, 2c) described *C. strausi* on the basis of a juvenile skeleton. Pidoplichko & Tarashchuk (1960, figs. 1–5) created *Macrocephalochelys pontica* on the basis of a partial skull and placed it in the Platysternidae. Mlynarski (1966, fig. 15) provided better figures of the type and noted similarities of the jaw to *Chelydra* (=*Chelydropsis*) but retained the species in *Trionyx*. Ckhikvadze (1971) referred *M. pontica* to *Chelydropsis*. Mlynarski (1968, 1976) placed the genus for *C. strausi* in quotations. Mlynarski (1976) noted that Ckhikvadze (1971) referred *M. pontica* to *Chelydropsis* and the Chelydridae but continued to retain *Ma. pontica* in the Platysternidae. Auffenberg (1974) transferred *T. grandis* to *Geochelone (Geochelone)*. Gaffney (1975, fig. 2D) continued to use *Macrocephalochelys* and placed it as the sister taxon of *Platysternon* in his subtribe Platysternina. Ckhikvadze (1980) synonymized *Tr. nopcsai, T. grandis, C. strausi*, and *Ma. pontica* under *Chelydropsis nopcsai*. Khosatsky & Redkozubov (1989, figs. 4–5) described a new skull and referred all of the above to *Ch. nopcsai* (Szalai).

DIAGNOSIS

Carapace reaches a length of 70 cm. The carapace is incompletely known but it has simple sulci (not doubled) and lacks costal-peripheral fontanelles in adults. Peripherals 8 and 9 are strongly serrate and supramarginals are present. The posterior margin of vertebral 5 is probably on or near the pygal-suprapygal suture.

DISTRIBUTION

Pliocene of Niedersachen West State, Germany (*C. strausi* of Schmidt 1966), southern Slovak Republic (*C.* cf. *decheni* of Mlynarski 1963, plates 23–25, 26, figs. 3–4, 1966, 1968),

Prichernomiskaya Nizmennost region (*Ma. pontica* of Pido-plichko & Tarashchuk 1960), Crimea region (*Ma. pontica* of Tarashchuk 1971, fig. 1), Ukrania, and Moldavia (*Ma. pontica* of Khosatsky & Redkozubov 1989, fig. 6), and Rumania (*Tr. nopcsai* of Szalai 1934 and Mlynarski 1966; *T. grandi* of Maca-rovici & Vancea 1960).

REMARKS

This species is poorly characterized and comprises various poorly characterized or differentiated taxa from the Pliocene swept together probably mostly on the basis of age. Many of the nominal taxa are poorly represented on the basis of shell material, making comparison with other species primarily represented by shells difficult. Tarashchuk (1971, fig. 1) referred a complete skull to *Ma. pontica* but only figured the dorsal and lateral views. The skull differs from that of *Ch. murchisoni* in the reduction in size of the tympanic opening and cheek emargination. Until these taxa are better sorted out and compared, I have adopted the undocumented but not unreasonable assertion of Ckhikvadze (1980) in sweeping all these records together under this species name.

Chelydropsis sp. indet.
DISTRIBUTION

Miocene (lower Helvetian, about 14 Ma) in the Loir-et-Cher Department of France (Broin 1977 as *Chelydropsis* sp. in the *Ch. sanctihenrici* group?), Tortonian (about 10 Ma) in Bohemia, Czech Republic (*C. argillarum* of Laube 1900 and Broin 1977), and Vallesian (about 10 Ma) in Baden-Württemberg, Germany (Schleich, 1986, fig. 3a,b). It is also reported from Neogene: Thracia, western Turkey (Schleich 1994, plate 1, figs. 5–7).

REMARKS

Specimens unidentified to species or of questionable validity but that otherwise appear to belong to the *Ch. murchisoni* group of Broin (1977) are included here. Laube (1900: 47, plate 2, fig. 3) named *C. argillarum* on the basis of partial juvenile carapace. Mlynarski (1976) placed the genus in quotation marks and noted that its systematic position is unclear. Ckhikvadze (1971) placed the species in *Chelydropsis* and Broin (1977) placed the species in quotations and questionably in the *Ch. murchisoni* group (=*Chelydropsis sensu strictu*). This species is specifically indeterminate.

CHELYDRINAE (GRAY 1870)
Taxonomic History

See Ernst, Chapter 1, for taxonomic history of this extant group.

Diagnosis

The jugal may* or may not be reduced and may* or may not be shorter than the quadratojugal but it is not excluded from the orbit by contact of the postorbital and maxilla (except in one specimen of *Macrochelys schmidti*). The cheek emargination may be large or reduced*. The temporal passage is not roofed over. The triturating surfaces of the maxilla and dentary are relatively narrow**. The pterygoid lacks* a posteriorly directed ridge on ventral surface and the pterygoid waist may be either wide or narrow**. The maxilla and premaxilla have a small to large hook**. The lower mandible has a small** and not noticeably elevated** coronoid process and the foramen dentofacialis majus anterior* to and near* the level of the coronoid. The carapace has a cephalic notch in dorsal view**. The costiform process of nuchal terminates medial** to rib insertion of the first costal. Suprapygal 1 is roughly quadratic** and posterior carapace margin is strongly serrated*. The posterior margin of vertebral 5 overlaps* the pygal. The hyoplastral buttress terminates on peripheral 4* or 5** and posterior buttress terminates on peripheral 7*. The carapace weakly to moderately tricarinate* and costal-peripheral fontanelles are usually present*. The plastron has both lateral* and central** fontanelles in adults. The epiplastron is narrow** and the gular scales are near or contacting entoplastron**. The entoplastron is anchor shaped**. There are four, three, or fewer** inframarginals and supramarginal scales may* or may not be present. The abdominal scale may** or may not be divided transversely.

Distribution

The fossil history of the group is confined to early Miocene to Recent of North America but extant species of *Chelydra* reach as far south as Columbia, South America (Ernst, Chapter 1). Undescribed fossils from the late Eocene (Duchesnean) may also belong to this group (Eaton et al. 1999b). Unpublished Tertiary records of *Chelydra*-like chelydrines from California, Oregon, and Nevada indicate a long unpublished record extending into the Oligocene.

Chelydra (Schweigger 1812)
Diagnosis

The jugal is not reduced and is longer than the quadratojugal. The skull has large posterodorsal and cheek emarginations. The maxilla and premaxilla hook is small and the pterygoid waist is wide. There are four, three, two**, or fewer ** inframarginals and supramarginal scales are absent. The abdominal scale is not transversely divided. The plastron is attached to carapace by ligaments*. The hyoplastral buttress terminates on peripheral 5*. The costal rib heads arise at the level of the vertebral-pleural scale sulcus** and vertebral 1 is equal or narrower** than vertebral 2–3 width.

The carapace is weakly to moderately tricarinate. (See Figs. 2.1C, 2.2F, and 2.3E.)

Taxonomic History

See Ernst (Chapter 1) for a key to and the taxonomic history of the three extant species. Only *C. serpentina* is reported in the fossil record. Gaffney (1975) included *Chelydra* in his tribe Chelydrini, but because this group now includes only the genus, it has been abandoned.

Chelydra serpentina (Linnaeus 1858)
TAXONOMIC HISTORY

See Ernst (Chapter 1) for the taxonomic history of the extant species. The two nominal Pleistocene species from Florida, *C. sculpta* Hay (1916, plate 4, figs. 7; plates 6, figs. 8–9; Hay, 1917, plate 3, fig. 1) and *C. laticarinata* Hay (1916. plate 6, figs. 6–7) are suppressed senior synonyms of *C. s. osceola* (see Ernst, Chapter 1).

DIAGNOSIS

See Ernst (Chapter 1) for characteristics to this extant species.

DISTRIBUTION

See Ernst (Chapter 1) for the distribution of the extant and fossil records of this species.

REMARKS

C. serpentina is well represented in the fossil record but it is confined to the United States and the genus is only represented by the extant species. Its record extends back to the late Miocene (Hemphillian, 5–9 Ma). The presence of *C. serpentina* in the Pleistocene of Nevada and Idaho shows that the species either crossed the continental divide into the Snake or Colorado River drainages or was there all along.

Chelydra sp. indet.
TAXONOMIC HISTORY

The anterior part of a plastron from late Miocene (Hemphillian) of the Edson Quarry, Sherman County, Kansas, was assigned to Chelonia by Adams & Martin (1931) and later to the Chelydridae by Hibbard (1934, 1939), to *Chelydra* by Galbreath (1948) as *Chelydra* sp., and finally to *C. serpentina* by Preston (1979). The specimen on which this is based has not been described in detail or figured and there is no justification presented for its assignment to *C. serpentina* except for the lack of any other currently recognized fossil species of *Chelydra*. Even its assignment to *Chelydra* needs to be verified.

Macrochelys (Gray 1856a)
Taxonomic History

Macrochelys and *Macroclemys* Gray 1856b are both used in recent literature (sometimes in the same paper) (Figs. 2.1D, 2.2G, and 2.3B, F), but see Ernst (Chapter 1) for the taxonomic history of this extant genus. Gaffney (1975) included *Macroclemys* (=*Macrochelys*) in his tribe Platysternini and *Macroclemys* in his subtribe Macroclemydina. Because the inclusion of *Platysternon* in the Chelydridae is rejected, his subtribe Macroclemydina is abandoned. Of the four recognized species, three are extinct and described on or known from skulls. Matthew (1924) named *Chelydrops stricta* but made no comparisons to *Macrochelys*. Whetstone (1978) is the most recent reviewer of the fossil forms but apparently overlooked *Chelydrops*. He concluded that *M. schmidti* is the sister taxon of *M. auffenbergi* and the extant *M. temminckii*. *Chelydrops stricta* is here included in *Macrochelys*, although the original and additional material is in need of further description and comparisons. *M. auffenbergi* is known from postcranial material and thus provides some nonshell characters for comparison with *M. temminckii*. Dobie (1968) suggested that *M. temminckii* gave rise to *M. auffenbergi* but this is strongly influenced by the reported record of *M. temminckii* from the Miocene by Zangerl (1945) based on a side of a rostrum (see below). If the record of the Miocene *M. temminckii* is confirmed, then there were two coeval species of the genus in the late Miocene (Hemphillian).

Osteological Diagnosis

The jugal is reduced* and shorter than quadratojugal*. The posterodorsal emargination is reduced* and cheek emargination greatly reduced**. The maxilla and premaxilla have a strong** hook and the pterygoid has a narrow** waist. There are three inframarginals and supramarginal scales are present*. The abdominal scale is divided transversely**. The plastron is attached to carapace by ligaments* or weakly gomphotic sutures. The hyoplastral buttress terminates on peripheral 4 or 5*. The costal rib heads arise well medial to vertebral-pleural scale sulcus and vertebral 1 is wider than vertebral 2–3. The carapace is strongly** tricarinate.

Distribution

Early Miocene (Hemingfordian, about 19 Ma) to Recent of the Atlantic and Gulf drainage of the United States.

Remarks

The fossil record of *Macrochelys* is represented by four species, three of which are extinct.

Macrochelys schmidti (Zangerl 1945)
TAXONOMIC HISTORY

Zangerl (1945:3, figs. 2–3) described and figured *M. schmidti* on the basis of a nearly complete juvenile skull. Whetstone (1978a, fig. 1A) later referred a virtually complete adult skull to the species as *M. schmidti*.

DIAGNOSIS

The skull is relatively shorter than other species with less constriction of the snout. There is no secondary triturating ridge on the maxilla and the lateral triturating ridge is straight. The jugal just extends to the orbit. The skull has a width of 106% of the basilar skull length. The width of the rostrum at the posterior rim of the orbit is 60% of the basilar skull length. The distance from the anterior margin of the jugal to the quadratojugal-postorbital junction is 37% of the basilar skull length. The distance from the anterior rim of the orbit to the nasal notch is 13% of the basilar skull length. The postcranial skeleton is unknown.

DISTRIBUTION

The species is only known from the early Miocene (early Hemingfordian) of the Marsland Formation of Box Butte and Dawes Counties of Nebraska.

REMARKS

This species is known only from skulls. Whetstone (1978) used the depression of the skull, broadly rounded cheek, and less elevated supraoccipital crest in his diagnosis of this species but these features are strongly, sometimes subtlety influenced by diagenetic compression. He compared several other dimensions of the skull and its parts to the basilar length. These indicate that the skull of *M. schmidti* is relatively broader across the quadrates and orbits than in *M. auffenbergi* and *M. temminckii* and has a relatively shorter snout and longer jugal. The jugal has only a point contact with the orbit in the type (Zangerl 1945, fig. 2) and apparently no contact in the referred specimen (Whetstone 1978, fig. 1A).

Macrochelys stricta (Matthew 1924)
TAXONOMIC HISTORY

Matthew (1924:208–210, fig. 63) figured and very briefly described the type *Chelydrops stricta* within a large paper dealing primarily with fossil mammals that was overlooked by subsequent North American students of fossil chelydrids (Zangerl 1945; Dobie 1968; Whetstone 1978). Broin (1977) is the first to note the similarity of *Chelydrops* to *Macrochelys*. It is here included in *Macrochelys*.

DIAGNOSIS

The snout is constricted*. There is a secondary triturating ridge* on the maxilla that is stronger than that of *M. auffenbergi*. The triturating ridges of maxilla are slightly concave* medially. The postcranial skeleton is unknown.

DISTRIBUTION

The species is only known from the middle Miocene (early Barstovian, about 15 Ma) from the Lower Snake Creek Fauna of the Olcott Formation of Sioux County, Nebraska.

REMARKS

The species is only known from two partial skulls, with the type consisting of the rostral part including the prefrontals, premaxillae, and maxillae. As known from Matthew's description, the large size, narrow and elongate snout, down-turned and pointed beak, indication of a depressed skull roof over the orbits, relatively small orbits, and construction of the palate distinguish the species from *Chelydra* and agree with *Macrochelys*. The sharp down-turned, constricted snout and exposure of the jugal in the orbit distinguish *M. stricta* from *Chelydropsis*. The presence of a second triturating ridge on the maxilla distinguishes *M. stricta* from *M. schmidti* and *M. temminckii*. The secondary triturating ridge is apparently stronger than in *M. auffenbergi* (Whetstone 1978) and the species is retained pending detailed comparisons. Lower mandibles in the Frick collection of the American Museum of Natural History (FAM 11384, 11612) from the same fauna remain undescribed.

Macrochelys auffenbergi (Dobie 1968)
TAXONOMIC HISTORY

Dobie (1968:59, fig. 1) named *Macroclemys auffenbergi* on the basis of a sample comprising a virtually complete skeleton except for the majority of the carapace and is the most completely known of the extinct species of the genus. Whetstone (1978, fig. 1B) provided additional information on the skull.

DIAGNOSIS

Snout is constricted* with a weak secondary triturating ridge* on the maxilla. The triturating ridges of maxilla are slightly concave medially. Skull has a width of 92% of the basilar skull length. The width of the rostrum at the posterior rim of the orbit is 45%* of the basilar skull length. The distance from the anterior margin of the jugal to the quadratojugal-postorbital junction is 26%* of the basilar skull length. The distance from the anterior rim of the orbit to the nasal notch is 15%* of the basilar skull length. The ratio of the distal width of the humerus to the length increases with an increase in size. The phalanges are relatively shorter than in *M. auffenbergi*. The distal anterior-posterior length of the costal 6 is longer than that of the costal 7.

DISTRIBUTION

The species is known only from the Late Miocene (early Hemphillian) of Alachua County, Florida.

REMARKS

M. auffenbergi reaches a carapace length of about 46 cm.

Macrochelys temminckii (Troost in Harlan 1835)
TAXONOMIC HISTORY

See Ernst (Chapter 1) for the taxonomic history of the extant species. Hay (1907:847 figs. 1–4; 1908a, figs. 282–285) described the Pleistocene species *M. floridana* on the basis of four isolated peripherals. Auffenberg (1957, fig. 2A–C) later placed *M. floridana* in synonymy with *M. temminckii*.

DIAGNOSIS

The snout is constricted with no secondary triturating ridge on maxilla. The triturating ridges of maxilla are slightly concave medially. Skull has a width of 95–115% of the basilar skull length. The width of the rostrum at the posterior rim of the orbit is 47–57%* of the basilar skull length. The distance from the anterior margin of the jugal to the quadratojugal-postorbital junction is 28–30%* of the basilar skull length. The distance from the anterior rim of the orbit to the nasal notch is 16–18%* of the basilar skull length. The ratio of the distal width of the humerus to the length decreases with an increase in size. The phalanges are relatively longer than *M. auffenbergi*. The distal anterior-posterior length of costal 6 is shorter* than that of costal 7 (see Figs. 2.1D, 2.2G, and 2.3B, F).

DISTRIBUTION

See Ernst (Chapter 1) for the distribution of this species. The type of *M. floridana* is from the Pleistocene (Rancholabrean) of the Peace Creek Beds of Hillsborough County, Florida.

REMARKS

Auffenberg (1957) (fig. 2A–C) noted that the scalation characters of the peripherals used to distinguish *M. floridana* are variable within *M. temminckii*.

Macrochelys species indet.
DISTRIBUTION

Early Miocene Runningwater Formation (Hemingfordian, about 16 Ma) of Box Butte County, Nebraska. Late Miocene of the lower Ash Hollow Formation of Brown County (Clarendonian, about 10 Ma, unpublished) and Knox County (late Hemphillian, about 6 Ma, Parmley 1992), Nebraska, and Bennett County (late Clarendonian, about 9 Ma, Zangerl 1945), South Dakota.

REMARKS

Zangerl (1945, fig. 4) referred a specimen from the late Miocene (late Clarendonian) of the Ash Hollow Formation of Bennett County, South Dakota, to *M. temminckii*, on the basis of the right half of the snout including the orbital bones. Zangerl noted that it exhibited no significant differences from the extant species, at least in the anterior part of the skull. He did not describe or figure the nature of the triturating surface. Additional material or reexamination of this specimen in comparison with subsequently named taxa is needed to confirm this identification. An undescribed pygal (FAM 9827) from the late Clarendonian of the Merritt Dam Member of the Ash Hollow Formation, Cherry County, Nebraska, may also belong to the same species.

The late Hemphillian record of *M. temminckii* reported by Parmley (1992) from Ash Hollow Formation, Knox County, Nebraska, is based upon four posterior peripherals that were not figured, compared with other species of *Macrochelys*, or described other than to note that the peripherals were deeply notched and smaller than the maximum size for *M. temminckii*. Because *M. auffenbergi* and probably other species of *Macrochelys* have deeply notched posterior peripherals, the assignment of the Ash Hollow specimens to *M. temminckii* is not presently justified.

An undescribed skull (FAM 11330) from the Hemingfordian of the Runningwater Formation, Box Butte County, Nebraska, is referable to *Macrochelys*, possibly *M. schmidti*. An undescribed and virtually complete skull (FAM 11556) comes from the basal Ash Hollow Formation of Brown County, Nebraska.

Chelydrini unident
Remarks

Holman & Sullivan (1981, fig. 1) reported on the medial moiety of a hypoplastron from the Valentine Formation (Barstovian), Cherry County, Nebraska, that they regarded as the earliest record of *Chelydra*. The arcuate splay and coarseness of the "teeth" of the medial suture and gentle slope of the free margin suggest that the specimen is part of the hyoplastron. They noted that the specimen is "markedly more massive and rugose" than the extant species of *Chelydra* but made no comparisons to *Macrochelys*. The medial

Table 2.1

Distribution of the character states of the genera and higher taxa of the Chelydridae*

Taxa	1	2	3	4	5	6	7	8	9	10	11	12	13	14	15	16	17	18	19	20
Protochelydrinae	0	0	?	0	0	0	0	0	0	?	?	0	0	0	0	0	0	0/1	0	0
Protochelydra	0	0	0	0	0	0	0	0	0	?	?	0	0	0	0	0	0	0	0	0
Denverus	?	?	?	?	?	?	?	?	?	?	?	?	?	?	?	0	0	1	0	0
Chelydropsinae	1	1	1	1	1	0/1	0	0	0	0	0	0	0	0/1	0	0/1	0	0/1	0/1	0
Chelydrasia	1	1	1	1	1	1	0	0	0	0	0	0	0	1	0	0	0	1	0/1	0
Chelydropsis	1	1	1	1	1	1	0	0	0	0	0	0	0	1	0	1	0	1	0	1
"Chelydrasia"	?	?	?	?	?	?	?	?	?	?	?	0	0	1	0	1	?	1	0	0
Chelydrinae	0	0/1	0/1	0/1	0	1	0	1	1	1	1	1	1	0	1	1	0/1	1	1	0/1
Chelydra	0	0	0	0	0	1	0	1	1	1	1	1	1	0	1	1	1	1	1	0/1
Macrochelys	0	1	0	1	0	1	1	1	1	1	1	1	1	0	1	1	0/1	1	1	0

Taxa	21	22	23	24	25	26	27	28	29	30	31	32	33	34	35	36	37	38	39	40
Protochelydrinae	0	0	0	0	0	0	0	0	0/1	0/1	0	0	0	0	0	0	0	0	0	0
Protochelydra	0	0	0	0	0	0	0	0	0/1	0	0	0	0	0	0	0	0	0	0	0
Denverus	0	0	0	0	?	1	?	?	?	1	?	0	?	1	0	0	?	?	?	1
Chelydropsinae	0	0/1	0/1	0/1	0	0/1	1	0/1	0/1	0/1	0/1	0	0	0/1	0	0/1	0/1	0	0	0
Chelydrasia	0	0	0	0/1	0	0/1	1	0	0	0	0	0	0	0/1	0	1	0/1	0	0	0
Chelydropsis	0	1	1	1	0	1	1	1	0	1	1	1	0	0/1	0	1	1	0	0	0
"Chelydrasia"	0	0	1	?	0	?	1	1	1	?	0	0	0	0	?	1	1	0/1	0	0
Chelydrinae	0/1	1	1	1	1	1	1	1	1	1	1	1	1	1	0/1	0	1	0/1	0/1	0
Chelydra	0	1	1	1	1	1	1	1	1	1	1	1	1	1	1	0	1	1	1	0
Macrochelys	1	1	1	1	1	1	1	1	1	1	1	1	1	1	0/1	1	1	0	0	0

* 0 = primitive; 1 = derived state.

† 1. Jugal reduced.
2. Jugal about equally as long as quadratojugal.
3. Jugal excluded from orbit by postorbital and maxilla.
4. Moderate cheek emargination.
5. Temporal passage roofed over.
6. Pterygoid without posteriorly directed ridge.
7. Pterygoid "waist" narrow.
8. Triturating surfaces of the maxilla and dentary relatively narrow.
9. Maxilla and premaxilla with hook.
10. Coronoid progress of lower mandible small and not greatly elevated.
11. Foramen dentofaciális majus situated anteriorly and near level of coronoid process.
12. Epiplastron relatively narrow.
13. Gular scales near or contacting entoplastron.
14. Gular area truncate anteriorly.
15. Entoplastron anchor shaped.
16. Hyoplastron terminates on peripheral 4 or higher.
17. Hyoplastron terminates on P5.
18. Hypoplastron terminates on peripheral 7 or lower.
19. Lateral plastral fontanelles present.
20. Inframarginals less than three.
21. Abdominal scale transversely divided.
22. Bridge region of plastron relatively narrow.
23. Xiphiplastron relatively narrow and with relatively straight lateral margin.
24. Plastron attached to carapace by ligaments.
25. Carapace with cephalic notch in dorsal view.
26. Carapace tricarinate.

27. Posterior margin of carapace distinctly serrated.
28. Caudal notch wider than one-half of pygal width.
29. Suprapygal 1 narrower than suprapygal 2.
30. Free perimeter dimension of anterior peripherals longer than deep.
31. Serrations of posterior peripherals relatively strong.
32. Costiform process terminates medial to costal rib 1.
33. Suprapygal 1 quadratic.
34. Costal-peripheral fontanelles usually present.
35. Costal rib heads arise at level of vertebral–pleural sulcus.
36. Supramarginal scales present.
37. Vertebral 5 overlapping pygal.
38. Vertebral 1 narrower than Vertebrals 2–3.
39. Pleural scale 1 overlaps nuchal.
40. Medial surface of bridge peripherals inflated.

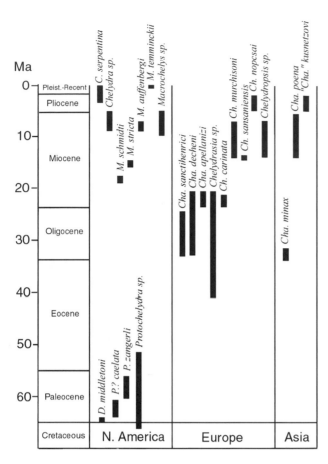

Fig. 2.4. Geographic and chronologic distribution of the taxa of fossil Chelydridae

moiety of the hypo- and hyoplastron of *Chelydra* and *Macrochelys* is strikingly similar but those of *Macrochelys* are more massive and rugose. The relative width of the fossil resembles *Chelydra* but the narrowness of the plastron in *Macrochelys* is a derived feature and is only known for *M. auffenbergi* and *M. temminckii*. The only other chelydrid from the Barstovian of central North America is *M. stricta*. Without additional information, it is premature to determine the genus.

SUMMARY

Of the six named genera of Chelydridae, *Protochelydra* (two species), *Denverus* (one species), *Macrochelys* (four species, including *Chelydrops* Matthew), and *Chelydra* (three species) are confined to the New World (North America excepting *Chelydra*) (Fig. 2.4). *Protochelydra* is the oldest of these and is known from the late Cretaceous (Lancian) to early Eocene (Wasatchian). Unnamed taxa in North America extend the chelydrid record into the early Late Cretaceous (Turonian). The fossil record indicates that the separation of the extant genera *Macrochelys* and *Chelydra* predates the latest Eocene (Duchesnean). *Chelydrasia* (six species) and *Chelydropsis* (three species) comprise the record in Eurasia starting in the Oligocene and unnamed taxa extend this record into the middle Eocene (Bartonian). *Chelydrasia* is essentially a wastebasket group for those Chelydropsinae lacking the derived features of *Chelydropsis*. *Chelydropsis sensu strictu* is probably derived from *Chelydrasia* in the Miocene. The known records support the origin of the Chelydridae in North America with spread to Eurasia in the late Paleocene or Eocene. *Macrochelys* and *Chelydra* are proposed as the sister group of *Chelydropsis* and *Chelydrasia* with *Protochelydra* and *Denverus* as the out-groups (Fig. 2.31 I, Table 2.1). Although *Denverus* exhibits some derived states present in later groups, these are variable in the Chelydropsinae and are regarded as especially informative at this level.

ACKNOWLEDGMENTS

I especially thank Eugene Gaffney, Jim Parham, Igor Danilov, and Pat Holroyd for looking up and providing literature on short notice. Bruce Erickson allowed access to unpublished material. I thank Eugene Gaffney for providing original photographs of *Ch. murchisoni* for comparison and Igor Danilov for providing translations of selected Russian papers. Eugene Gaffney and Jim Parham reviewed the manuscript and provided useful criticisms. University of California Museum of Paleontology (UCMP) contribution no. 1774.

3

An Introduction to the Skull of *Chelydra serpentina*, with a Review of the Morphology Literature of *Chelydra*

EUGENE S. GAFFNEY

*C*HELYDRA IS A TURTLE that is common in North America, and this is probably the main reason that it has been used for morphologic studies since Agassiz (1857). In contrast to chelonioids, which were of phylogenetic interest early in the twentieth century, *Chelydra* has usually been studied as a representative of all turtles or of terrestrial cryptodires. As a result of this, the work on *Chelydra* is very uneven. The head is relatively well known, but the postcranium has not been subjected to systematic description and illustration. Parts of the morphology are known, but much is not.

This chapter has two purposes: (1) to present a summary description of the *Chelydra* skull and (2) to review the literature describing any facet of *Chelydra* morphology. The skull of *Chelydra* has never been the subject of a description per se, so a brief one is presented here. The skull of *Chelydra* provides a useful introduction to turtle skull anatomy because it is a convenient size. *Chelydra* skulls are available in most universities and museums, and good literature exists for determining soft part anatomy and development. The cranial morphology of *Chelydra* is illustrated in Gaffney (1972), some of which is repeated here, and Gaffney (1979). Soft parts are illustrated in Rieppel (1990), Nick (1912), and Soliman (1964). Although the limbs of *Chelydra* are not well described in the literature, Gaffney (1990) illustrates the osteology of the limbs and girdles of its closest living relative, *Macroclemys* (=*Macrochelys*), as well as of *Podocnemis* and *Proganochelys*. The anatomical terms describing the skull are those adopted by Gaffney (1972, 1979), both of which contain glossaries of these terms. The phylogenetic context of *Chelydra* is emphasized in the descriptive section. For more information on the other taxa in turtle phylogeny, see Gaffney & Meylan (1988), Gaffney (1990) for *Proganochelys,* Gaffney et al. (1987) for

Kayentachelys, Gaffney (1983, 1996) for *Meiolania,* and Gaffney (1975, 1979) for *Platysternon.*

GUIDE TO THE CRANIAL OSTEOLOGY OF *CHELYDRA SERPENTINA*
Dermal Roofing Elements
Prefrontal

The prefrontal in *Chelydra* (see Figs. 3.1, 3.2, and 3.4) lies at the anterodorsal margin of the skull, forming the dorsal margin of the apertura narium externa (Fig. 3.1) and the roof of the fossa nasalis due to the absence of nasals. Nasal bones form much of these features in *Proganochelys* and cryptodires like *Kayentachelys.* As in most eucryptodires the prefrontal in *Chelydra* is a large element forming the anterodorsal margin of the fossa orbitalis (Fig. 3.1). The ventral surface of the prefrontal forms part of the sulcus olfactorius (Fig. 3.2), the groove for the olfactory (I) nerve. A ventromedial process forms part of the foramen orbitonasale (Figs. 3.2 and 3.3), an opening between the fossa nasalis and the fossa orbitalis. A ventral process characteristic of cryptodires and usually lacking in pleurodires contacts the vomer and palatine.

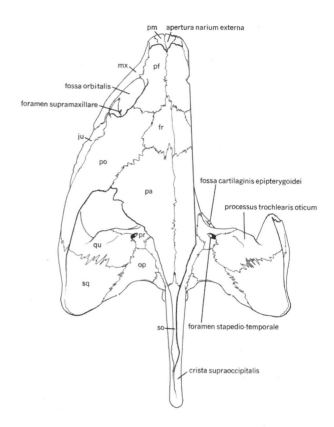

Fig. 3.1. *Chelydra serpentina,* AMNH 107385, dorsal view of skull with most of right side removed to show otic chamber. Adapted from Gaffney (1972).

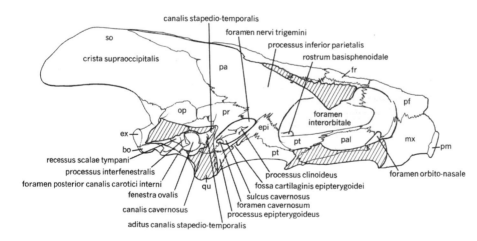

Fig. 3.2. *Chelydra serpentina,* AMNH 9249. (*upper*) Lateral view of skull with a parasagittal section on the right side removed to show structures in the cavum acusticojugulare. Adapted from Gaffney (1972). (*lower*) Sagittal section of skull. After Siebenrock (1897), adapted from Gaffney (1972).

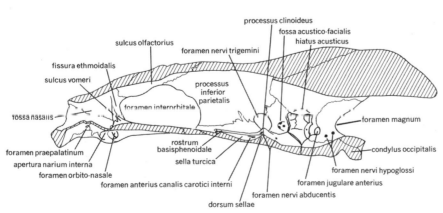

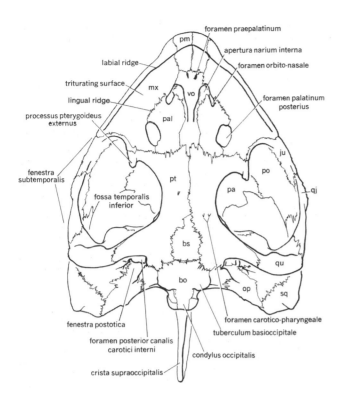

Fig. 3.3. *Chelydra serpentina*, AMNH 5305, ventral view of skull. Adapted from Gaffney (1972).

Frontal

The frontal in *Chelydra* is a small, flat element and is retracted from the orbital margin (Fig. 3.1) in contrast to the larger prefrontal that forms part of the orbit in most cryptodires. The ventral surface forms part of the sulcus olfactorius (Fig. 3.2) which contains cranial nerve I, the olfactorius.

Parietal

The parietal in *Chelydra* is a relatively large element made up of a horizontal plate and a parasagittal, vertical plate. The horizontal plate (Figs. 3.1 and 3.3) forms the posteromedial part of the skull roof and forms the anteromedial margin of the temporal emargination, which, in *Chelydra,* is more extensive than the primitive condition seen in *Proganochelys* and *Kayentachelys* but less extensive than seen in most Testudinoidea and Trionychoidea.

The vertical plate is the processus inferior parietalis (Fig. 3.2) and forms the side wall of the braincase. Anteriorly the processus is shallow where it grades into the sulcus olfactorius, but it deepens to form the posterior margin of the foramen interorbitale (Fig. 3.2; filled with cartilage in life, tra-

versed by the optic II, oculomotor III, trochlearis IV, and the opthalamic branch of the trigeminal V nerves). The processus inferior parietalis forms the dorsal margin of the foramen nervi trigemini (Fig. 3.2), which contains the mandibular and maxillary branches of the trigeminal (V) nerve. Between the foramen nervi trigemini and the foramen interorbitale (Fig. 3.2), the parietal contacts the epipterygoid bone, an element found in *Proganochelys* and cryptodires but lost in pleurodires.

Jugal

The jugal in *Chelydra,* as in other turtles, has a lateral cheek plate and a medial process (Figs. 3.1, 3.3–3.5). The lateral plate in *Chelydra* forms part of the posterior margin of the orbit and extends posteriorly to form the dorsal margin of the cheek emargination. The jugal in *Macrochelys* and *Platysternon* is much smaller, forming little or none of the orbital and cheek margins.

The medial jugal process is exposed in the posterior wall of the fossa orbitalis, the anterior wall of the fossa temporalis inferior (Fig. 3.3; containing much of the jaw musculature), and the posterior margin of the triturating surface (Fig. 3.3).

Quadratojugal

The quadratojugal in *Chelydra* (Figs. 3.3 and 3.4) and most turtles is a flat, C-shaped element forming the posterior part of the cheek and curving around the semicircular cavum tympani (Fig. 3.4) of the quadrate.

Squamosal

The squamosal in *Chelydra* (Figs. 3.1, 3.3–3.5), and nearly all other turtles, is a cone-shaped bone fitting around the antrum postoticum (Figs. 3.4 and 3.5) of the quadrate. The squamosal projects posterodorsally and its medial surface bears jaw muscle attachments and its posteroventral surface bears the depressor mandibulae attachment (Schumacher 1973).

Postorbital

The postorbital in *Chelydra* (Figs. 3.1, 3.3, and 3.4) is a large, flat bone covering the fossa temporalis and forming the posterior orbital margin. In most Testudinoidea and Trionychoidea the postorbital is much smaller because of the larger temporal emargination in these groups. In *Macrochelys* and *Platysternon* the postorbital is larger than in *Chelydra. Kayentachelys,* plesiochelyids, and chelonioids also have a large postorbital, the presumed primitive cryptodiran condition.

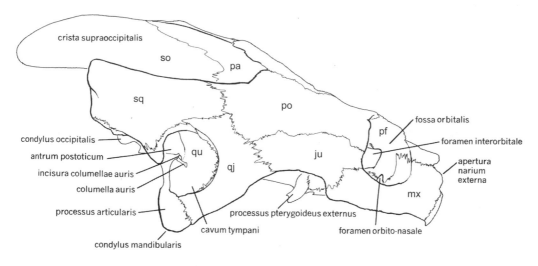

Fig. 3.4. *Chelydra serpentina*, lateral view of skull. Adapted from Gaffney (1972).

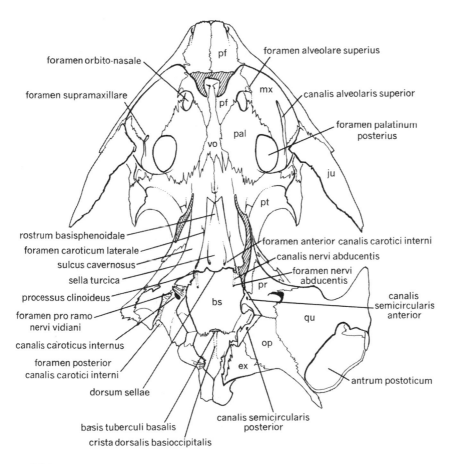

Fig. 3.5. *Chelydra serpentina*, AMNH 107386, dorsal view of skull with skull roof removed, hatching indicates cut surfaces, right squamosal, left quadrate, prootic, opisthotic, and squamosal removed. Adapted from Gaffney (1972).

Palatal Elements
Premaxilla

The premaxilla in *Chelydra* (Figs. 3.1–3.3), and other turtles lies at the anterior margin of the skull and bears part of the horny beak. The dorsal surface of the premaxilla forms the floor of the fossa nasalis (Fig. 3.2) and the ventral margin of the apertura narium externa (Fig. 3.1). The ventral surface of the premaxilla forms the acute labial ridge (Fig. 3.3) anteriorly and a flat or slightly concave triturating surface posteriorly. *Chelydra* has a slight hook on the labial ridge, which is much more pronounced in *Macrochelys* and *Platysternon* but absent in many turtles.

Maxilla

The maxilla (Figs. 3.1–3.4) lies at the anterolateral edge of the skull and consists of a vertical plate and a horizontal plate. In *Chelydra* and other turtles the vertical plate has a dorsal process forming the anterior orbital margin and the lateral wall of the fossa nasalis. The ventral margin of the vertical plate is the labial ridge, more shallow in *Chelydra* than in *Macrochelys*.

The horizontal part of the maxilla forms most of the triturating surface (Fig. 3.3), the nutrient-rich portion of the skull that bears the horny rhamphotheca or beak. The triturating surface varies a great deal in turtles but the surface in *Chelydra* is relatively generalized and not very different from the primitive condition in forms such as *Proganochelys* and *Kayentachelys*. The medial margin of the triturating surface is a variably developed ridge, the lingual ridge (Fig. 3.3), which in *Chelydra* is low and indistinct, but it is deeper and heavier in *Macrochelys*.

Medially the horizontal plate of the maxilla forms the lateral margins of the apertura narium interna (Fig. 3.3), the foramen orbito-nasale (Figs. 3.2 and 3.3), and the foramen palatinum posterius (Fig. 3.3). The apertura narium interna is the palatal opening of the nasal choanae, the tube or tubes that begin in the fossa nasalis and extend posteriorly. In *Chelydra* the apertura is relatively short in contrast to turtles like chelonioids, which have secondary palates and a more posterior apertura narium interna. The foramen orbito-nasale is an opening between the fossa nasalis and fossa orbitalis and is usually formed by prefrontal, palatine, and maxilla.

The foramen palatinum posterius (Fig. 3.3) is an opening between the palatal surface and the fossa orbitalis in cryptodires and between the palate and the sulcus palatinopterygoideus in pleurodires.

Vomer

The vomer (Figs. 3.2, 3.3, and 3.5) is an unpaired, elongate element lying in the midline of the palate, between the paired apertura narium interna (Fig. 3.3). The vomer in most turtles is roughly dumbbell shaped with the anterior and posterior portions expanded and a narrow connecting bar separating the choanal passages. The vomer is exposed on both the dorsal and ventral surfaces of the palate.

Palatine

The palatine (Figs. 3.2, 3.3, and 3.5) is a flat bone in the anterior portion of the palate, forming much of the palatal roof and part of the floor of the fossa orbitalis. Laterally it forms part of the foramen palatinum posterius (Fig. 3.3) and foramen orbito-nasale (Fig. 3.3). Dorsally it forms the roof of the apertura narium interna. In *Chelydra* the palatine is flat, but in *Macrochelys* it is more arched, forming a deeper trough for the choanal passage.

Palatoquadrate Elements
Quadrate

The quadrate (Figs. 3.1–3.6) is a complex element involved with several organ systems in the turtle skull and containing features diagnostic for turtles in general and cryptodires in particular.

In lateral view the quadrate forms the roughly kidney-shaped cavum tympani (Fig. 3.4), a concave surface with the tympanic membrane attached around its margin (Baird 1970). In living turtles the quadrate forms a wall between the more lateral cavum tympani and more medial cavum acustico-jugulare (Fig. 3.6). In *Proganochelys* the quadrate is C-shaped with a shallow, barely formed cavum tympani, in contrast to the more extensive fully formed cavum tympani of *Chelydra* and other Casichelydia.

The cavum tympani (Fig. 3.4) has two important elements: the incisura columellae auris (Fig. 3.4) and the antrum postoticum (Figs. 3.4 and 3.5), both clearly developed in *Chelydra*. The incisura columellae auris is the canal or groove in the quadrate traversed by the stapes or columella auris (Fig. 3.4). It is variable in turtles, widely open in *Proganochelys* but completely closed into a canal in *Chelydra*. The incisura may or may not also contain the eustachian tube; it is separated from the incisura by bone in *Chelydra*. The antrum postoticum is a cone-shaped extension of the cavum tympani, extending posterodorsally and formed by quadrate and squamosal.

Ventral to the cavum tympani is a block of bone, the processus articularis (Fig. 3.6), that bears the articulation with the lower jaw, the condylus mandibularis (Figs. 3.4 and 3.6).

The quadrate in turtles has a series of medial processes that strongly connect the quadrate with the braincase. The quadrate forms the lateral half and the prootic forms the medial half of the foramen stapedio-temporale (Figs. 3.1 and 3.7), the canalis stapedio-temporalis (Fig. 3.2), and the aditus canalis stapedio-temporalis (Fig. 3.2), structures that contain the stapedial artery as it passes from the cavum acustico-jugulare, over the stapes, and into the fossa temporalis su-

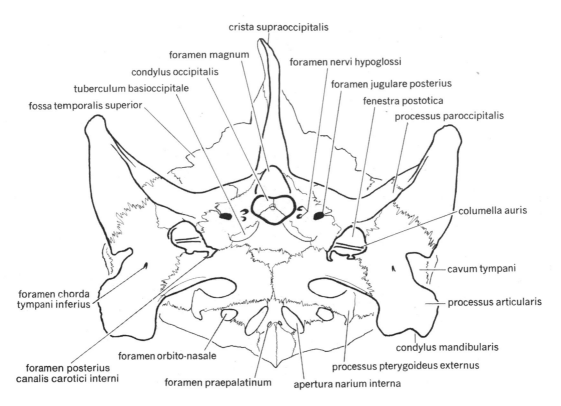

Fig. 3.6. *Chelydra serpentina*, AMNH 67015, posterior and slightly ventral view of skull. Adapted from Gaffney (1972).

perior (Albrecht 1976). This canal and associated foramina are very stable in Casichelydia but differ from *Proganochelys*, in which the quadrate-prootic contact is much less extensive with no definitive canal formed.

The quadrate and prootic together form the definitive derived character of cryptodires, the processus trochlearis oticum (Fig. 3.1). Cryptodires and pleurodires both have a trochlear process that reorients the main adductor jaw muscle tendon to change its line of action (Schumacher 1973; Gaffney 1975, 1979; Gaffney & Meylan 1988; Rieppel 1990). This process is developed on the anterodorsal edge of the prootic and quadrate, and it bears a synovial capsule and cartilage for the articulation of the adductor tendon or boden aponeurosis.

The quadrate-braincase (i.e., prootic and opisthotic) connection in turtles bridges the cranio-quadrate space (Goodrich 1930; Gaffney 1975, 1979). This space contains the stapedial artery, the lateral head vein, and the hyomandibular branch of the facial nerve. Pleurodires and cryptodires differ in the connections across the cranioquadrate space. *Chelydra* clearly has the cryptodiran condition with the soft structures contained together rather than separated by bone as in pleurodires. In ventral view the quadrate in *Chelydra* shows the broad articulation with the pterygoid, diagnostic of the Selmacryptodira and absent in *Proganochelys*, pleurodires, and *Kayentachelys*. *Chelydra* and other cryptodires have no medial process of the quadrate, as found in pleurodires.

In posterior view the quadrate forms the lateral margin

of the fenestra postotica (Figs. 3.3 and 3.6), the posterior opening of the cavum acustico jugulare. This area is variably ossified in turtles and represents the remnants of the cranio-quadrate space that contains various soft structures.

Pterygoid

The pterygoid of turtles is a complex element, which in *Chelydra* (Figs. 3.3 and 3.7) shows the eucryptodiran condition of the carotid circulation, the cryptodiran bracing of the palato-quadrate, and the casichelydian akinesis.

In ventral view (Fig. 3.3) the pterygoid in *Chelydra* is a simple flat element that shows the basic morphology of amniotes with a quadrate process and transverse flange (processus pterygoideus externus; Fig. 3.3). The open interpterygoid vacuity and movable basisptergoid articulation of the pterygoid in *Proganochelys* and primitive amniotes is altered in *Chelydra* and other Casichelydia to an akinetic condition with a completely closed interpterygoid vacuity and suturing of the pterygoid and braincase.

The dorsal surface of the pterygoid (Figs. 3.5 and 3.7) in *Chelydra* shows the characteristic features of the eucryptodiran pterygoid and is a good introduction to many of the basicranial features in turtles. The dorsal surface of the pterygoid bears a roughly parasagittal, vertical plate, the crista pterygoidea, that articulates dorsally with the processus inferior parietalis (Fig. 3.2) and (in cryptodires) the epipterygoid, to form the lateral wall of the braincase. This wall separates the medial cavum cranii from the lateral fossa

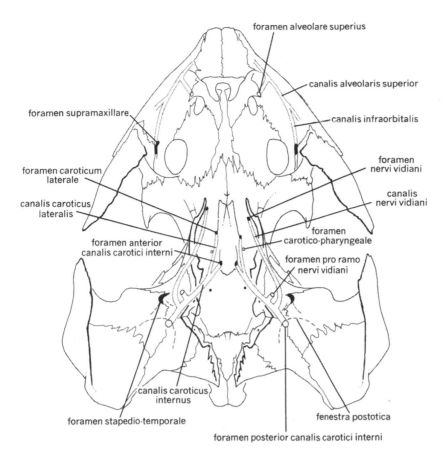

Fig. 3.7. *Chelydra serpentina,* composite dorsal view of skull showing internal canals (dotted). Foramina visible on dorsal surfaces shown solid; foramina hidden in dorsal view are open circles. Diameter of some canals exaggerated for clarity. See also Albrecht (1976). Adapted from Gaffney (1972).

temporalis inferior (Fig. 3.3). The area medial to the crista pterygoidea in turtles encloses a cavum epiptericum, a space anatomically between the primary braincase and the palatoquadrate (see Goodrich 1930 and Gaffney 1979, p. 106, for more on the cavum epiptericum).

Structures medial to the crista pterygoidea (Fig. 3.5) are mostly related to the arterial and venous circulation and the cranial nerves. Just medial to the crista pterygoidea is a trough, the sulcus cavernosus (Fig. 3.5), which represents the anterior extension of the cranioquadrate space and contains the lateral head vein. The posterior part of the sulcus cavernosus is enclosed dorsally by the prootic and is called the canalis cavernosus. The anterior opening of the canalis cavernosus, where it becomes the sulcus cavernosus, is the foramen cavernosum (Fig. 3.2). The trigeminal nerve (V) ganglion overlies the sulcus cavernosus just anterior to the foramen cavernosum. The maxillary (V2) and mandibular (V3) branches exit the cavum cranii via the foramen nervi trigemini (Fig. 3.2), a notch in the crista pterygoidea roofed by the prootic and parietal in *Chelydra*. In most cryptodires the mandibular artery also exits via the foramen nervi trigemini but in pleurodires it does not. In the medial wall of the canalis cavernosus is the foramen pro ramo nervi vidiani (Fig. 3.7), a short canal formed by prootic and pterygoid,

that connects the canalis caroticus internus (Fig. 3.7) and the canalis cavernosus. It transmits the vidian nerve (the palatine branch of VII) and a small branch of the internal carotid artery (Albrecht 1967, 1976). The canalis nervi vidiani (Fig. 3.7) is a small canal usually formed within the pterygoid that contains the vidian nerve and extends anteriorly to open near the anterior margin of the pterygoid.

The pterygoid bone (Fig. 3.7) is a key indicator of the arterial circulation in the skull (Albrecht 1967, 1976). The internal carotid artery enters the skull at the posterior edge of the pterygoid at the foramen posterius canalis carotici interni (Fig. 3.7). The more primitive cryptodiran condition, as seen in *Kayentachelys* and Paracryptodira (i.e., baenids) is a more anterior position (Gaffney 1975, 1979, 1983, 1996; Gaffney & Meylan 1988). The internal carotid artery extends anteromedially via the canalis caroticus internus (Fig. 3.7) to enter the basisphenoid midway along the length of the pterygoid. Just as the canalis caroticus internus enters the basisphenoid it gives off an anterior branch, the canalis caroticus lateralis (Fig. 3.7), usually formed between the pterygoid and the basisphenoid. The canalis caroticus lateralis contains the palatine artery and opens anteriorly at the foramen caroticum laterale (Fig. 3.7), also formed between the basisphenoid and pterygoid in *Chelydra*.

Epipterygoid

The epipterygoid (Fig. 3.2) is a small, platelike bone that lies in the side wall of the braincase between the parietal and pterygoid. The bone does not ossify in pleurodires and is fused to the pterygoid in most baenids.

Braincase Elements
Supraoccipital

The supraoccipital (Figs. 3.1–3.4, 3.6) is an unpaired median element lying in the posterodorsal region of the skull. It forms the dorsal margin of the foramen magnum (Fig. 3.6), the dorsal portion of the cavum labyrinthicum, and the crista supraoccipitalis (Figs. 3.1–3.4, 3.6). The crista supraoccipitalis is a vertically oriented plate extending posteriorly to serve as the attachment site for the M. adductor mandibulae externus pars profundus (Schumacher 1973).

The supraoccipital has paired ventrolaterally projecting processes that form the roof of the inner ear, the cavum labyrinthicum, and contain the recessus labyrinthicus supraoccipitalis. The recessus labyrinthicus supraoccipitalis contains the membranous common crus. The indentations or canals formed by the dorsal portions of the anterior and posterior semicircular ducts (Baird 1960, 1970) are the canalis semicircularis anterior and canalis semicircularis posterior, respectively. A notch or foramen in the medial wall of the supraoccipital portion of the cavum labyrinthicum is the foramen aquaducti vestibuli, which carries the endolymphatic duct from the otic sac in the cavum cranii to the sacculus in the cavum labyrinthicum (Baird 1960, 1970). The hiatus acusticus (Fig. 3.2) is usually limited dorsally by the supraoccipital.

Exoccipital

The exoccipital in *Chelydra* (Figs. 3.2, 3.3, 3.5, and 3.6) is an irregularly shaped, paired bone lying lateral to the foramen magnum (Fig. 3.6), forming part of the condylus occipitalis (Fig. 3.6), and part of the posterior wall of the cavum acustico-jugulare. The exoccipital forms the posterior wall to the recessus scalae tympani which contains the periotic sac (Baird 1960), the vagus (X) nerve, and the vena cerebralis posterior. As exposed in the side wall of the cavum cranii the exoccipital forms the posterior margin of the foramen jugulare anterius (Fig. 3.2) and all of both foramina nervi hypoglossi (Fig. 3.2). The exoccipital also forms the exit foramina for these, the vena cerebralis posterior (vena jugularis in mammals) leaving via the foramen jugulare posterius (Fig. 3.6). The hypoglossal nerves (XII) exit via the foramina nervi hypoglossi (Fig. 3.6). The vagus (X) and accessory (XI) nerves follow the vena cerebralis posterior (Soliman 1964; Nick 1912).

Basioccipital

The basioccipital in *Chelydra* (Figs. 3.2, 3.3, and 3.6) is a roughly triangular, unpaired bone lying at the posteroventral edge of the skull, below the foramen magnum (Fig. 3.6). It forms the ventral third of the condylus occipitalis (Fig. 3.6), lying between the exoccipitals. Posteroventrally it forms the small tuberculum basioccipitale (Fig. 3.6). On its dorsal surface it is exposed in the floor of the cavum cranii (Fig. 3.2) just anterior to the foramen magnum.

Prootic

The prootic (Figs. 3.1, 3.2, 3.5, and 3.7) is an irregular, blocklike bone, forming the anterior ossification of the inner ear and intimately connected to palatoquadrate and basicranial elements. The prootic forms the anterior part of the otic chamber, and with the quadrate, forms the processus trochlearis oticum (Fig. 3.1) and foramen stapedio-temporale (Fig. 3.1). Anteromedially, the prootic forms the posterior margin of the foramen nervi trigemini (Fig. 3.2).

As exposed in the side wall of the cavum cranii, the prootic forms the anterior margin of the hiatus acusticus (Fig. 3.2) and the posterior margin of the foramen nervi trigemini (Fig. 3.2). In the center of the prootic is a depression, the fossa acustico-facialis (Fig. 3.2), which contains the ganglion vestibulare (Soliman 1964). Three openings are in the fossa: two foramina nervi acustici (branches of VIII) and the foramen nervi facialis (VII). The foramina nervi acustici penetrate into the cavum labyrinthicum and the foramen nervi facialis opens into the canalis cavernosus.

The prootic forms the anterior portion of the inner ear, called the cavum labyrinthicum, and contains the recessus labyrinthicus prooticus that houses the ampullae of the anterior and horizontal semicircular canals. The lateral surface of the prootic forms the anterior margin of the fenestra ovalis (Fig. 3.2) which holds the footplate of the stapes. The three bones that ossify around the inner ear, the prootic, opisthotic, and supraoccipital, and the fenestra ovalis, are very conservative in amniotes and differ very little among turtles.

Opisthotic

The opisthotic forms the posterior part of the inner ear, the roof of the cavum acustico-jugulare, and structures of the perilymphatic system. Anteriorly it forms the posterior half of the cavum labyrinthicum, the recessus labyrinthicus opisthoticus. A ventral process, the processus interfenestralis (Fig. 3.2), forms the posterior margin of the fenestra ovalis (Fig. 3.2). The posterior surface of the processus interfenestralis forms the anterior wall for the recessus scalae tympani, which contains part of the perilymphatic system. More medially the processus interfenestralis forms the fenestra perilymphatica which contains the periotic sac of the inner ear (Baird 1960). The medial surface of the opisthotic forms the posterior edge of the hiatus acusticus (Fig. 3.2) and the anterior edge of the foramen jugulare anterius (Fig. 3.2).

The glossopharyngeal nerve (IX) runs through a series of foramina in the opisthotic at the base of the processus interfenestralis. Posterolateral to the processus interfenestralis, the opisthotic has a large extension, the processus paroccipitalis (Fig. 3.6), that lies behind the quadrate and, along with that bone, forms the roof of the cavum acustico-jugulare. The opening entering the cavum acustico-jugulare is the fenestra postotica (Fig. 3.6) through which the stapedial artery (arteria stapedialis), the lateral head vein (vena capitis lateralis), the vagus nerve (X), and the hyomandibular branch of the facial nerve (VII) enter the skull. *Chelydra* has the condition common in eucryptodires, but in some turtles the fenestra postotica may be subdivided or it may be larger and communicate with the foramen jugulare posterius (Fig. 3.6).

Basisphenoid

The basisphenoid (Figs. 3.2, 3.3, 3.6, and 3.7) is a triangular, median bone, lying in the center of the basicranium. Its ventral surface is smooth in *Chelydra*, but its dorsal surface shows several structures. Anteriorly the rostrum basisphenoidale (Figs. 3.2 and 3.5) is formed by the ossified, anteriorly converging trabeculae. Between the bases of the trabeculae is the concave depression for the pituitary, the sella turcica (Figs. 3.2 and 3.5). At the posterolateral corners of the sella turcica is the paired foramen anterius canalis carotici interni (Figs. 3.2, 3.5, and 3.7), the entry of the internal carotid artery into the cavum cranii.

The dorsum sellae (Figs. 3.2, 3.5, and 3.7) is a raised, slightly overhanging area forming the posterior margin of the sella turcica. The processus clinoideus (Figs. 3.2 and 3.5) extends anterodorsally from each side of the dorsum sellae. The abducens nerve (VI) penetrates the base of each processus clinoideus through the canalis nervi abducentis (Fig. 3.5). The posterior half of the basisphenoid is a shallow depression that meets the hiatus acusticus laterally (Fig. 3.2).

The Lower Jaw
Dentary

The dentary (Figs. 3.8 and 3.9) is the most anterior and most prominent bone in the lower jaw, and meets the other jaw bones posteriorly in a complex series of squamous and interleaving sutures. The dentary bears nearly all the triturating surface, the area covered in life by the horny rhamphotheca. As in the maxilla, the triturating surface of the dentary (Fig. 3.8) has a lateral labial ridge and medial lingual ridge, both very low in *Chelydra*. Accessory ridges or pits seen in some other turtles are absent and the triturating surface of *Chelydra* is flat, curving anteromedially to a small symphyseal projection. In *Macroclemys* the symphyseal projection or hook is much larger.

The medial surface of the dentary has a longitudinal groove, the sulcus cartilaginis meckelii (Fig. 3.9), which contains the remnant of Meckel's cartilage (Kunkel 1912). Posteriorly, the sulcus is continuous with the fossa meckelii (Fig. 3.9). The foramen alveolare inferius (Fig. 3.9) also occurs on the medial surface of the dentary usually near the foramen intermandibularis medius (Fig. 3.9). The foramen alveolare inferius leads into the canalis alveolaris inferior (Fig. 3.9), the largest internal canal in the dentary, transmitting nerves and

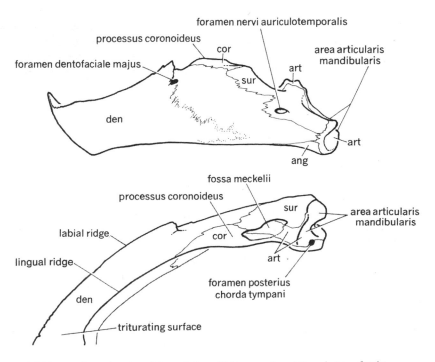

Fig. 3.8. *Chelydra serpentina*, AMNH 67015, lower jaw. (*upper*) Lateral view of left ramus. (*lower*) Dorsal view of right ramus. Adapted from Gaffney (1972).

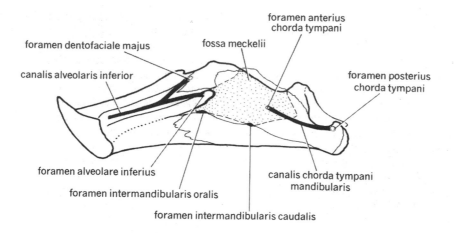

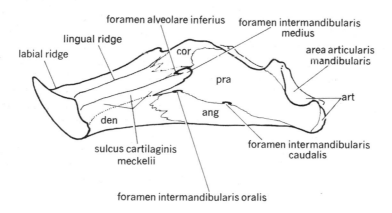

Fig. 3.9. *Chelydra serpentina,* AMNH 67015, lower jaw. (*upper*) Medial view of right ramus showing internal structures. (*lower*) Medial view of right ramus. The fossa meckelii is stippled, internal canals are solid, hidden foramina are open circles, and visible foramina are solid. Adapted from Gaffney (1972).

blood vessels (Soliman 1964; Gaffney 1979). The foramen dentofaciale majus (Figs. 3.8 and 3.9) is an opening on the lateral surface of the dentary, usually found near the posterior limits of the rhamphotheca, that also leads anteriorly into the canalis alveolaris inferior (Fig. 3.9).

Angular

The angular (Figs. 3.8 and 3.9) is a long and slender sheet of bone that curves around the posterior half of the mandible from the anteromedial surface to the posterolateral surface and forms much of the posteroventral part of the jaw. The angular forms the ventral margins of two foramina, the more anterior foramen intermandibularis oralis (Fig. 3.9) and the more posterior foramen intermandibularis caudalis (Fig. 3.9), that penetrate into the fossa meckelii from the medial surface of the jaw (see Prearticular).

Surangular

The surangular (Fig. 3.8) is a roughly platelike bone that forms most of the lateral wall of the fossa meckelii above the dentary. The fossa meckelii (Fig. 3.9) is the space behind the processus coronoideus (Fig. 3.8) and anterior to the jaw

articulation. It is open dorsally and communicates anteriorly with the sulcus cartilaginis meckelii (Fig. 3.9). The upper opening is usually formed by the coronoid, articular, and prearticular, as well as the surangular. The fossa meckelii contains nerves and blood vessels entering the lower jaw from the skull (Soliman 1964; Albrecht 1967, 1976; Gaffney 1979) and muscle fibers of the M. pseudotemporalis and M. adductor mandibulae (Schumacher 1973).

The foramen nervi auriculotemporalis (Fig. 3.8) penetrates the surangular from the lateral jaw surface into the fossa meckelii, containing a branch of the mandibular (V3) nerve (Soliman 1964).

Coronoid

The coronoid (Figs. 3.8 and 3.9) is a thick, roughly triangular bone lying midway along the jaw ramus at the highest point of the jaw in lateral view. Its apex is the processus coronoideus (Fig. 3.8), not as large in *Chelydra* as in forms like *Platysternon* (Gaffney 1975, 1979). Fibers of the M. adductor mandibulae externus, and more importantly, the main external adductor tendon or bodenaponeurosis, attach on the processus coronoideus (Schumacher 1973; Gaffney 1979; Rieppel 1990). Posteriorly the coronoid borders the fossa meckelii, and anteriorly it reaches the posterior border of the triturating surface.

Articular

The articular (Figs. 3.8 and 3.9) is an irregular blocklike bone lying at the posterior end of the jaw. It is the only lower jaw element that ossifies from cartilage; it is the posterior end of Meckel's cartilage (Kunkel 1912). The dorsal portion of the articular bears most of the area articularis mandibularis (Figs. 3.8 and 3.9), the surface that articulates with the condylus mandibularis of the quadrate. The posteroventral surface of the articular in *Chelydra* lacks a retroarticular process, but it has a concavity for the attachment of the M. depressor mandibulae (Schumacher 1973; Rieppel 1990).

The articular forms the foramen posterius chorda tympani (Fig. 3.8; see Prearticular).

Prearticular

The prearticular (Fig. 3.9) is a large, flat sheet covering most of the posteromedial surface of the lower jaw in *Chelydra*. It also forms most of the medial wall of the fossa meckelii (Fig. 3.9). The prearticular forms the posterior edge of the foramen intermandibularis medius (Fig. 3.9), the large anterior opening of the fossa meckelii containing a branch of the mandibular (V3) nerve as well as the meckelian cartilage and blood vessels (Soliman 1964). In the prearticular-angular suture are the two foramina that contain further subdivisions of the mandibular nerve, the foramen intermandibularis oralis (Fig. 3.9) and the foramen intermandibularis caudalis (Fig. 3.9; Soliman 1964; Gaffney 1979).

The chorda tympani nerve, a branch of the facial (VII) nerve, enters the lower jaw via the foramen posterius chorda tympani just medial to the area articularis mandibularis in the articular (Fig. 3.8). The foramen posterius chorda tympani opens into the canalis chorda tympani (Fig. 3.9), found in the suture between the prearticular and articular. The nerve continues anteriorly to enter the fossa meckelii.

A GUIDE TO THE MORPHOLOGY LITERATURE OF *CHELYDRA SERPENTINA*
Cranial Osteology

The skull of *Chelydra* has never been the subject of a description per se, but it is nonetheless relatively well known from figures in Gaffney (1972, 1979), Nick (1912), Soliman (1964), and Rieppel (1990). Nick (1912) and Soliman (1964) contain a series of thin sections of juvenile *Chelydra* heads and show soft structures as well as bones. Gaffney (1972) illustrates adult skulls primarily for the identification of structures and some of these figures are reproduced here. Gaffney (1979) is a summary description of all turtle skulls and con-

tains numerous comparisons and observations of *Chelydra* as well as figures and references to other aspects of turtle cranial morphology.

The external features of *Chelydra* skulls have been figured by Sobolik & Steele (1996, fig. 4; dorsal view only), Gray (1856b), Boulenger (1889, fig. 3, repeated in Wermuth & Mertens 1961), Ashley (1955, figs. 6, 7, 10, 12), Ernst & Barbour (1972, p. 21), Gaffney (1975, figs. 1–5; 1979, figs. 217, 218, 224), and Obst (1986, p. 136).

Internal structures of the skull in *Chelydra* are most extensively described in Gaffney (1979, fig. 10–quadrate, fig. 11–ethmoid region, fig. 16–ear section, fig. 18–basicranium, fig. 20–pterygoid, fig. 27–pterygoid, fig. 33–canals and foramina, fig. 37–braincase, fig. 39–braincase, fig. 42–braincase, fig. 49–opisthotic and exoccipital, fig. 50–ear region, fig. 51–otic chamber, fig. 64–frontal section, fig. 76–median section, fig. 81–median section, fig. 87–ear region, figs. 111–112–lower jaw). Other sources are Albrecht (1976), who describes cranial canals, and Siebenrock (1897).

Postcranial Osteology

Despite the prominence of shell morphology in turtle systematics, no detailed description of the *Chelydra* shell exists in the literature. The most used shell figure is probably that in Boulenger (1889, fig. 4). Other useful figures include Sobolik & Steele (1996, figs. 60–63, disarticulated shells), Mlynarski (1976, fig. 66–1, a disarticulated nuchal bone with lateral processes), Zangerl (1953, fig. 63, ventral view of first and second thoracic vertebrae), Williams & McDowell (1952, figs. 5, 6, dorsal view of anterior plastral lobe), Richmond (1964, fig. 1, dorsal view of plastron), and Medem (1977, figs. 5–7, carapace; figs. 2, 8, plastron).

The vertebral column of *Chelydra* is represented by good cervical figures in Williams (1950, fig. 8), redrawn in Hoffstetter & Gasc (1969, fig. 8). Although there is no paper specifically devoted to *Chelydra* vertebrae, Hoffstetter & Gasc (1969) is a good review for turtles with Gaffney (1996) providing discussions of vertebral characters including chelydrids.

Caudal vertebrae of *Chelydra* have only been figured in Newman (1906, fig. 58) in lateral view. Descriptions are in Hoffstetter & Gasc (1969) and Gaffney (1996).

The shoulder girdle of *Chelydra* is figured in Sobolik & Steele (1996, figs. 9, 10), Gaunt & Gans (1969, plate 2) with a general discussion in Walker (1973). Gaunt & Gans (1969) show the positions of the girdles in the shell. The pelvis of *Chelydra* is figured in Sobolik & Steele (1996, figs. 14–16), Ruckes (1929, fig. 4), and Zug (1971, figs. 9, 10). Zug (1971) describes and compares the *Chelydra* pelvis with other cryptodires. Walker (1973, fig. 19) also figures a *Chelydra* pelvis in the context of a more general comparison among turtle pelvic girdles.

The limbs (excluding manus and pes) of *Chelydra* are figured in Sobolik & Steele (1996, figs. 6–13). The carpus and tarsus of *Chelydra* are figured and described in Burke & Alberch (1985, fig. 2) in the context of development and homology among tetrapods as well as other turtles. Rabl (1910, plate 5, fig. 6; plate 3, fig. 9) also figures the carpus and tarsus of *Chelydra* in a more specific context of comparisons with other turtles. The femoral head of *Chelydra* is figured in Zug (1971, fig. 11). All the limb and girdle bones of *Macroclemys* are figured in Gaffney (1990).

The hyoid skeleton of *Chelydra* is figured in Hildebrand et al. (1985, figs. 12–10, 13–1), but see also Spindel et al. (1987, fig. 2) for the hyoid of *Macroclemys*.

Muscular System

The jaw musculature of *Chelydra* is the best known among turtles thanks to the work of Rieppel (1990, figs. 1–11) who described its development and adult morphology. In this paper Rieppel rejected the idea that turtle jaw muscles were consistent with a diapsid origin, as he later (deBraga & Rieppel 1997; Rieppel & Reisz 1999) argued that turtles were related to diapsid placodonts. Schumacher (1973) is a general review of all turtle jaw musculature including references to his earlier works. Nick (1912) and Soliman (1964) figure *Chelydra* thin sections showing the jaw muscles.

The postcranial musculature of *Chelydra* was somewhat diagrammatically described for the forelimb and hindlimb by Sieglbauer (1909, plate 8, fig. 10; plate 9, figs. 21, 22; plate 10, figs. 25–28). Walker (1973), however, reviewed the limb musculature of all turtles including chelydrids.

Gaunt & Gans (1969), in their study of respiration in *Chelydra,* figured and described many of the muscles that attach on the internal surface of the carapace and plastron, and showed the positions of the limb girdles. Some elements of the neck and hyoid musculature of *Chelydra* are indicated in Hildebrand et al. (1985, fig. 12–10). The jaw and hyoid musculature of *Macroclemys* is described in Schumacher (1973; see earlier references), and the hyoid and its musculature are described for *Macroclemys* in Spindel et al. (1987). Gaunt & Gans (1969, fig. 1) described the glottis, hyoid, and associated muscles in *Chelydra*. The axial musculature of turtles in general was reviewed by Gasc (1981). Bojanus (1819), Ashley (1955), and Thomson (1932) should be seen for detailed descriptions in other turtles.

Nervous System

Along with chelonioids, the cranial nerves of *Chelydra* are the best known among turtles thanks to the work of Soliman (1964). Eleven thin sections document the thorough description of the cranial nerves (ibid., fig. 3) in Soliman's paper. Sections in Nick (1912) and Rieppel (1990) also show the cranial nerves. Hanson (1919, plates 1, 2) described the anterior cranial nerves of *Chelydra*. The brain of *Chelydra* was first described by Agassiz (1857, plates 23, 25, labels on pages

576, 577), and later by Humphrey (1894). The only studies of which I am aware on the postcranial nerves in *Chelydra* are in Sieglbauer (1909, figs. 2, 4, plates 7, 9) who described the brachial and pelvic plexi, and some of the limb nerves in *Chelydra* as well as in other turtles.

Other Organ Systems

The gastrointestinal and other organ systems of *Chelydra* have never been described in detail. Agassiz (1857, plate 25, figs. 3, 3a) described and figured the dissection of an entire *Chelydra* hatchling as well as other organs in embryonic and hatchling *Chelydra*. For information on the nonskeletal, nonmuscular organ systems, it is necessary to go to works describing other turtles, Bojanus (1819), Thomson (1932), and Ashley (1955), as there is nothing in the literature specifically on *Chelydra* that I am aware of.

The penis of *Chelydra* is described and figured by Zug (1966, fig. 1), who used it as the generalized penis including structural features seen in other turtles.

Development

Beginning with Agassiz (1857), *Chelydra* has been the subject of many developmental studies. Agassiz (1857), despite its publication more than 100 years ago, is still useful for the later stages of development. Descriptions and figures of circulation and various organs in the embryo and hatchling still have not been repeated.

Yntema (1964, 1966, 1968, 1970a, 1978) established criteria for stage recognition in *Chelydra* embryos and demonstrated its use in determining processes of shell and limb development. Meier & Packard (1984) used *Chelydra* as an example of a "primitive amniote" (ibid., p. 310) in their studies of neural crest and its relationship to cranial segmentation. Packard & Packard (1989) also used *Chelydra* embryos for biochemical studies.

The embryogenesis of the head in *Chelydra* is not covered in its entirety in any paper, but important aspects have been published. For wider scope coverage of the chondrocranium in turtles, see Bellairs & Kamal (1981), who also figure (figs. 76, 77, 78) the chondrocranium in *Chelydra,* based on DeBeer (1926, 1937) and Bellairs (1949). Rieppel (1976, 1977) also described parts of the *Chelydra* chondrocranium. Rieppel (1993) is the most important paper for descriptions of the development of the bony skeleton, but in particular, the skull (see also Rieppel 1990), in *Chelydra*. He figured the ossification of the dermatocranium and described the ossification in the whole skull. The inner ear, middle ear, and stapedial development in *Chelydra* are described by Toerien (1965a, 1965b).

Chelydra figures prominently in an important series of developmental papers by Burke and associates: Burke (1989a, 1989b), Burke (1991), and Gilbert et al. (2001). These involve descriptive and experimental methods to determine shell origin and growth. Burke & Alberch (1985) described the carpus and tarsus in adult and developing *Chelydra*.

ANATOMICAL ABBREVIATIONS

ang	angular	pal	palatine
art	articular	pf	prefrontal
bo	basioccipital	pm	premaxilla
bs	basisphenoid	po	postorbital
cor	coronoid	pr	prootic
den	dentary	pra	prearticular
epi	epipterygoid	pt	pterygoid
ex	exoccipital	qj	quadratojugal
fr	frontal	qu	quadrate
ju	jugal	so	supraoccipital
mx	maxilla	sq	squamosal
na	nasal	sur	surangular
op	opisthotic	vo	vomer
pa	parietal		

4

Molecular Insights into the Systematics of the Snapping Turtles (Chelydridae)

H. BRADLEY SHAFFER

DAVID E. STARKEY

MATTHEW K. FUJITA

SOME PROBLEMS IN SYSTEMATICS and phylogenetics are relatively easy to work out, and some are just plain difficult. There are many reasons why this may be the case, but the result is often the same—a few taxa seem to provide endless discussion, debate, and challenge for systematists, and frustration for taxonomists and comparative biologists who require well-resolved phylogenies. Among turtles, one of the most difficult problems has been the content and phylogenetic relationships of the family Chelydridae.

The relatively recent use of molecular markers in phylogenetics has greatly increased our knowledge of evolutionary relationships among organisms, from intraspecific genealogies to the tree of life. Molecular evidence is often most useful when it provides new insights into contentious phylogenetic questions that were previously intractable with more traditional sources of data. Although molecular similarities cannot replace strong morphological differences (any more than morphological similarity can override molecular divergence), new molecular data often force us to critically evaluate conclusions based on sparse sampling or poorly defined morphological characters. Protein electrophoresis (allozymes), chromosome morphology, and immunological cross-reactivity provided the first molecular data for phylogenetic analysis, and they continue to provide important insights into phylogenetic relationships. In general, however, these techniques were developed as a proxy for direct observations at the DNA sequence level. Because DNA sequencing has become increasingly straightforward for both mitochondrial (mtDNA) and nuclear (nDNA) genomes, DNA sequence data have grown to be the major source of molecular characters for phylogenetic inference. Optimal models of DNA evolution can be estimated for a set of sequences and incorporated into phylogenetic methods such as maximum likelihood and Bayesian analysis. The combination of model-based phylogenetic methods, ease of data acquisition, and a virtually limitless source of characters has rendered mtDNA and nDNA sequences the data of choice for many vexing phylogenetic issues.

In this chapter, we review molecular data that are relevant to three questions in the systematics of the Chelydridae. The first is the relationship of the alligator snapping turtle, *Macrochelys temminckii,* to *Chelydra.* The alligator snapping turtle is geographically restricted to the central and southeastern United States, and some debate exists on whether *Macrochelys* plus *Chelydra* form a monophyletic group.

Second, we consider the relationships and classification of the genus *Chelydra.* As currently recognized, the single species *Chelydra serpentina* consists of four subspecies. *Chelydra s. serpentina* occurs commonly across the United States from the East Coast as far west as the Rocky Mountains, *C. s. osceola* is largely restricted to Florida, *C. s. rossignoni* is found in scattered localities from southern Mexico to northern Central America, and *C. s. acutirostris* ranges from southern Central America to northern South America. Although some have argued that the Florida snapping turtle, *C. s. osceola,* deserves species recognition, others have doubted the validity of even subspecific status. Neither chromosomal nor serological studies appear to distinguish *serpentina* from *osceola,* raising further questions about the validity of *osceola* as a distinct evolutionary form. More recently, mtDNA and allozyme data from a relatively limited sampling of the genus *Chelydra* have shed some light on the validity of these four subspecies, although controversy remains over the interpretation of these results.

Finally, we consider molecular insights regarding the relationships of the big-headed turtle, *Platysternon megacephalum,* because previous studies have disagreed about its phylogenetic position. Skeletal morphology as well as locomotor characters suggest to some researchers that *Platysternon* is allied with *Chelydra* and *Macrochelys,* whereas chromosomal, electrophoretic, and immunological precipitation data place it with Emydidae, Geoemydidae (= Bataguridae), and Testudinidae. Existing mtDNA data either weakly support a *Chelydra/Platysternon* relationship or are ambiguous, and combined mtDNA/morphological evidence strongly support a *Chelydra/Platysternon* sister group relationship (Shaffer et al. 1997). However, new nDNA data do not support this relationship (Krenz et al., unpublished). At least from a molecular perspective, the placement of *Platysternon* remains problematic, although new data reviewed below are shedding light on this problem.

THE NEW WORLD CHELYDRIDS

We address two primary questions concerning the relationships of the New World snapping turtles. First, are *Chelydra* and *Macrochelys* sister taxa, as is generally implied in the literature? Second, is evolutionary diversification within the species *C. serpentina* and *M. temminckii* reflected in their current taxonomy?

Gaffney (1975) developed the intriguing hypothesis that *Platysternon* and *Macrochelys* are sister taxa, with *Chelydra* sister to that pair. Given the overall similarity in geographic range, morphology, and general biology of *Chelydra* and *Macrochelys* compared with *Platysternon,* this is a somewhat counterintuitive hypothesis, yet surprisingly little follow-up work has focused on the sister-group relationship of *Chelydra* and *Macrochelys.* Starkey (1997) provided mtDNA data strongly supporting the sister-group relationship of *Chelydra* and *Macrochelys,* as did Whetstone (1978b) on the basis of morphological characters. Starkey (1997) is the only completed molecular study of which we are aware that addresses this point, but his results are unambiguous. Based on a broad taxonomic sample of the living turtles of the world, he recovered a sister-group relationship of the New World chelydrids (*Chelydra* and *Macrochelys*) with very high statistical confidence (bootstrap proportion [BP] =100; Starkey, unpublished result). Unpublished data from our lab for a ~1–kb intron from the nuclear R35 neural transmitter gene also demonstrates, with similarly high bootstrap support, a *Chelydra–Macrochelys* sister-group relationship. We consider the combined mtDNA and nDNA results to be biologically reasonable and statistically extremely strong. Thus, for the remainder of this chapter we use the term "New World Chelydridae" to refer to this apparently monophyletic group.

At a lower phylogenetic level, the taxonomic content of *Chelydra* has been much more contentious. Phillips et al. (1996) studied 19 individual snapping turtles (10 *C. s. serpentina,* 5 *C. s. osceola,* 2 *C. s. acutirostris,* 2 *C. s. rossignoni*) using restriction endonuclease fragment patterns of mtDNA and isozyme variation. mtDNA fragment analysis uncovered two main groups, one in the United States (including *C. s. serpentina* and *C. s. osceola*) and a second comprising non-U.S. taxa (*C. s. rossignoni* and *C. s. acutirostris*). For the mtDNA, there were no diagnostic differences found between the fragment patterns in *C. s. serpentina* and *C. s. osceola,* and sequence divergence estimates between the two were extremely low, ranging from 0.0 to 0.3%. In contrast, *C. s. rossignoni* and *C. s. acutirostris* were strongly differentiated from the U.S. taxa in their mtDNA. Estimates of percent sequence divergence calculated from the restriction enzyme data ranged from ~3.0 to 4.5% for *C. s. acutirostris–C. s. serpentina* comparisons, and ~4.5 to 5.6% for *C. s. rossignoni–C. s. serpentina.* *C. s. acutirostris* and *C. s. rossignoni* were less divergent from each other, with pairwise divergences ranging from 1.7 to 2.6%.

Although limited in scope both in terms of number of loci examined and turtles sampled (see Sites and Crandall, 1997, for a critique), the allozyme data presented by Phillips et al. (1996) were generally consistent with their mtDNA results. They found no unique alleles in eight allozyme loci ex-

Table 4.1

Genetic distances among currently recognized subspecies
of *Chelydra serpentina*.

	acutirostris	rossignoni	osceola	serpentina
acutirostris	—	5.45	8.39	8.39
rossignoni	0.319	—	6.03	6.03
osceola	0.347	0.008	—	0.23
serpentina	0.430	0.028	0.054	—

Note: Above the diagonal are the average uncorrected pairwise distance comparisons derived from mtDNA control region sequences described in the text. Below the diagonal are Nei's *D*-values calculated from eight allozyme loci reported in Phillips et al. (1996).

amined between *C. s. serpentina* and *C. s. osceola,* demonstrating that the two are not diagnosably distinct for these loci. One of the eight loci (*Pnp*) showed a reasonably large frequency difference that may imply some differentiation between *C. s. serpentina* (frequency of the "a" allele = 0.67, *f*(b) = 0.33) and *C. s. osceola* (*f*(a) = 0.20, f(b) = 0.80), although it is difficult to make too much of this with the sample sizes available. Consistent with the mtDNA data, *C. s. acutirostris* was fixed for unique alleles for two of eight allozyme loci (*M-Icdh* and *S-Icdh*) based on the two individuals examined. However, the two *C. s. rossignoni* were virtually identical to the *C. s. serpentina* and *C. s. osceola* sampled for these eight loci. Thus, Nei's (1978) genetic distances (calculated by us using the data presented in Phillips et al. 1996) are quite large between *C. s. acutirostris* and the other three taxa (0.43 to 0.32; Table 4.1), but low among the remaining three taxa (0.008–0.05).

Walker et al. (1998) built on these results by examining mtDNA sequence variation among 66 *Chelydra* from ten states in the Southeast. They sequenced a 409–bp fragment of control region (CR) mtDNA, and found this fragment to be identical in 60 of the 66 individuals (91%) examined. The few variants differed from the common haplotype by a single base pair transition, or the single transition plus an additional single base pair indel event. Walker et al. (1998) provided excellent geographic coverage of both southern *C. s. serpentina* and *C. s. osceola,* as well as a transect across a phylogeographic transition zone in western Florida that is consistently deep for many other vertebrate taxa. This lack of variation within *Chelydra* across the southeastern United States is unique for turtles with similar distributions in the same region, leading these authors to agree with Phillips et al. (1996) that *C. s. serpentina* and *C. s. osceola* together represent a single evolutionary unit.

These studies point to a significant lack of diversity among common snapping turtles from the United States, with Phillips et al. (1996) hypothesizing a significant disjunction between non-U.S. and U.S. taxa. Although these results may appear straightforward, both studies have limits to their resolution. The fundamental conclusions of Phillips et

al. are based on shared restriction fragment patterns, and not mapped restriction sites, which provide more accurate estimates of sequence divergence. Their allozyme data, while suggesting that *C. s. acutirostris* is a differentiated taxon, are based on very small samples of individuals (two each of the non-U.S. taxa) and loci (eight allozyme loci). Walker et al. (1998) provide compelling sequence evidence from the rapidly evolving CR that *C. s. serpentina* and *C. s. osceola* are virtually identical in the southeastern United States, but only limited population sampling for *C. s. serpentina* across its entire range and no data for the non-U.S. taxa.

To address the question of diversity within and among the United States taxa more completely (Shaffer, Starkey, and Fujita, unpublished data), we have sampled common snapping turtles from throughout their U.S. range (50 localities across 20 states). We analyzed 428 base pairs of CR mtDNA that is essentially the same fragment as that studied by Walker et al. (1998). The results of our study are consistent with the previous two studies in that we found almost no variation within U.S. common snapping turtles, with only a single new haplotype among the 50 or more turtles sampled. This new haplotype differed from the most common haplotype of Walker et al. (1998) by a single base pair insertion-deletion (indel) event, resulting in a maximum sequence divergence of 0.23% across a broad representation of *C. s. serpentina* and *C. s. osceola* (Table 4.1, Fig. 4.1). We also sequenced the same individuals of *C. s. rossignoni* and *C. s. acutirostris* used in Phillips et al. (1996), as well as two new samples of *C. s. rossignoni* from Mexico, one from Honduras, and a single new *C. s. acutirostris* from Panama. At the sequence level, we found even deeper divergences (ranging between ~5.4 and 8.4%) between these two taxa and the U.S. snapping turtles (Table 4.1). Thus, a relatively consistent molecular picture across several laboratories and data sets appears to be emerging. In the United States a single, virtually invariant lineage of *Chelydra* is broadly distributed, with no evidence of differentiation between the subspecies *C. s. serpentina* and *C. s. osceola*. This result is supported with excellent geographic sampling for mtDNA sequence data, thin sampling for allozymes, and currently no information for nDNA sequences. However, the Central American *C. s. rossignoni* and the South American *C. s. acutirostris* are deeply differentiated from each other and from the U.S. snapping turtles (Fig. 4.1), in particular, for mtDNA sequence data.

Unlike the wide-ranging genus *Chelydra*, *M. temminckii* has a limited distribution in the southeastern United States. Roman et al. (1999) assessed variation among alligator snapping turtles utilizing a 420–base-pair fragment of CR mtDNA. They found considerably more variation within alligator snapping turtles than codistributed common snapping turtles, with estimates of pairwise diversity ranging from 1.7 to 2.4%. Furthermore, this variation was geographically structured, with 8 of 11 haplotypes restricted to single rivers, and major lineages identified in the western

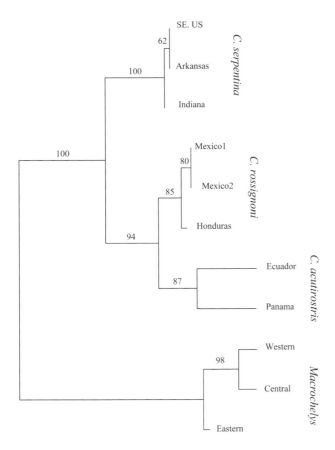

Fig. 4.1. Phylogram showing relationships of all unique sequences found in *Chelydra* and *Macrochelys,* based on 428 (462 including primers) base pairs of the mtDNA CR. The tree is a maximum likelihood reconstruction using the model HKY+Γ, with a–ln L = 1131.65. The following parameters were calculated from the data set using Modeltest 3.0 (developed by D Posada and KA Crandall in 2001): t_i/t_v = 2.5462, Γ = 0.3921, a = 0.2988, c = 0.2137, g = 0.1293, and t = 0.3582. Branch lengths are proportional to levels of mtDNA sequence divergence. Numbers above nodes are bootstrap proportions based on 1,000 bootstrap pseudo-replications. *C. s. osceola* is not shown because it was identical with the most common *C. s. serpentina* sequence.

(Mississippi), central (Apalachicola), and eastern (Suwannee) portions of the range. Roman et al. (1999) noted that the level of differentiation that they observed is similar to that found among species of map turtles in the same region and suggested that three evolutionarily significant units (ESUs) exist within *Macrochelys*. However, they did not recommend taxonomic recognition of these ESUs.

Why might the level of diversity be so different within *Chelydra* versus *Macrochelys* in the United States? They are both extremely aquatic turtles that can attain great size and age (Ernst et al., 1994). The range of the common snapping turtle extends much further north than that of the alligator snapper, and at least some of this range must represent post-Pleistocene expansion into previously glaciated habitat. While such a recent range expansion may explain the lack of genetic variation in the northern range of *Chelydra,* it does not explain the extraordinarily low levels found in the south where both *Chelydra* and *Macrochelys* coexist. Although both turtles are normally aquatic, common snappers display a

much greater propensity for overland movement, which may result in greater gene flow among aquatic habitats. This difference in terrestrial ecology may explain the low levels of among-drainage differentiation in *Chelydra* compared with *Macrochelys*. However, even complete panmixia would not explain the lack of genetic diversity within and among populations of U.S. *Chelydra.* Two possible explanations are a massive, recent range expansion across the entire U.S. *Chelydra* population, or a selective sweep on the mitochondrial genome. Selective sweeps, where natural selection for a single haplotype leads to a specieswide fixation of that favored haplotype and a drastic reduction in genetic variation (Galtier et al. 2000; Przeworski 2002), is a particularly important mechanism for mtDNA because the entire molecule is a single recombinational unit. Differentiating between range expansion and selective sweeps requires data from nuclear genes, and we are currently collecting these data in our laboratory.

IS *PLATYSTERNON* A CHELYDRID?

More than 50 years after Williams's (1950) synthetic analysis of the higher-order phylogeny of the living turtles, the relationships of the big-headed turtle *Platysternon megacephalum* remain as controversial as ever. In many ways, the placement of this enigmatic species requires the full resolution of cryptodire relationships, since it appears to be an ancient lineage with no particularly close living relatives. Although we remain uncertain of the relationships of this key species, recent molecular evidence seems to be shedding light on parts of this deep phylogenetic puzzle, and the hope for future resolution is promising.

Three recent studies at the DNA sequence level have provided new insights on the possibility of a sister-group relationship of *Platysternon* with the New World Chelydridae, as proposed by Gaffney (1975) based on morphological evidence. Shaffer et al. (1997) examined two mitochondrial genes across 23 taxa spanning the breadth of living turtle families. They used 892 base pairs of the cytochrome *b* gene to provide resolution for the shallower nodes of the tree, whereas 325 base pairs of the slowly evolving 12S ribosomal gene were aimed at resolving deeper nodes within the Testudines. Both alone and in combination, the two genes provided little insight into the relationships of *Chelydra* and *Platysternon* to each other or to other turtles, with all relevant bootstrap proportions (BPs) <50 (see fig. 4a–c in Shaffer et al. 1997). Although statistical support was poor, the combined gene analysis placed *Chelydra* and *Platysternon* together, suggesting that weak molecular support for this relationship may exist. Alternatively, the weak support of these two relatively isolated lineages may reflect a "long branch attraction" problem in the molecular data. When combined with a data set of 115 morphological characters, the final molecules-plus-morphology analysis strongly supported (BP = 94) a sister-group relationship for *Chelydra* and

Platysternon, although this relationship was driven largely by the morphological data.

In a more detailed taxonomic sampling strategy, Starkey (1997) examined phylogenetic relationships among 72 recognized genera, and 230 species/subspecies of living turtles for a 991–base-pair fragment of the mitochondrial ND4 gene plus three adjacent tRNAs. Like the mitochondrial results in Shaffer et al. (1997), Starkey (1997) found little resolution for the relationships of *Platysternon* and the New World Chelydridae, apparently due to saturation of this relatively rapidly evolving mitochondrial gene. Two key results did emerge from Starkey's work, however. First, *Platysternon* is extremely divergent from all other turtles, and was reconstructed as a long, isolated branch near the base of the Cryptodira. Second, *Platysternon* and the New World Chelydridae showed no evidence of a sister-group relationship. Rather, the New World Chelydridae is sister (with a reasonable support, BP = 72) to Kinosternidae plus Dermatemydidae, and *Platysternon* is sister (BP <50) to the remaining cryptodires. Starkey's data are similar to Bickham and Carr's (1983) hypothesized relationships of *Platysternon* as close to the clade consisting of Testudinidae, Bataguridae (=Geoemydidae), and Emydidae (Testudinoidea of Shaffer et al. 1997), although Starkey's hypothesis also includes the marine turtles in this latter clade.

Both Shaffer et al. (1997) and Starkey (1997) used mtDNA for their sources of molecular data, and found little support at the deep nodes where the resolution of *Platysternon* relationships appear to reside. A recent analysis of 2,793 base pairs of the nuclear RAG-1 gene (Krenz et al. unpublished) for essentially the same set of taxa as in Shaffer et al. (1997) is a major addition to this literature. RAG-1 is a slowly evolving, single-copy, protein-coding gene that has proved very useful as a phylogenetic tool for deep relationships of birds and mammals, and provides a clean phylogenetic signal for the relationships of *Platysternon* to other turtle clades. Although Krenz et al. (unpublished) could not say with certainty the precise placement of *Platysternon* or the New World Chelydridae, they were able to reject (with reasonably strong statistical support) the hypothesis that the two form a monophyletic group. Based on RAG-1, *Platysternon* appears to be allied to the Testudinoidea, whereas *Chelydra* is close to a clade consisting of Kinosternidae, *Dermatemys,* and the marine turtles. This result is very similar to that of both Starkey (1997) and Bickham and Carr (1983), and is certainly not in conflict with the molecular results of Shaffer et al. (1997). However, the RAG-1 data are still not capable of toppling the morphological evidence suggesting a sister-group relationship of the New World Chelydridae and *Platysternon,* and the combined evidence tree still marginally supports the morphological view with weak bootstrap support (Krenz et al. unpublished).

To summarize the available molecular data, we constructed a supertree based on the molecular results of Bickham and Carr (1983), Shaffer et al. (1997), Starkey (1997), and

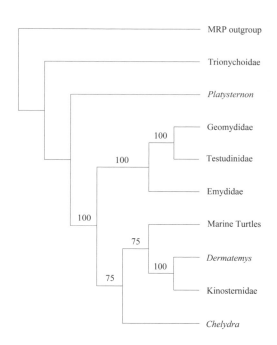

Fig. 4.2. A majority rule consensus of four equally parsimonious supertrees of some of the major clades of living turtles. For input trees, we used chromosomal data (Bickham & Carr 1983, their fig. 3), combined mtDNA from cytochrome *b* and 12S (Shaffer et al. 1997. their fig. 4c), mtDNA from ND4 plus adjacent tRNAs (Starkey 1997; new analysis based on the 23 taxa in Shaffer et al. 1997), and RAG-1 (Krenz et al. 2005; their fig. 5). We used the matrix representation with parsimony (MRP) supertree consensus method calculated in the program RadCon 1.1.5, and constructed a maximum parsimony tree in PAUP*. Numbers refer to the percentage of four equally parsimonious supertrees in which a given node was found. Trionychoidae is the monophyletic group including Trionychidae plus Carettochelyidae (Shaffer et al., 1997). Note that *Platysternon* and *Chelydra* are not sister taxa.

Krenz et al. (unpublished). The resulting supertree, which is a summary of the topological relationships of the four input trees, is presented in Fig. 4.2. The key result is the distant relationship of *Platysternon* and *Chelydra* that consistently appears in most molecular analyses.

CLASSIFICATION RECOMMENDATIONS AND FUTURE RESEARCH DIRECTIONS

Within the genus *Chelydra,* existing mtDNA and allozyme data point to extremely weak differentiation between the U.S. subspecies *C. s. serpentina* and *C. s. osceola.* With less than a quarter of a percent sequence divergence in the rapidly evolving control region, and a minimal Nei's D of 0.05 (Table 4.1), no molecular evidence supports the recognition of these two taxa. Our broad sampling of mtDNA, combined with the published, detailed population level analysis of southeastern U.S. populations all argue for a surprising lack of variation in U.S. *Chelydra.* Whether this eliminates *C. s. osceola* as a "good" subspecies (whatever that may mean) that is distinct from *C. s. serpentina* is still open to debate, and the final decision must include a careful analysis of morphological variation (Feuer 1971). However, the molecular data imply that *C. s. osceola* and *C. s. serpentina* are not well-differentiated evolutionary lineages.

The situation for *C. s. acutirostris* and *C. s. rossignoni* is quite different. MtDNA and allozyme data both demonstrate deep differentiation between *C. s. acutirostris* and all other *Chelydra,* with diagnostic characters and an overall level of differentiation that is frequently found among well-differentiated turtle species. The same is true for *C. s. rossignoni* based on mtDNA sequence data, but not allozymes. Given that these taxa have allopatric distributions, we favor species designations based on character differentiation rather than inferred interbreeding. Based on both mitochondrial and allozyme evidence, *C. s. acutirostris* appears to qualify as a good phylogenetic species. One reasonable solution is to withhold judgment on the status of *C. s. rossignoni* pending additional nuclear DNA evidence, given the lack of allozyme differentiation between it and *C. s. serpentina*. Alternatively, the relatively deep mtDNA differentiation and allopatric distribution both argue that *C. s. rossignoni* is a separate evolutionary lineage, and therefore is best viewed as a separate species as well.

Based on currently available molecular evidence, we favor recognizing a monotypic, widespread *C. serpentina* across the continental United States and southern Canada, and abandoning *C. s. osceola* as an evolutionary entity. We recognize that there may be morphological evidence indicating that the Florida populations are somewhat differentiated, and when a full morphological analysis is complete, there may be sound reasons to recognize *C. s. osceola* at either the specific or subspecific levels. However, the available molecular data do not indicate any substantial differentiation between these taxa. We further recommend elevating the Central and South American forms to full species status as *C. acutirostris* and *C. rossignoni*. This was the recommendation of Phillips et al. (1996) based on incomplete sampling, and it has been substantiated and enhanced, at least for mtDNA, with additional samples. We concur with Sites and Crandall (1997) that additional sampling of *C. acutirostris* and *C. rossignoni* would be extremely valuable, and we provide limited additional sampling in Fig. 4.1. In total, the molecular evidence currently in hand is consistent with the three-species interpretation advocated by Phillips et al. (1996).

At the family level, the molecular evidence available strongly supports the interpretation of *Chelydra* and *Macro-*

chelys as sister taxa, and their placement in a monophyletic Chelydridae is clearly warranted. The situation for *Platysternon* is less clear, and the resolution of the molecular and morphological conflict over the relationship of *Platysternon* requires additional data and analysis. However, the available molecular evidence points to *Platysternon* as allied with the Testudinoidea, not Chelydridae. The prudent choice at this point is probably to place *Platysternon* in a monotypic family Platysternidae pending additional information. If and when the phylogenetic placement of *Platysternon* stabilizes, then we would favor relegating it to a more inclusive family that reflects its phylogenetic relationships, rather than placing it in a monotypic family that simply emphasizes its uniqueness.

Both for understanding relationships within *Chelydra*, and in particular for the resolution of the placement of *Platysternon*, multigene nuclear data are clearly vital. For relationships within *Chelydra*, the key issue centers on the validity of *rossignoni* and *acutirostris* as distinct from *serpentina*. The possibility that *osceola* is a real evolutionary lineage still should be considered, although it appears less likely based on mtDNA and allozymes. We are currently sequencing several nuclear introns, and they should shed light on this set of problems. At the deep phylogenetic level of *Platysternon* relationships, the RAG-1 data imply that nuclear exons may be informative, and unpublished intron data from our lab also appears promising. All available evidence currently points to *Platysternon* and Chelydridae as critical phylogenetic taxa that will require considerable sequence data before we fully understand their relationships to each other and other turtle lineages. In addition, if the nonmonophyly of these two taxa is even more strongly supported with additional molecular evidence, understanding the apparent homoplasy in several morphological characters that have previously been interpreted as homologies emerges as a fascinating problem in morphological evolution.

ACKNOWLEDGMENTS

We thank Chris Phillips, Richard Vogt, and Peter Meylan for tissues, Peter Meylan for comments on the manuscript, and members of the Shaffer lab group for input and discussion. This material is based on work supported by National Science Foundation grants 9727161 and 0213155 and the University of California Agricultural Experiment Station.

PART 2 PHYSIOLOGY, ENERGETICS, AND GROWTH

5

Respiratory and Cardiovascular Physiology of Snapping Turtles, *Chelydra serpentina*

NIGEL H. WEST

THE SNAPPING TURTLE is of special interest to ecologists as the most widely distributed chelonian species in North America. It ranges across all of the United States and southern Canada east of the Rocky Mountains and as far south as Ecuador (Carr 1952). The species undoubtedly is highly adaptable and is capable of inhabiting most types of freshwater environments within its range, although some slight preference may exist for muddy bottoms in which individuals can hibernate and/or lie in ambush for prey. Snapping turtles are very cold tolerant and have been observed to be active under ice. They are highly aquatic and bask only occasionally in the southern portion of the range and somewhat more in the north. Although most of the time is spent submerged, snappers do occasionally make long overland journeys (Dillon 1998).

Surprisingly little is known about the systemic physiology of this highly adaptable species and there have only been a few concerted physiological studies at any level of organization. Perhaps most evidence has been accumulated on the respiratory and cardiovascular physiology of adult *Chelydra,* in relation to its highly aquatic habit.

RESPIRATORY PHYSIOLOGY
Gas Exchange in Ovo

The gas-exchange requirements of *Chelydra serpentina* in ovo are met entirely by diffusion across the egg shell. The eggs of *Chelydra* possess an outer mineral layer and an inner shell membrane of nearly equal thickness. The outer mineral layer gradually exfoliates and represents a small fraction of the thickness of the eggshell at hatch (Packard 1980). Packard (1980) recognized that the conductance of respiratory gases and water through flexible eggshells is a function of the degree of hydration of the egg, the structure of the flexible membrane, and progressive exfoliation of the mineral layer as the egg swells. Feder et al. (1982) measured the separate contributions to the conductance of O_2 and H_2O through the eggshell membrane and the mineral layer by comparing both intact and decalcified half-shells. They also measured the effect of eggshell membrane hydration on conductance by progressive desiccation. As expected, decalcified shells lost water at higher rates than intact shells. Oxygen conductance was increased by dehydration in both intact shell fragments and those lacking a mineral layer. Rather unexpectedly, experimental removal of the mineral layer by 5% EDTA frequently resulted in a *decrease* in the O_2 conductance of eggshell fragments. Feder et al. (1982) suggest that this anomalous result may be due to the adjacent pillar structure of the mineral layer normally holding pores in the eggshell membrane in an open configuration. These pores appeared to close when the mineral layer was removed by treatment with EDTA. In normal development, the gradual exfoliation of the mineral layer serves to match O_2 supply to increasing demand of the embryo by progressively increasing the oxygen conductance of this diffusion-based system of oxygen delivery.

Lung Phospholipids and the Onset of Air-breathing

In common with other air-breathing vertebrates, lung compliance in *C. serpentina* is increased by a surfactant system composed of phospholipids, neutral lipids, and proteins, which act to lower surface tension within the lungs. The monolayer of surfactant, by reducing the negative fluid pressure in the liquid lining of the alveoli, also reduces the net influx of interstitial fluid into the alveolar gas space (Orgeig et al. 1997). Surfactant is secreted by alveolar type II cells into the lungs of all air-breathing vertebrates. Here it serves to reduce the work of breathing by reducing surface tension forces at the wall of gas-exchange units by lining the interface between alveolar gas and water. Copious amounts of pulmonary surfactant were found in crocodiles, snakes, lizards, and turtles (Daniels et al. 1996). Thyroid hormones and glucocorticoids both contribute to the maturation of the surfactant system in mammals (Sullivan et al. 2001). In particular, thyroid transcription factor-1 (TTF-1) is a regulator of the gene expression of surfactant proteins in mammals. Johnston et al. (2002) used immunohistochemistry to determine the expression of TTF-1 in both the conducting airways and gas-exchange areas of the lung in embryonic, pipped, and hatchling snapping turtles. TTF-1 expression was detected in embryonic snapping turtles and declined in hatchlings. The same cells within the respiratory epithelium stained for both TTF-1 and the surfactant protein, SP-B, suggesting a role for TTF-1 in regulating the expression of surfactant protein. The onset of air-breathing coincided with the maximum secretion

of total phospholipid. An increased ratio between disaturated and total phospholipid (DSP/PL) occurred at pipping (from about 16 to 25%) as did a decrease in the cholesterol/total phospholipid ratio (CHOL/PL) from about 7% to 2%. In adults of three species of turtles these ratios ranged from 27 to 30% (DSP/PL) and 6 to 10% (CHOL/PL), respectively (Daniels et al. 1995). Thus, the maturation of the phospholipid content coincided with the onset of air-breathing in the *Chelydra* hatchlings. Both saturated and total phospholipids increased dramatically at the initiation of lung ventilation upon pipping, presumably to increase lung compliance and to reduce the work of breathing.

Gas Exchange in Adult Chelydra

Snapping turtles are highly aquatic, raising the possibility that bimodal gas exchange represents a significant adaptive mechanism in this species. Warm-acclimated turtles that were free to submerge obtained 5% of their oxygen consumption from water at 20°C. This increased to 11% at 4°C (Gatten 1980), reflecting both the lower metabolic rate and the increased oxygen capacitance coefficient in water at the lower temperature. Thus, cutaneous oxygen uptake may assume increased significance in winter, when snapping turtles are denied access to air by ice, or are hibernating in mud. Boyer (1963) found no relationship between oxygen consumption and inspired oxygen levels in turtles breathing gas mixtures ranging from normoxic (21% O_2) to severely hypoxic (2% O_2), suggesting that aquatic gas exchange alone may be sufficient to support aerobic metabolism in submerged turtles during winter when metabolic rate is low.

Gatten (1980) measured aerial and aquatic oxygen uptake for 12 days in an unrestrained, cold-acclimatized snapping turtle at 4°C. This animal did not breathe for 202 consecutive hours of 288 hours of measurement. Under these conditions, aquatic oxygen uptake was 31% of the total, compared with aquatic oxygen uptake of 11% of the total for the warm-acclimated animals. This suggests that cold acclimatization may influence cutaneous gas exchange by physiological mechanisms, such as vasodilatation and capillary recruitment, which increase cutaneous oxygen conductance. Turtles overwintering under ice may take up dissolved oxygen via the pharynx, skin, or cloaca, whereas those overwintering in mud can only utilize the pharynx for aquatic gas exchange (Gatten 1980).

Bagatto & Henry (1999) found the overall respiratory gas exchange ratio (R.E., the ratio of carbon dioxide production volume and oxygen utilization volume in unit time) of freely diving snapping turtles to be 0.76, partitioned into a lung R.E. of 0.55 and a cutaneous R.E. of 5.4. Thus, the preferred route for CO_2 loss was aquatic, not pulmonary, whereas the oxygen pathway was largely pulmonary. Aquatic oxygen consumption represented only 4% of total in *Chelydra* acclimated to 25°C. Aquatic carbon dioxide exchange in these animals was 30% of the total, reflecting the relatively higher capacitance coefficient of CO_2 in water, and the maintenance of a significant transcutaneous gradient for CO_2, even in the presence of air-breathing. Bagatto & Henry (1999) point out that the reduced plastron area in the snapping turtle has resulted in an increased cutaneous surface area exposed to water, but the skin is relatively thick. Combined with the relative capacitance coefficients of the respiratory gases in water, this may impose more severe diffusion limitations on the transcutaneous diffusion of O_2 than that of CO_2. Because *C. serpentina* live in waters that frequently become hypoxic, these authors consider that may have been little advantage in significant adaptations for aquatic O_2 uptake, providing air-breathing was possible (Bagatto & Henry 1999).

The standard metabolic rate (SMR) of submerged cold-acclimated (10°C), unrestrained snapping turtles (with access to air) at 10, 20, and 30°C was not significantly different from SMR in air at the same temperatures. SMR meets metabolic maintenance needs, without accounting for growth or activity. In warm-acclimated animals (25°C), voluntary diving resulted in reduction in oxygen consumption compared with that in air—estimated to be a reduction of about 16% at the acclimation temperature (Gatten 1980). The slight reduction in SMR of warm-acclimated turtles may reflect a slightly decreased maintenance cost of metabolism during submersion. It has been suggested that this allows a greater proportion of self-contained oxygen stores to be used in aerobic muscular activity (Gatten 1978), increasing the scope for underwater exercise in warm-acclimated turtles at 25°C.

Warm-acclimated turtles in water and cold-acclimated turtles in air have similar SMRs at 10, 20, and 30°C, but lower rates than warm-acclimated individuals in air, suggesting that both cold adaptation and submergence lower SMR, but in a nonadditive fashion (Gatten 1980). No lactate concentrations have been measured, but the fact that voluntary submergence in turtles with access to air results, at most, in only a small reduction in oxygen consumption suggests that metabolism during voluntary submergence under these conditions is aerobic, with little or no reliance on anaerobic mechanisms.

On the other hand, a potential for anaerobic metabolism is implied by the fact that *C. serpentina* can occupy eutrophic bodies of water that become anoxic during winter (Reese et al. 2002). Certainly, snapping turtles have been reported to overwinter in hibernacula buried in anoxic mud (Brown & Brooks 1994). Turtles submerged in normoxic water at 3°C for several months with no access to air showed only a small lactate accumulation, suggesting that metabolism was overwhelmingly aerobic (Reese et al. 2002). The slight metabolic acidosis was fully compensated by increased loss of CO_2 by extrapulmonary routes, resulting in a slight decrease in the partial pressure of carbon dioxide in the blood (Pco_2) and an unchanged blood pH (Reese et al. 2002). In contrast, *C. serpentina* submerged in anoxic water at 3°C are forced to depend on anaerobic metabolism. Lac-

tate accumulated progressively over 100 days of submergence, titrating plasma bicarbonate levels from 51.5 to 4.9 mmol/liter. Arterial pH fell from 8.1 to 7.5 over the first 25 days, simultaneously with an increase in blood Pco_2, which returned to control values by day 50. Blood pH decreased more slowly to 7.1 during the remaining 75 days. In *Chrysemys picta belli*, it is estimated that the shell and bone buffer about 75% of total body lactic acid (Jackson et al. 2000a). Calcium and magnesium carbonates in bone are transported into blood and buffer the protons derived from lactate. Reese et al. (2002) calculate that *C. serpentina* can survive four months of anoxia at 3°C, which, while less than the anoxia resistance of *Chrysemys* (Reese et al. 2000), enables *C. serpentina* to overwinter in hypoxic bodies of water such as eutrophic ponds.

Respiratory Chemosensitivity in Chelydra

The oxygen demands of adult snapping turtles at 20°C are largely (95%) met by intermittent air ventilation (Gatten 1980). Adult snapping turtles in nature are primarily bottom dwellers, although they are occasionally found basking. Under laboratory conditions at 20–25°C undisturbed turtles exhibited nonventilatory periods of some 18 minutes that were interrupted by short breathing periods (4–5 min) in which about five breaths were taken (West et al. 1989). Earlier, Boyer (1966) found nonventilatory periods of 16 minutes in undisturbed, normoxic turtles. In shallow water the only movement in undisturbed turtles is repetitive neck extension, which also anticipates the start of the breathing period. The partial pressure of arterial oxygen (Pao_2) and arterial pH (pH_a) rise as a result of repetitive pulmonary ventilation cycles and the partial pressure of arterial carbon dioxide ($Paco_2$) falls (Fig. 5.1). Pao_2 and pH_a then fall and $Paco_2$ rises during the succeeding nonventilatory period until the next period of ventilation is initiated.

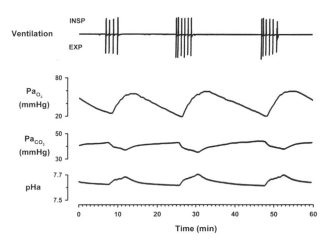

Fig. 5.1. The pattern of ventilation recorded by a pneumotach and the corresponding changes in Pao_2, $Paco_2$, and pH_a in a 2.3-kg snapping turtle. The animal was unrestrained except for an arterial cannula and was held in a tank at 22–24°C. Adapted from West et al. 1989.

Central and peripheral chemoreceptors have been demonstrated in chelonians (Davies & Sexton 1987; Hitzig & Jackson 1978; Ishii et al. 1985). If feedback from such receptors is an important determinant of the initiation of breathing periods, then manipulation of blood gases should affect the duration of the nonventilatory period. Turtles that were made hyperoxic, or hypoxic, by changing the fraction of oxygen in inspired gas from 10 to 30% initiated breathing periods over a wide range of Pa_{O_2} values. At the end of the nonventilatory period, when the breathing period was initiated, Pa_{O_2} varied from nearly 80 mm Hg above the mean value (breathing hyperoxic gas) to nearly 40 mm Hg below the mean value (breathing hypoxic gas). This wide variation of Pa_{O_2} at the initiation of breathing periods suggests that drive from oxygen-sensitive chemoreceptors does not play a determining role in the initiation of these periods in the snapping turtle. Pa_{CO_2} at the termination of nonventilatory periods showed a less than 5 mm Hg variation about the mean value, suggesting that it may be the predominant chemical drive initiating breathing bouts. Also, a tendency for nonventilatory periods terminated at higher Pa_{CO_2} values to contain more breaths, but the Pa_{O_2} at the end of the nonventilatory period had no significant effect on the number of breaths in the succeeding period of ventilation.

West et al. (1989) interpreted these findings to indicate that a Pa_{CO_2} threshold (or the corresponding pH_a) was an important chemical stimulus in determining the initiation of ventilation bouts in relatively undisturbed C. serpentina. This conclusion, although reasonable at the time, needs to be reexamined in the light of more recent findings in freshwater turtles. Isolated brainstems from the red-eared slider, Trachemys, show bursts of respiratory motor activity in hypoglossal nerve rootlets when superfused with a Hepes solution bubbled with 5% CO_2/95% O_2 (Johnson et al. 2002). In these preparations, respiratory activity persists after synaptic inhibition within the brainstem, suggesting not only that respiratory rhythm generation is intrinsic to the brainstem, but also that some neurons in the respiratory network have potential pacemaker properties under these conditions. In a similar preparation, Johnson et al. (1998) showed that the frequency of episodes of respiratory motor bursts increased as superfusate pH decreased in the range 8.1–7.4, demonstrating central pH/Pco_2 chemosensitivity. Respiratory motor discharge was insensitive to hypoxia, and a robust respiratory motor discharge was produced even after 2 h of superfusion with 5% CO_2/95% N_2 at 22°C. This result appears to be in accord with findings in intact Chelydra that Pa_{O_2} did not appear to be significant in initiating periods of breathing (West et al. 1989).

The normal breathing pattern in turtles, including Chelydra, is then almost certainly the result of complex interactions between central rhythm-generating processes in the brain and feedback from central, and possibly peripheral, chemoreceptors, as well as from other respiratory-related receptors such as pulmonary stretch receptors. In the anuran amphibian Bufo marinus, breathing is intrinsically periodic even under conditions in which the values of arterial blood gases and pH and lung volume are clamped at constant values (West et al. 1987). It is currently unknown whether such intrinsic respiratory periodicity exists in intact C. serpentina or indeed any chelonian. Certainly, this would appear to be a potentially profitable area of experimentation, in particular, in species with unicameral lungs that could be unidirectionally ventilated to clamp blood gases at constant values of Pco_2 and Po_2 as is possible in anuran amphibians (West et al. 1987).

More recently, Frische et al. (2000) investigated the regulation of respiration and oxygen transport by the blood in snapping turtles freely moving in an experimental tank at 25°C and exposed to hypoxia. The experimental protocol consisted of a stepped reduction of inspired oxygen levels from 21% O_2 through 15%, 10%, 5%, followed by return to normoxia. Each level of inspired oxygen was maintained for 24 h. The interval between breathing bouts in normoxia was comparable to that found by West et al. (1989) at 19 ± 4 min, but decreased to 4 ± 2 min when the animals breathed 5% O_2. A similar shortening of the nonventilatory period was found by Boyer (1963) in snapping turtles exposed to hypoxia in a dry, open-circuit respirometer. At 5% O_2, the number of breaths in a bout of breathing fell from three to five breaths in normoxia to a single breath in each bout, and the ventilation pattern became more regular (Frische et al. 2000). The studies of West et al. (1989) and Frische et al. (2000) concur that there is no, or little, chemoreceptor influence on the number of breaths per breathing bout at milder levels of hypoxia (10–15% O_2).

CARDIOVASCULAR
Cardiac Frequency in Ovo

The potentially wide range of temperatures experienced by reptilian embryos may have a profound influence on the cardiovascular physiology of these embryos, in particular, influencing the cardiovascular system's role as a conveyer of nutrients and respiratory gases to developing tissues. In Chelydra, Birchard & Reiber (1996) found that eggs incubated at 29°C had a higher heart rate (69–70 beats/min) for the first month than those incubated at 24°C (52–57 beats/min). Q_{10} was relatively low (1.6) during this phase of development. By day 42, heart rate was the same at both incubation temperatures and there was a slight reduction in heart rate just before, and a significant drop following, hatching. Heart rates, at any given temperature, were lower late in incubation than those observed earlier in incubation (Birchard 2000). Heart rate is still sensitive to acute temperature change after day 42; for example, 59-day embryos at 24°C still show a significant increase in heart rate when warmed to 29°C. The decrease in heart rate just before and after hatching may be related to changes in metabolic demand and the increasing efficiency of the lung as a gas-exchange organ. It is likely that it is also the result of the development of both afferent and efferent neural cardiovascular control mechanisms that result in a net increase in vagal (parasympathetic) tone to the heart.

Cardiorespiratory Interactions in Chelydra.

Changes in heart rate accompany changes in the intermittent pattern of breathing in adult *Chelydra serpentina*, in particular, those associated with the shortening of nonventilatory periods in response to hypoxia. It is agreed that average heart rate increases in hypoxia (Boyer 1963, 1966; Frische et al. 2000). Frische et al. (2000) found that heart rate averaged about 9 beats/min during normoxia at 25°C, but increased to 18 beats/min when animals were breathing 5% O_2. This was due largely to an increase in heart rate during nonventilatory periods to 16 beats/min, although no significant change occurred in heart rate during ventilatory periods. Boyer (1966) and Gaunt & Gans (1969) found an anticipatory increase in heart rate just before the ventilatory period. A small anticipatory increase in heart rate occurred as the head stretched toward the water surface, followed by a larger increase in heart rate as breathing started. This suggests that feed-forward as well as chemoreceptor and/or mechanoreceptor feedback mechanisms are involved in cardiorespiratory control in this species.

The relative increase in heart rate during the short nonventilatory periods at 5% O_2 suggests, in the view of Frische et al. (2000), that considerable gas exchange between lung gas and pulmonary blood may occur between breaths in *Chelydra*. The significance of the changes in heart rate in relation to cardiac output and intracardiac shunting during ventilatory and nonventilatory periods in resting *Chelydra* is a subject for speculation, as systemic and pulmonary blood flows have not been directly measured in this species. In resting green turtles (*Chelonia mydas*), tachycardia occurs during periods of ventilation and is associated with an increase in cardiac output and the ratio of pulmonary/aortic blood flow (West et al. 1992). Such tachycardia is almost certainly a result of feed-forward mechanisms, rather than feedback from pulmonary stretch receptors sensing lung volume (West et al. 1992; Herman et al. 1997). Frische et al. (2000) speculate that the relatively high heart rate (and presumably, cardiac output) during nonventilatory periods in severe hypoxia in *Chelydra* may be related to a high pulmonary blood flow during these periods. They suggest that this would assist in maintaining arterial blood oxygenation by maintaining pulmonary perfusion. In acute preparations of *Trachemys scripta*, increases in cardiac output above 40 ml·min^{-1}·kg^{-1} resulted in incremental increases of cardiac output, composed almost entirely of systemic venous return being directed to the lungs (Ishimatsu et al. 1996), thereby progressively reducing the proportion of cardiac output bypassing the lungs (R-L shunting).

The effects of ventilation and relative systemic and pulmonary blood flow on blood oxygen transport have been modeled for freshwater turtles (Wang & Hicks 1996). The model shows that, in normoxia, the pressure of arterial oxygen (Pao$_2$) is hardly increased by increases in ventilation, although reduction or elimination of systemic venous blood bypassing the lungs (R-L shunt) is effective in protecting Pao$_2$ (Wang & Hicks 1996). In hypoxia, on the other hand, the effect of increasing ventilation assumes a greater importance than reducing shunt flow in ameliorating a fall in the arterial pressure of oxygen (Pao$_2$). Frische et al. (2000) modeled the influence of increases in ventilation, pulmonary blood flow, and elimination of the R-L shunt on Pao$_2$ and arterial saturation in hypoxia (5% O_2) using the model of Wang & Hicks (1996). Increasing ventilation alone only produced a Pao$_2$ of 8.1 mm Hg and a hemoglobin saturation of 9.7%. Increasing ventilation and pulmonary blood flow and the elimination of the R-L shunt were estimated to produce a Pao$_2$ of 10.9 mm and a hemoglobin oxygen saturation of 15%. This rose to 28% at the same Pao$_2$ when the rise in pH$_a$ due to increased ventilation (respiratory alkalosis), and the resulting left shift in P$_{50}$ resulting from a Bohr factor of -0.95 (West et al. 1989) was taken into account. Thus a large left-shift of the P$_{50}$ (partial pressure of blood oxygen at half-saturation) from 29 to 18.6 mm Hg was significant in protecting hemoglobin oxygen saturation in hypoxia.

No changes occurred in hematocrit, hemoglobin concentration, hemoglobin synthesis, or ATP concentration in *C. serpentina* after 4 days of exposure to hypoxia (5% O_2) (Frische et al. 2000), indicating either that there are no molecular or red cell adjustments to hypoxia in *C. serpentina*, or alternatively that this period of hypoxia was too short to induce them. Similarly, Reese et al. (2002) also found no changes in hematocrit in *C. serpentina* submerged in anoxic water for up to 100 days, although hematocrit in animals submerged in normoxic water increased from 16% to 30% by the tenth day of submergence. Plasma pH$_a$ rose from 7.50 to 7.72 in hypoxia, probably reflecting a respiratory alkalosis.

Hypoxia tolerance in turtles, despite a lack of adjustments at the molecular level, in particular, unchanging triphosphate concentrations in response to hypoxia and anemia (Wang et al. 1999), probably reflects the effectiveness of the cardiorespiratory adjustments outlined above. Given the sedentary nature of *C. serpentina*, and its predictable ventilatory responses to hypoxia, this species may represent an ideal model in which to explore experimentally shunting of pulmonary and systemic blood flow.

FUTURE DIRECTIONS

The development of new technologies makes a convergence of ecology and physiology possible and profitable. Studies on the physiological ecology of the snapping turtle are now feasible with the aid of technologies such as time-depth recorders (TDRs) and enable us to answer questions at the interface between ecology and physiology. For example, do the submerged nonventilatory periods recorded in the laboratory (West et al. 1989) reflect the respiratory pattern of the animal in its natural environment? Furthermore, is this pattern influenced by time of day or changes in the thermal or the aquatic gaseous environment? For example, Gordos & Franklin (2002) recently used TDRs to compare the diving performance in two species of freshwater turtle, *Rheodytes leukops* and *Emydura macquarii*, respectively, with a high and low potential for aquatic respiration. This was reflected in differences in mean and maximum dive length for the two species, with *R. leukops* having longer periods of submergence.

Another potentially fascinating area of research is at the boundary between respiratory physiology and behavior. Periodic air ventilation in *C. serpentina* is the last step in a complex behavioral sequence leading to accessing the air/water interface. In nature, such behavior interacts with other behavioral suites such as those involved in the feeding repertoire. Van Sommers (1962) showed that immature *Pseudemys scripta* held in a sealed, water-filled chamber could be trained to lever-press to secure access to a recess in the chamber that was filled with air for periods of 5–10 s. The response could be extinguished if not reinforced by the air reward. The most usual pattern of response was for the turtle to place its head in the recess and then to lever-press for the recess to be filled with air. The response was reduced if 10% CO_2 was added to air, approximately doubled if the ambient temperature was increased by 10°C, and increased if the diet was enriched. These findings have not been replicated in any species, although such an approach seems useful in investigating periodic ventilation in turtles. This is, after all, a complex behavioral as well as a physiological phenomenon. It would be interesting to follow up on these findings. The sedentary nature of *C. serpentina*, combined with its tolerance of a wide range of temperatures and trophic environments, suggests that it would be a useful model species for such work.

Reproductive Physiology of the Snapping Turtle

IBRAHIM Y. MAHMOUD

ABDULAZIZ Y. A. ALKINDI

MORPHOLOGICAL CHANGES associated with the gonadal cycles have been studied thoroughly in several freshwater turtles, including the snapping turtle. Some of these studies were based on gravimetric and histological investigations (Ernst & Barbour 1972; Moll 1979; Licht 1984). Other studies examined seasonal steroid levels (Callard et al. 1978; Lewis et al. 1979; Licht et al. 1980; Kuchling et al. 1981; Licht 1982, 1985a,b; McPherson et al. 1982; Silva et al. 1984; Mahmoud et al. 1985a, 1985b, 1992; Mendonca & Licht 1986a; Mesner et al. 1993). Lastly, a few studies investigated the relationship between plasma gonadotropins and plasma steroid levels (Licht et al. 1985a; Mendonca & Licht 1986a, 1997).

The common snapping turtle (*Chelydra serpentina*) extends over a wide range throughout North America, with different geographical populations associated with specific ecological and climatic conditions. These conditions can influence the pattern of gonadotropin (Gn) release, gonadal events, and the production of sex steroids, mainly estradiol (E_2), progesterone (Pro), and testosterone (T). Moreover, ecological factors and geographical distribution may influence the duration and the pattern of the gonadal cycle in turtles (Licht 1982). These factors are crucial in influencing specific physiological adjustments relative to hormonal profiles and gonadal events. Maintaining a high hormonal profile throughout the short gonadal cycle, as is the case in snapping turtles from northern latitudes, may be a valuable physiological adaptation. Snapping turtles in northern latitudes have short ovarian and testicular cycles (95–100 days) with a thermoactivity range (TAR) of 18–26°C from middle or late May to early September. Most of the freshwater turtles from more southern latitudes have a longer gonadal cycle, which begins in March and ends in November because of the warmer temperatures and longer photoperiodicity. In the extreme southern-most region of its range snapping turtle may have an acyclical pattern with reproductive activity occurring during most of the year (Licht 1982).

The timing of male and female reproductive cycles of snapping turtles in northern populations (e.g., Wisconsin) is closely related to temperature (Mahmoud & Licht 1997). Temperature may alter gonadal response to circulating steroids and gonadotropins (Licht et al. 1985a; Mahmoud & Licht 1997). A rise in temperature during spring and summer appears to be the major factor that stimulates gonadal growth, spermatogenesis, vitellogenesis, and elevation in hormonal levels. Conversely, lower temperatures in the fall cause a drop in hormone levels and, consequently, testicular regression and ovarian quiescence. It is also possible that a decrease in temperature can lower the sensitivity of the gonads to gonadotropin, which is concomitant with testicular regression (Licht 1975; Mendonca & Licht 1986a). In snapping turtles, gonadal quiescence begins when water temperature falls below 18°C in the fall, and hormonal levels subsequently drop. Conversely, a slight rise above 18°C in the spring can stimulate a slow rise in hormonal levels, sufficient to stimulate gonadal growth.

Snapping turtles in northern latitudes do not typically bask and exhibit minimal if any behavioral thermoregulation, which tends to complicate comparative thermal relationships among freshwater species. Behavioral thermoregulation is a common practice in most freshwater turtles, serving to keep their body temperature within a TAR by basking when water temperature changes. Thus, in snapping turtles, body temperature remains stable and reflects the water temperature at all times.

In determining the TAR of the snapping turtles in Wisconsin, field observations such as feeding, swimming, and mating were recorded (Mahmoud & Licht 1997). Also, water and cloacal temperatures were recorded. Field data indicated the values for cloacal temperature were similar to and highly correlated ($P = 0.001$) with water temperature, consistent with the apparent lack of basking behavior in this species (Mahmoud & Licht 1997); however, we cannot rule out the possibility of limited behavioral thermoregulation by this species. Snapping turtles were observed to surface and feed under floating mats of plants such as algae or water lilies where water temperature is usually higher than the surroundings. Field observations indicated that snapping turtles were active within a temperature range of 18–26°C but they became inactive above or below this range. Therefore, it was concluded that the TAR for snapping turtle in Wisconsin is 18–26°C (Mahmoud & Licht 1997).

This chapter addresses morphological and biochemical gonadal events in both male and female snapping turtles relative to the dynamics of hormonal levels, including sex steroids and gonadotropins. Histological and ultrastructural changes were related to growth and development. The hormonal levels in both male and female turtles were monitored during captivity to evaluate the degree of induced stress throughout the cycle. In addition, experimental procedures relevant to luteolysis and induced ovulation are described and compared with natural conditions.

OVARIAN CYCLE

The snapping turtle is a single clutch (monoclutch) species, thus ovulation occurs only once during the reproductive season. The ovarian cycle in snapping turtles from northern populations begins in May and terminates in early September. The cycle can be divided into three phases: preovulatory–ovulatory (mid May–mid June), postovulatory or quiescence (mid June–mid July), and growth or vitellogenic (mid July–early September) (Fig. 6.1). At the onset of the cycle, the ovaries are at their maximum mass and contain large preovulatory follicles (18–22 mm in diameter). During ovulation, all the preovulatory follicles enter the uterine horns within hours (24–48 h). After ovulation is completed, the ovaries are at a minimum mass and the follicles contained within are, in general, 5 mm or smaller in diameter. The ovaries remain in this condition throughout the egg retention period (~2 weeks). Vitellogenesis begins one week after oviposition in July. At this time luteolysis is already in progress while the eggs are in the oviduct (Mahmoud & Licht 1997). The follicles begin to grow rapidly and become fully mature preovulatory follicles by middle to late August. Preovulatory follicles remain in this condition throughout the winter and undergo ovulation the following spring. During this period the ovaries contain steroidogenically active corpora lutea that remain active while the eggs are in the oviduct and continue to be active for a short period after oviposition (Mahmoud & Licht 1997).

Luteolysis starts while the eggs are still in the oviduct with a slow regression in the size of the corpora lutea until they finally disappear from the ovarian stroma during early vitellogenesis. These follicles are numerous and are found in different sizes and numbers throughout the year. A subpopulation of these follicles undergoes rapid growth to become preovulatory follicles at the end of the vitellogenic phase. During growth the developing follicles reach 10–14 mm during late July–early August, and by mid August to early September they reach maturity size.

OVULATION

Field observations revealed that ovulation occurred when water temperature rose above 18°C during May and early June. This is based on 23 turtles captured and immediately palpated behind the plastral bridge for the presence of eggs (Mahmoud & Licht 1997).

In captivity six of seven turtles ovulated between May and June. The turtles were captured in September with fully grown preovulatory follicles and were kept in an outdoor tank in 4 ft water until the following spring. They did not accept food between September and April as water tempera-

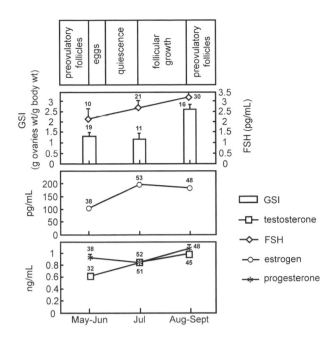

Fig. 6.1. Mean levels of ovarian gonadosomatic index (GSI), T, FSH, plasma E$_2$, and Pro during phases of ovarian cycle for female *Chelydra serpentina* in Wisconsin 1989–1993. Vertical lines represent 1 standard error (SE); numbers are sample sizes. Radioimmunoassay (RIA) procedures for steroids were the same as those used by Mahmoud et al. (1989). Plasma FSH was measured by RIA based on rabbit antiserum raised against *Chelonia mydas* (Licht & Papkoff 1985). Adapted from Mahmoud & Licht 1997.

ture was below their TAR. After April, the turtles accepted food when the temperature rose above 18°C, and eggs were detected in the oviduct shortly after. After ovulation, circulating E$_2$ and Pro levels were significantly higher than before ovulation in these captive turtles (Mahmoud & Licht 1997). This suggests that the active corpora lutea begin to synthesize sex hormones (mainly Pro) which are essential for the development and maintenance of eggs in the oviduct.

Female Hormonal Levels Based on Freshly Caught Turtles from Natural Populations

Snapping turtle hormone levels were compared with the gonadosomatic index (GSI) (g gonad wt/g body wt = $wt \times 100$). The GSI was used to minimize body weight effects (Mahmoud & Licht 1997). Fig. 6.1 presents the main events of the ovarian cycle. Follicle-stimulating hormone (FSH) was at low levels during ovulation and postovulation but afterward rose for the rest of the summer. However, no significant differences existed among the samples taken during the cycle. There was a significant correlation between GSI and FSH but FSH was not significantly correlated with the steroids (P <0.01). It appears that FSH is essential for follicular growth as FSH began to rise after June before vitellogenesis (Fig. 6.1).

During the postovulatory period (mid June–mid July), the ovaries were at a minimum mass, as E$_2$ levels significantly increased for the rest of the cycle in association with the rise

in FSH and subsequent follicular maturation and growth (Fig. 6.1) (Mahmoud & Licht 1997).

The elevation in E$_2$ levels through the postovulatory period is comparable to that observed in other turtle species. In loggerhead turtles (*Caretta caretta*), approximately 4–6 weeks prior to migration from feeding grounds to mating and nesting, circulating E$_2$ levels increased significantly and remained high during the follicular growth (Wibbels et al. 1990, 1992). In freshwater turtles, E$_2$ remained significantly elevated throughout the period of follicular growth (Callard et al. 1978 on *Chrysemys picta;* McPherson et al. 1982; Mahmoud et al. 1985b; Mendonca & Licht 1986a on *Sternotherus odoratus;* Mahmoud & Licht 1997 on *C. serpentina*). Specifically in *C. picta,* the growing follicles secrete all three steroids under the influence of higher gonadotropin levels and it was suggested that a rise in E$_2$ values is associated with vitellogenesis and sexual behavior (Callard et al. 1978). McPherson et al. (1982) stated that the first peaks of estradiol in *S. odoratus* are associated with renewal of vitellogenic activity in the summer and before the formation of clutches.

In the snapping turtle, Pro levels were high in June (Fig. 6.1) because of active corpora lutea while the eggs were still in the oviduct. No follicular growth occurred at this time. After oviposition there was a slight drop, but Pro rose again at the onset of vitellogenesis and follicular growth in July (Fig. 6.1). There were significant differences among Pro samples throughout the cycle (Mahmoud & Licht 1997). Moreover, we cannot rule out the possibility that a brief Pro surge occurred shortly after ovulation that may have been missed during blood sampling.

An elevation in FSH accompanied by a rise in Pro during the preovulatory period is similar to the surges in luteinizing hormone (LH) and FSH levels observed in *C. caretta* (Wibbels et al. 1992). They suggested that these hormones might facilitate follicular growth, ovulation, egg production, and nesting. Although FSH values have a similar profile in both studies, Pro values in *C. caretta* were approximately 25 times higher.

LH, an important gonadotropin, is undetectable throughout the ovarian cycle. Since radioimmunoassay (RIA) showed high cross-reaction with snapping turtle LH, failure to measure LH suggests very low circulating levels during the ovarian cycle (Mahmoud & Licht 1997). However, the interpretation of these data is limited by a lack of direct evidence for detection of circulating LH in snapping turtles. Moreover, the role of LH in reptiles is still unclear except for preovulatory surges that occur in association with FSH and elevated Pro in several species of sea turtles (Licht et al. 1979, 1980; Licht 1982; Wibbels et al. 1992). LH release in the snapping turtles may surge briefly and may have been missed when blood samples were taken (Mahmoud & Licht 1997). These rapid hormonal changes are associated with mating behavior, egg production, ovulation, nesting, and vitellogenesis. Even though we were unable to detect LH in snapping turtles, some increase in FSH was detected in the ovulating turtles.

FSH, previously undetected in other female freshwater turtles, was observed for the first time at a detectable level in a female snapping turtle by Mahmoud & Licht (1997). The significance of such a finding is still unclear, however, as only a few studies have been conducted on gonadotropins thus far.

Testosterone levels were lower in mid May to June, but rose significantly in August in association with follicular growth and vitellogenesis (Fig. 6.1). Such association may suggest the involvement of this steroid in these processes. Under natural conditions T may also be involved in ovulation, as well as in sensitizing the follicles and corpora lutea to LH stimulation (Wibbels et al. 1992). In snapping turtles, a strong correlation exists between E_2 and T during vitellogenesis and follicular maturation.

PREOVULATORY FOLLICLES

Full-sized mature (preovulatory) follicles are ready for ovulation in spring. On the free side (unattached to ovarian stroma) of each follicle there is an elongated slit, the aperture of rupture for the passage of follicle during ovulation. This slit will become the central cavity of the corpus luteum. The theca consists of an outer portion (theca externa) and inner portion (theca interna) which contain blood vessels (Cyrus et al. 1978). The follicular granulosa cells are compressed between the zona pellucida and the theca interna. The granulosa cells are steroidogenic and very active during this stage. Microvilli originate from the granulosa cells and the oocyte. The granulosa cells are elongated with large nuclei. The endoplasmic reticulum is indistinct with numerous microfilaments. The mitochondria are elongated with lamellar cristae. There are very few fat droplets, but they increase significantly during luteinization (Cyrus et al. 1978).

The morphological changes relative to follicular growth during the ovarian cycle have been investigated in several chelonian species (Moll 1979). In another study on ovarian follicles, mature female snapping turtles from Wisconsin were collected throughout the year to examine size and condition of follicles during phases of the ovarian cycle (Mahmoud et al. 1985). Follicle growth was compared with changes in plasma estradiol levels and plasma levels of calcium, total protein, inorganic phosphate, and cholesterol. In addition, the activity of Δ^5-3β-hydroxysteroid dehydrogenase (3β-HSD), an important enzyme in conversion of Δ^5-3-hydroxysteroids to Δ^4-3-ketosteroids in the granulose follicular and theca cells of snapping turtle, was examined in relation to follicle size and plasma estradiol levels.

The results revealed that plasma E_2 levels coincide with size and development of follicles. The E_2 levels rose during preovulatory phase in spring while the follicles were at their maximum size (18–22 mm) with steroidegenically active corpora lutea. Shortly after oviposition, follicular growth commences with luteolysis already underway. By early August the follicles had increased to 10 mm and E_2 levels increased significantly (P <0.05). As E_2 levels continued to rise, the follicles reached their maximum preovulatory size by late August to early September, ready for ovulation the following spring.

McPherson et al. (1982) reported that high estrogen levels are associated with vitellogenic periods during autumn and from spring to early summer in *S. odoratus*. The high E_2 levels during preovulatory period in spring in snapping turtles are perhaps necessary for not only the final maturation of follicles prior to ovulation but also for building up the glandular tissues in the reproductive tract. Our observations with a scanning electron microscope (Cyrus et al. 1982) indicate that the lining of the oviduct is highly developed during preovulatory phase.

Plasma calcium, total protein, and inorganic phosphate rose significantly (P <0.05) during late vitellogenesis. However, albumin, which is a major component of total protein, remained largely unchanged during the cycle with only a slight rise during vitellogenesis.

In other chelonians, estrogen causes an increase in plasma calcium (Clark 1967; Callard et al. 1978) which is directly related to the production of vitellogenin since the latter is a calcium-binding protein complex. The pattern of plasma phosphate resembles that of calcium, which is essential for the production of vitellogenin (Clark 1967), and total plasma protein is associated with the rise in vitellogenin in the western painted turtle (*C. picta bellii*) during vitellogenesis (Callard et al. 1978).

In the snapping turtles, cholesterol levels remain relatively constant throughout the cycle, except during vitellogenesis when there is a slight rise. The lack of a significant increase may indicate that cholesterol is continuously used as a sterol pool during the cycle. This is contrary to the multi-clutch western painted turtle, where there is a significant rise in cholesterol during early spring followed by a drop prior to ovulation, which suggests follicular uptake (Callard et al. 1978).

The histochemical reaction for enzyme 3β-HSD follows the same pattern as E_2 relative to follicular events. It becomes strong during the preovulatory phase, mainly in the granulose follicular cells. Preovulatory follicles examined in spring and late summer have relatively large quantities of the enzyme compared with preovulatory follicles examined in January. Postovulatory follicles examined in July have less of this enzyme than is found in large follicles.

It appears that vitellogenesis and development of follicles in this species are stimulated by rising E_2 levels. The high intensity of 3β-HSD, which indicates ovarian steroidogenesis, also correlates with follicular growth and E_2.

In another aspect of this study (Mahmoud et al. 1985b), snapping turtles that had already oviposited their eggs (mid June) were collected from the field in central Wisconsin and were injected intraperitoneally 24 h after capture with estradiol benzoate at 48-h intervals. The controls were given only ethanol:propylene-glycol. The E_2-injected turtles had significantly higher ($P <0.001$) levels of serum cholesterol, calcium, and phosphate over the controls. In the experimental group, plasma proteins also increased, but not significantly. The increase in these parameters did not take place until the fourth injection. The slow response to E_2 might be related to the condition of the ovaries because this species is a monoclutch species where all the follicles of mature size had already ovulated at the time the experiment was conducted. Moreover, the ovaries were in quiescent phase with small follicles (5–7 mm). The rise in the plasma parameters after E_2 injections again suggests that E_2 is essential for follicle maturation and vitellogenesis which coincides with the natural condition (Mahmoud et al. 1985b).

The relationship of steroid regulation to follicular growth and glandular development has also been monitored through hormone–receptor interactions during the ovarian cycle. Mahmoud et al. (1980, 1986) detected high-affinity progesterone-binding components in the oviductal lining of the snapping turtle. The physiological chemical properties of these compounds are similar to those of western painted turtle receptors (Ho and Callard 1984). Sexually mature female snapping turtles were collected from the field during the following phases: preovulatory and vitellogenic (low Pro, high E_2, 18- to 22-mm follicles); early postovulatory (high Pro, low E_2, 5- to 7-mm follicles, active corpus luteum [CL]); and inactive (hibernation: low Pro and E_2, 18- to 22-mm follicles). Some turtles were sacrificed immediately for receptor assays and RIA of Pro and E_2. Others were kept for at least one week and injected with estradiol-17β dissolved in 7:3 (v/v) propylene glycol-ethanol solution (0.01 μg/g body wt) to test the effect of E_2 priming on receptor levels. These turtles were also used for receptor assays. Results suggest the presence of two high-affinity Pro binding components: R-sites specific for Pro binding and Q-sites for binding both Pro and corticosteroids (Mahmoud et al. 1985b). During the preovulatory and vitellogenic phases the binding components were detectable without E_2 priming. In contrast, during the early postovulatory and hibernation phases the components were undetectable despite E_2 priming. This supports previous studies suggesting that Pro may depress the synthesis of its own receptors (Vu Hai et al. 1977). Evidently E_2 priming during early postovulatory phase cannot override the effect of a large output of Pro from the corpus luteum. During hibernation the ovaries are inactive and binding components were undetectable. During the preovulatory and vitellogenic phases Pro levels were low and thus Pro receptors were produced.

Progesterone is essential for stimulating glandular activity such as albumen and shelling secretion while the eggs are still retained in the oviduct. A previous study on this species revealed that Pro has an inhibitory effect not only on its own receptors but also on the growth of follicles when the corpora lutea are still steroidogenically active (Mahmoud et al.1985b). Follicular growth commences immediately after luteolysis.

Steroidogenic Activity of Corpus Luteum Based on Ultrastructural Features

After ovulation the thecal wall of the preovulatory follicle collapses and becomes thicker and folded. Eventually the folds disappear and the follicular wall becomes a disc-shaped corpus luteum. Under the influence of luteinization, the granulosa follicular cells of the preovulatory follicle are transformed to granulosa lutein cells, which become the steroidogenic cells of the corpus luteum (Cyrus et al. 1978).

After the collapse of the thecal layers, the thecal connective tissue with their blood vessels surrounding the "aperture of rupture" invaginates, resulting in the entrapment of the lutein cells between the two layers. The lutein area is avascular as with other turtles (Rahn 1938; Altland 1951). Septae from the internal theca invade the granulosa, which becomes the blood supply of the lutein cells.

The lutein cells have abundant tubular and cisternal agranular endoplasmic reticulum (AER) throughout, which is frequently interconnected. The tubular AER is very dense and closely spaced, whereas the cisternal AER is loose throughout (Fig. 6.2).

The cisternal ER is frequently organized into concentric whorls or elaborate folded membrane arrays. The whorls in some areas are interconnected with one or more adjacent whorls. The folded membrane arrays and the whorls are often associated with lipid droplets or mitochondria. In the mammalian corpus luteum, an abundance of smooth endoplasmic reticulum is arranged in concentric whorls, and active mitochondria are closely associated with lipid droplets (Fig. 6.2). These features are also found in snapping turtle granulosa lutein cells, and cholesterol synthesis presumably takes place in the lipid droplets (Klicka & Mahmoud 1973). Cholesterol undergoes side-chain cleavage before entering the steroidogenic pathway.

Pregnenolone synthetase, the enzyme responsible for this cleavage, probably resides in the mitochondrion (Dorfman & Unger 1965). The close association between lipid droplets and mitochondria suggests a cholesterol–mitochondrion interaction. Lipid droplets may supply the mitochondria with sterol precursor as well as a source of energy. A mechanism for progesterone synthesis in luteal tissue involving the transport of lipid between the AER and mitochondrion was suggested by Bjersing (1967).

Steroidogenic Activity of Corpus Luteum Based on Biochemical Analysis

Klicka & Mahmoud (1972, 1973) reported that corpus luteum homogenate is capable of converting pregnenolone to progesterone and cholesterol to progesterone. In another

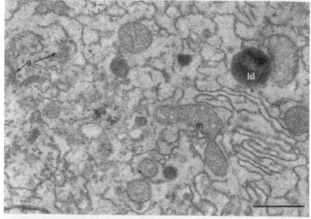

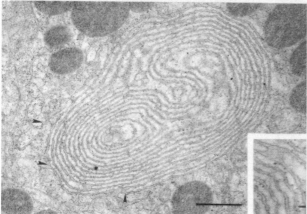

Fig. 6.2. (*upper*) Branched cisternae with some areas joined to each other to form a network dividing the cytoplasm into irregular islands. Note the folded array of cisternae in the lower right corner of the photograph and the Golgi complexes (g) in the upper left. ld is a lipid droplet. Bar = 1 μm. (*lower*) An elaborate cisternal whorl, a common feature found in active corpora lutea. Note the radial interconnections between the cisternae (*inset*) and the continuity between the cisternae of the whorl and other cisternae (arrows). Bar = 1 μm. (Inset: bar = 0.1 μm). Adapted from Cyrus et al. 1978.

weeks of their life span, with very low production in the last three weeks. The sharp decline in steroid production in July may be related to the disruption of the AER. Cyrus et al. (1978) found that fibrosis of the corpus luteum in the snapping turtle is accompanied by a massive disruption of AER. This may lead to a reduction in the enzyme Δ^5-3β-steroid dehydrogenase, which is associated with the AER.

In mammals, Δ^5-3β-steroid dehydrogenase is associated with AER and is responsible for converting pregnenolone to progesterone. The histochemical localization method of Bara and Anderson (1973) for Δ^5-3β-steroid dehydrogenase in frozen sections of corpora lutea of snapping turtles (unpublished data) revealed that the enzyme was abundant in the corpus luteum in June but absent in July.

Degeneration of Corpus Luteum (Luteolysis)

The invasion of the granulosa cells by connective tissue from the thecal septae as well as the connective tissue from the central cavity marks the onset of the process of luteolysis. The CL remains steroidogenically active for about 5–6 weeks, but it begins to regress in size shortly after oviposition. As a result, most of the granulosa luteum cells are isolated into islands surrounded by connective tissue. After that, a further regression takes place until all the corpora lutea disappear from the ovarian stroma.

Luteolysis is a gradual process associated with a general disruption of AER and mitochondria. The tubular AER becomes extremely vesiculated, but no significant change occurs in the number of lipid droplets of the luteum cells compared with the active corpus luteum (Cyrus et al. 1978).

study (Mahmoud et al. 1980), homogenates of corpora lutea were incubated with [4–^{14}C] pregnenolone. Two sets of conversions were investigated: (1) homogenates from active corpora lutea in June (4–7 mm diameter from six turtles) with well-developed endoplasmic reticula, and (2) homogenates from inactive corpora lutea in July (less than 4 mm diameter from six turtles) with general disruption of endoplasmic reticula and mitochondria.

Homogenates from June corpora lutea showed a positive correlation with respect to time of incubation and tissue concentration. Longer incubation time and/or higher tissue concentrations yielded greater synthesis of progesterone (Fig. 6.3). The maximal rate of progesterone synthesis occurred with 160 mg equivalents of tissue incubated for 20 min. The conversion percentages of corpora lutea of less than 4 mm diameter were extremely low and were not included in the plotted data set (Mahmoud et al. 1980).

The conversion study demonstrated that the corpora lutea are capable of producing steroids during the first 2–3

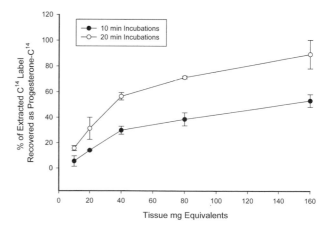

Fig. 6.3. Conversion of pregnenolone to progesterone by different amounts from homogenates of 4.0- to 7.0-mm corpora lutea from six June turtles during two different incubation periods. Results are presented as means ± SE. Longer incubation time and/or higher tissue concentrations yielded greater synthesis of progesterone. The maximal rate of progesterone synthesis occurred with 160 mg equivalents of tissue incubated for 20 min. The conversion percentages for corpora lutea smaller than 4.0 mm (early July) were extremely low and were not plotted. Adapted from Mahmoud et al. 1980.

Role of Arginine Vasotocin and Prostaglandin F$_{2a}$ in Oviposition and Luteolysis

Oviposition was not induced by administration of exogenous arginine vasotocin (AVT) to early gravid snapping turtles with steroidogenic CL and high Pro levels. In contrast, 12 days later when luteolysis was underway and Pro was at low levels, administration of AVT induced oviposition.

Controls injected with saline on the third day (high Pro levels) or on the 12th day (moderate Pro levels) after ovulation did not oviposit immediately. Instead, they laid their eggs 15–23 days after ovulation when there was a significant drop in Pro compared to days 3 and 12. In addition, in the control turtles, the eggs retention time in the uterus was similar to turtles under natural conditions (Mahmoud et al. 1988).

In another study, Mahmoud et al. (1987) investigated the influence of progesterone, estradiol-17β and AVT on in vitro contractility of oviductal segments during the ovarian cycle. Mature snapping turtles were collected in Wisconsin during the following phases: preovulatory, early postovulatory (active CL), late postovulatory (inactive CL), and vitellogenic. In the first experiment (Exp. 1) turtles were sacrificed 24–48 h after capture. Three segments were taken from each oviduct: anterior, middle, and posterior. Each segment was attached to a physiograph to record the amplitude, frequency and latency of contraction. Some segments were treated with AVT, others only saline. In a second experiment (Exp. 2), turtles were injected daily with either progesterone or estradiol-17β (0.01 μg/g) for one week. After surgical removal, oviductal segments were not treated with AVT but physiograph measurements were the same (for details, see Mahmoud et al. 1987).

Segments with AVT (Exp. 1) showed increased contractility (amplitude and frequency) and decreased latency between contractions, except when the CLs were steroidogenically active (early postovulatory phase). The lack of responsiveness to AVT during early postovulatory phase suggests that high Pro and possibly other luteal steroids or factors may inhibit AVT influence during egg retention periods. Thus, high Pro levels inhibit the contractility of the oviduct while eggs are present so the shelling process can take place.

In Exp. 2, turtles treated with Pro had decreased contractility during early postovulatory phase but no measurable change during late postovulatory and vitellogenic phases. In the estradiol-17β group, increased contractility was recorded during all phases except early postovulatory phase.

The data presented here support the hypothesis that Pro and other luteal factors decrease uterine contractility during active CL phase regardless of AVT presence (Jones & Guillette 1982; Guillette & Fox 1985). In other investigations, administration of exogenous Pro inhibited uterine contractions in vitro in *C. picta bellii* (Callard & Hirsch 1976) and *C. serpentina* (Mahmoud et al. 1984, 1987). Jones et al. (1982) reported that when corpora lutea are active, AVT has no effect on the in vitro contractility of the uterus in the lizard *Anolis carolinensis*. However, AVT was effective 24 h after deluteinization. Figler et al. (1986) reported that the endogenous AVT in sea turtles (*Lepidochelys olivacea* and *C. caretta*) increased 200- to 250-fold from base levels during oviposition and then decreased to lower levels after nesting exercise.

The increase in AVT may be caused by the degeneration of CL, which triggers a sensory response from the ovary to the hypothalamus for the AVT secretion. Early removal of CL (deluteinization) is further evidence of the luteal influence in blocking AVT release during the time of egg retention.

In reptiles, a correlation exists between steroidogenic activity of CL and egg retention. Roth et al. (1973) and Cuellar (1979) reported that deluteinization induced premature oviposition in lizards. Klicka & Mahmoud (1977) observed that bilateral ovariectomy has the same effect on gravid *C. picta bellii*. Arslan et al. (1978) concluded that luteal steroid levels (progesterone, estradiol, and testosterone) peak during the middle of gravidity in lizard *Uromastix hardmicki* and fall before oviposition. Similarly, in snapping turtles, circulating Pro peaks in early postovulatory phase and then falls just before oviposition (Mahmoud & Licht 1997).

A single administration of prostaglandin F$_{2\alpha}$ (PGF) to snapping turtles with active CL induced luteolysis (Mahmoud et al. 1988). This was confirmed by a decrease in endogenous Pro concentrations within 30 h, and also by the loss of ultrastructural steroidogenic features such as smooth endoplasmic reticulum and mitochondria with tubular cristae. Saline-injected turtles maintained the same Pro levels. In addition, PGF triggered the invasion of a hyaline material from the theca to the granulosa within 48 h of treatment. Hyaline invasion was also reported during natural luteolysis in this species (Cyrus et al. 1978). It was suggested that the hyaline material might interfere with cellular transport and thus enhance luteolysis. Moreover, in control turtles with active CL, hyaline material was not present. The administration of PGF to the lizard *A. carolinensis* caused a significant decline in lipid vacuoles and lysosomes as well as loss of AER (Guillette et al. 1984).

The process of luteolysis in the snapping turtle under natural conditions is a gradual process and may take 2–3 weeks after ovulation; thus, PGF accelerated the luteolysis process. Furthermore, luteolysis might be triggered by prostaglandins of ovarian or oviductal origin. The high plasma levels of luteal steroids may prevent luteolysis by inhibiting synthesis of prostaglandins.

THE EFFECT OF STEROIDS ON OVIDUCTAL MORPHOLOGY

The histological and ultrastructural changes in the reproductive tract (oviduct) of the snapping turtle during active phases of the ovarian cycle were related to steroid hormones such as progesterone, estradiol, testosterone, and follicle-stimulating hormone (AlKindi et al. unpublished data). The increase in hormone levels coincided with an increase in the number and growth of endometrial and luminal epithelial glands and an increase in the secretory activity. For example, while circulating steroid levels were high, direct in situ observations were made of freshly ovulated reproductive tracts of snapping turtles in different stages of gravidity. Albumen deposition in uterine tubes and eggshell membrane formation in oviducts were evident. These events suggest a close relationship between the presence of high plasma steroid levels and endometrial glandular activity. Overall, the high hormone levels may play a major role in development, maintenance, and secretory activity of the oviductal glands.

TESTICULAR CYCLE

The cycle commences in mid to late May when the water temperature exceeds 18°C in Wisconsin. In the beginning, the testes are small and flaccid, at a minimum mass. There is a significant increase in testicular mass between mid July and late August, coincident with spermiogenesis and spermiation. The seminiferous tubules in association with testicular development reach their maximum mass by August. The regression of the testes in the fall is caused by lowering temperatures and it is possible that a drop in temperature can reduce the sensitivity to gonadotropin (Licht 1975; Mendonca & Licht 1986a). Accelerated testicular regression in the snapping turtle occurs with significant drop in temperature below the TAR for a few days. On the other hand, early phases of spermatogenesis in the spring begin under low temperature and with low FSH levels, which are nevertheless sufficient to stimulate steroidogenesis in Leydig cells.

The epididymis attains a maximum mass in late August to early September as the sperm begin to migrate into epididymal luminae during the fall mating period. The epididymi remain in large mass until the following spring and begin to regress gradually, reaching minimum mass by early July.

Testicular growth and spermiation have been investigated in several different chelonian species throughout the temperate zone: notably on *C. picta* (Christiansen & Moll 1973; Mitchell 1985a; Moll 1973) and *S. odoratus* (Risley 1938; Klicka & Mahmoud 1972; McPherson & Marion 1982; McPherson et al. 1982; Licht et al. 1985a; Mitchell 1985b; Mendonca & Licht 1986b). These investigators reported that the peak of growth and spermiation occurs during the summer and terminates in the fall. This is followed by testicular regression. In the southern populations, spermatogenesis and growth begin earlier and terminate later than in northern populations.

In Wisconsin snapping turtles, maximal testicular mass peaks during the second half of July through August coincident with spermiogenesis and spermiation. However, testicular recrudescence in snapping turtles from Tennessee commences during early spring with maximal testicular mass and spermiation attained in October (White and Murphy 1973).

Seasonal steroidogenesis in chelonian testes has been well documented in several species. These studies described morphological, histochemical, and hormonal levels on several genera: *Chrysemys* (Licht 1972; Lofts & Tsui 1977; Tsui & Licht 1977; Dubois et al. 1988); *Pseudemys* (Licht 1972; Tsui & Licht 1977); *Kinosteron* (Licht 1972); *Trionyx* (Lofts & Tsui 1977); *Gopherus* (Tsui & Licht 1977); *Chelonia* (Licht et al. 1979); *Caretta* (Wibbles et al. 1987, 1990); and *Chelydra* (Mahmoud et al. 1985a; Mesner et al. 1993; Mahmoud & Licht 1997).

The presence of steroidogenic ultrastructural features such as smooth endoplasmic recticulum (SER) and mitochondria with tubular cristae during the testicular cycle in Leydig and Sertoli cells reflects the changes in hormone levels (Mesner et al. 1993 in turtles; Mahmoud et al. 1985, and Mahmoud & Licht 1997 on *C. serpentina*; Dubois et al. 1988 on *C. picta*). At the end of the testicular cycle in August, Leydig cells in snapping turtles become robust and well developed with tubular SER and mitochondria with tubular cristae. Reactions for 3β-HSD and cholesterol are strong. Throughout the winter the SERs become vesiculated with other features remaining unchanged, and the reactions become weak. In spring, under the influence of the rising T and FSH, the tubular SER reappears with a depletion of cholesterol-rich lipid droplets and moderate reactions for 3β-HSD, signifying the beginning of steroidogenic activity (Mahmoud et al. 1985a).

In early spring Sertoli cells remain inactive as they are still in winter phase. However, at the onset of spermatogenesis during late spring, the Sertoli cells become active under the influence of T and FSH and commence steroidogenic activity through the summer (Mahmoud et al. 1985a; Mahmoud & Licht 1997). At the end of the testicular cycle, Sertoli branches withdraw from the center of the lumen and there is a general disruption of ER, a decrease in mitochondria, and an increase in lipid droplets. 3β-HSD and cholesterol are present at moderate levels.

The accumulation of cholesterol-rich lipid in Leydig and Sertoli cells of reptiles is indicative of a decline in steroidogenesis. Conversely, the depletion of these compounds indicates an increase in steroidogenesis (Lofts 1972; Lofts & Berns 1972; Lofts & Tsui 1977). In chelonians, depletion of cholesterol-rich lipid in Sertoli cells occurs during spermatogenesis, suggesting conversion of cholesterol into androgens (Lofts 1972). However, in another study on the snapping turtle, there was no depletion of lipid droplets in Leydig cells at any time (Mahmoud et al. 1985a). Thus, the accumulation or depletion of lipid droplets cannot be used as an indicator of steroidogenesis in Wisconsin snapping turtles. In Sertoli cells, however, the accumulation and depletion of lipid droplets are similar to those observed in other chelonians.

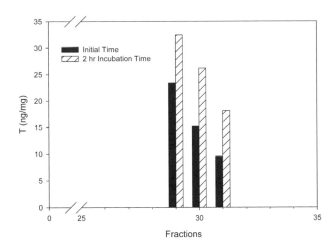

Fig. 6.4. Testosterone levels of Leydig cells isolated on Percoll gradient. Fractions either assayed for T at initial time (filled bars) or at 2 h incubation time (hatched bars). Note the presence of T in fractions 28–31 only. Also, following 2 h incubations, Leydig cells were found to produce significant amounts of T (P <0.05) compared with controls. Adapted from Mesner et al. 1993.

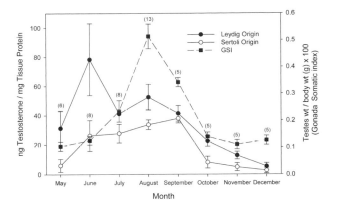

Fig. 6.5. Seasonal changes in Sertoli and Leydig cell culture for T levels and gonadal somatic index in snapping turtles from natural populations. Peak in Sertoli cell T synthesis was in conjunction with maximal testicular mass, which is associated with spermiogenesis and spermiation late July to early September. Leydig cell T levels peaked late May to early July coincident with early mating and onset of spermatogenesis. Leydig cells were the predominant sites of T synthesis with the greater difference in spring (P <0.05). Adapted from Mesner et al. 1993.

Minced testes of several reptilian species produce androgen when incubated with mammalian or nonmammalian gonadotropins (Gns). The response to FSH or LH was positive regardless of the seasonal condition of the testes. The fact that turtle testes respond rapidly to Gns at any time during the year suggests that the steroidogenic ultrastructural features reappear at any time of the year when stimulated with Gns and thus have the potential to produce androgens regardless of the season.

Mesner et al. (1993) measured the testicular steroidogenic activity in Sertoli cells (intratubular site) and in Leydig cells (intertubular site) of the snapping turtle after enzymatic separation of Leydig cells from the seminiferous tubules and then purification by isosmotic Percoll gradient centrifugation enabling the steroidogenic activity in the two sites to be measured separately. The cells in fractions 29–31 were identified as Leydig cells whereas the Sertoli cells remain in the tubules (Fig. 6.4). When active, both cell types had ultrastructural steroidogenic features such as smooth tubular endoplasmic reticulum and mitochondria with tubular cristae, as well as staining positively for 3β-HSD activity. Moreover, following incubation, cells from both sites were capable of producing a significant amount of testosterone (T), above the control levels prior to incubation (Fig. 6.4). There was a strong correlation between T levels in both sites and testicular growth and spermatogenesis.

During the testicular cycle both sites were active in producing a large amount of T. Leydig cells were active at the beginning of the cycle in producing large amounts of T and remained active throughout the cycle. They appear to be the major source of T in this species (Fig. 6.5), whereas Sertoli cells were inactive during early May prior to testicular recrudescence when the testes were at a minimum mass. At the onset of spermatogenesis, T levels in Sertoli cells start to rise gradually in conjunction with testicular growth, spermiogenesis, and spermiation in late July to early September (Fig. 6.5).

Moreover, there is evidence of a temporal separation of function in the two sites, which is based on asynchronous changes in ultrastructural steroidogenic features as well as hormonal levels of the two cells. This was clearly demonstrated in the snapping turtles when the two sites were separated by the enzyme method (Mesner et al. 1993). This is consistent with the concept of a temporal separation of androgen secretion and spermatogenesis (Licht 1982). This study reveals a biphasic cycle of T synthesis for Leydig cells and a monophasic cycle for Sertoli cells (Fig. 6.5) in direct agreement with changes in the ultrastructural steroidogenic features reported previously during the testicular cycle in the snapping turtle (Mahmoud et al. 1985a).

The continuous steroidogenic activity of the interstitial site (Leydig cells) throughout the male cycle in the snapping turtle is in contrast to other chelonians, but this condition may exist in other reptiles with short gonadal cycles from the northern temperate regions. The hormone levels that remain high throughout the cycle may account for the nonseasonality in mating in Wisconsin snapping turtles. Such a pattern may be related to a short gonadal cycle with rapid sperm maturation, a feature critical to ensure fertilization. Turtles with longer gonadal cycles in the southern latitudes mate in spring or fall, in association with higher temperatures and longer photoperiods. This is coincident with the availability of active sperm in the epididymis and elevated T levels (Kuchling et al. 1981; McPherson et al. 1982; Licht 1982).

MALE HORMONAL LEVELS BASED ON FRESHLY CAUGHT TURTLES FROM NATURAL POPULATIONS

A summary of male hormone levels in relation to gonadosomatic index (GSI) is illustrated in Fig. 6.6. FSH was at low levels during the early phase of spermatogenesis (mid May to June), but during spermatocytogenesis in July, FSH levels rose significantly in relation to the increase in 1° and 2° spermatocytes and the appearance of some spermatids (Mahmoud & Licht 1997). The FSH levels remained high for the rest of the cycle and differences in FSH levels were significant over the entire sampling period. There was a gradual rise in T levels at the beginning of the cycle, and by July T levels had peaked and remained high for the rest of the cycle. Significant differences occurred in T levels among groups, but there was no significant correlation for the entire cycle either between FSH and T levels or between FSH and GSI (Mahmoud & Licht 1997).

STRESS

Reptiles kept in captivity for an extended period and subjected to overcrowding, extreme temperatures, and/or nutritional deficiency exhibit reproductive failure, lack of growth, weight loss, and immune system suppression. They eventually become chronically stressed with high susceptibility to disease and mortality (for review, see Lance 1994). Acute stress caused by removal from natural conditions, serial bleedings, and sudden exposure to extreme temperatures can also alter immune and neuroendocrine systems from their natural state (Lance 1994). We have observed snapping turtles respond similarly when held for extended periods in Wisconsin biological supply houses.

Stress can activate the hypothalomopituitary-adrenal (HPA) axis resulting in an increase in catecholamines, glucocorticoids, glucose, and other blood components (Chrousos & Gold 1992). Such hormonal elevation can suppress immunity and growth.

In loggerhead sea turtles (*C. caretta*), plasma corticosterone concentration increased significantly following acute captivity stress but declined 6 h later (Gregory et al. 1996). Also, the ordeal of nesting can cause a temporary elevation in catecholamines (AlKindi et al. 2001). In stranded breeding green turtles maximal measured concentrations of corticosterone were similar for both heat (cloacal temperature, <6 h effect) and capture stressors (8 h). The small increase in plasma corticosterone was not associated with a decrease in plasma sex steroids (Jessop et al. 2000). In other studies on reptiles, restrainment followed by serial bleedings of juvenile alligators resulted in temporary elevation of catecholamines and corticosterone (Lance & Elsey 1999). Juvenile alligators implanted subcutaneously with pellets containing corticosterone experienced growth and immu-

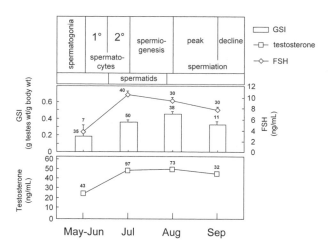

Fig. 6.6. Mean levels of testicular GSI, plasma T, and FSH during phases of spermatogenic cycle for male *Chelydra serpentina* in Wisconsin 1989–1993. Vertical lines represent 1 SE; numbers are sample sizes. See Fig.1 for the procedures.

nity suppression even though the alligators were not under stress (Morici et al. 1997). The effect of stress on gonadal conditions and hormone levels has been well documented in reptiles (see Licht 1984; Moberg 1985; Greenburg & Wingfield 1987). Recent studies show that stress caused by capture, confinement, and repeated blood samplings in both sexes triggers rapid changes in plasma concentrations of sex steroids, gonadotropins, and corticosteroids in bullfrogs (Licht et al. 1983; Mendonca et al. 1985), turtles (Licht et al. 1985a; Mendonca & Licht 1986a; Mahmoud et al. 1989), alligators (Lance & Elsey 1986; Mahmoud et al. 1996), and tuatara (Cree et al. 1990a, b). These studies suggest that stress can alter the hormonal levels from the natural conditions.

DEGREE OF STRESS IN RELATION TO DIFFERENT PHASES OF THE GONADAL CYCLE
Male Turtles

To examine the effects of stress on gonadal condition and circulating hormone levels, snapping turtles were placed under the influence of stressful conditions such as captivity, confinement, serial bleedings, and changes in temperature (Mahmoud & Licht 1997). Sexually mature turtles (20–25 kg) were collected from the field during two phases of the testicular cycle: spermatocytogenesis (mid June) with low T levels and regressed testes, and spermiogenesis (late July) with high T levels and enlarged testes (Mahmoud & Licht 1997).

To assess the effect of stress on hormonal levels, field-captured turtles were immediately divided into three groups and kept in water tanks during the course of the 5-day experiment. Each group was exposed to water temperature of either optimum (20–26°C) or above or below optimum TAR. After 24 h, the group in the optimum tank was divided and transferred to tanks above or below optimum while the others were transferred to an optimum-temperature tank. There was a gradual, but significant decline in T levels ($P < 0.05$) during the first 24 h of study in turtles captured in June and July (Fig. 6.7A and B), regardless of temperature changes and phase of the cycle. In June, however, the turtles demonstrated a significant decline in T concentrations after 5 days of capture despite exposure to various temperatures. Levels of T continued to decline whether the turtles were exposed to temperatures within or outside the TAR (Fig. 6.7A).

July turtles also showed a significant decline in T during the 5-day experiment (Fig. 6.7B). However, a lesser decline in T levels was noted in turtles exposed to temperatures within the TAR than in turtles exposed to extreme temperatures. Under all varied conditions of temperature, the T levels remained significantly higher during exposure to TAR. It ap-

pears that stress due to short-term captivity and frequent bleedings caused a rapid drop in T levels in freshly captured male turtles during June and July. The turtles were more sensitive to stress caused by bleedings and temperature changes in June than in July since there was a significant and rapid decrease in T values when the turtles were exposed to optimum temperatures. The response to stress varies in intensity during the testicular cycle, and thus the degree of stress depends on the hormonal levels and the condition of the gonads.

In another investigation (Mahmoud et al. 1989), male snapping turtles were collected from the field in Wisconsin during June, July, and August and females were collected in June only. Immediately after capture, blood samples were drawn at intervals of 0 (initial), 6, 24, 48, 72, and 168 h. Both sexes exhibited an initial rise in serum steroid levels followed by a decline. In June, male turtles exhibited no significant change in T levels from the initial values as a group. Males captured in July exhibited the greatest increase in T followed by a rapid decline by 48 h. August males showed no elevation in T levels at the beginning, but declined significantly at a relatively slow rate after 7 days of captivity.

It appears the effects of stress on circulating sex steroid levels vary among different phases of the testicular cycle. This may be attributed to the ability of the testes to synthesize different seasonal concentrations of androgen under natural conditions (Duvall et al. 1982; Licht et al. 1985a; Mahmoud & Licht 1997). Thus, the degree of male response to stress depends on the steroidogenic condition of the testes.

Female Turtles

Gravid and nongravid snapping turtles collected in June showed no significant differences between individuals from the initial levels of Pro and E_2 after 7 days of capture. Nongravid were vitellogenic and had significantly heavier ovaries than did gravid females. As in males, during the course of the experiment, captivity induced some change in serum sex steroid values. The initial increase in serum was almost identical in both gravid and nongravid females. Serum Pro concentrations peaked 24 h after capture for both groups. Nongravid females exhibited a significant rise in Pro 6 h after capture ($P < 0.05$) whereas gravid females took 24 h. After 7 days of captivity, Pro values were not statistically different from initial values. However, a different pattern of response was observed for E_2 when the two groups were compared. Gravid females showed a significant rise in E_2 values after 24 h ($P < 0.01$) whereas nongravid females showed no change from initial values. However, plasma E_2 concentrations after 7 days were approximately equal to those obtained initially in both groups. Unlike males, captivity did not significantly depress serum steroid levels below the initial values by the end of the 7-day captivity.

The mechanism controlling the changes in steroid levels is not well understood but several hypotheses, singly or interactively related, can be suggested: (1) captivity stress

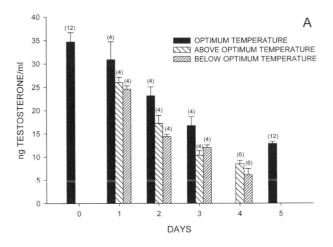

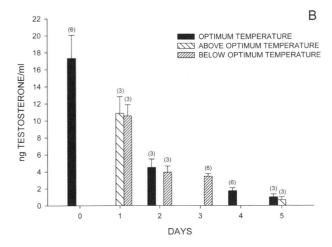

Fig. 6.7. (A) Serum testosterone (T) levels of male snapping turtle, *Chelydra serpentina*, exposed to temperature within the thermoactivity range (TAR: 20–26°C) or to extreme water temperatures (14–17 or 29–32°C) for 5 days during June (low T, spermatocytogenesis). The T levels continued to decline whether the turtles were exposed to TAR or to extreme temperatures (P < 0.05). Sample sizes in parentheses (mean ± SE). (B) Serum testosterone (T) levels of male snapping turtle, *Chelydra serpentina*, exposed to temperature within the TAR or to extreme water temperatures (14–17 or 29–32°C) for 5 days during July (high T, spermiogenesis). Like June turtles, all groups experienced a decline in T levels during the course of the experiment (P < 0.005) but T levels remained higher in turtles exposed to TAR temperatures (P < 0.05). Sample sizes in parentheses (mean ± SE). Adapted from Mahmoud & Licht 1997.

causes the release of ACTH and gonadotropins, which stimulates steroidogenesis in gonad and adrenal glands; (2) stress changes the steroid degradation process; (3) stress inhibits the uptake of steroids at the target tissue, thus remaining in circulation for a longer period; (4) stress changes the phase relationships of daily hormone rhythms.

EFFECT OF CAPTIVE STRESS AT DIFFERENT PHASES OF GONADAL CYCLE

Healthy captive snapping turtles were observed between 1980 and 1988 at the Lemberger Company, Oshkosh, WI (Mahmoud & Licht 1997). The turtles were kept in aquatic tanks, fed frequently, and sacrificed at different times of the summer for commercial meat sold by the company. When sacrificed, the ovaries and the testes were examined and records were kept on the status of the gonads at different phases of the cycle. The time of capture and the length of captivity were also recorded. The gonadal condition of the healthy captive turtles used in this study is presumed to be the same as that of turtles under natural conditions.

Interesting data were obtained on the effect of stress on gonads at different phases of the cycle. Male turtles ($n = 35$) caught in spring and sacrificed in June and July, had regressed testes. Apparently, the testes did not undergo spermatogenesis or growth when the turtles were under captivity stress. However, male turtles ($n = 56$) captured in late June to early July and sacrificed in mid to late August had well developed testes with testicular growth and spermiation similar to the natural condition (Mahmoud & Licht 1997).

In another experiment (Mahmoud & Licht 1997), captive female turtles ($n = 37$) caught in early spring with full preovulatory follicles and then sacrificed in May through late June did not undergo ovulation while in captivity. Female turtles ($n = 46$) that had already oviposited their eggs and turtles ($n = 38$) that had not oviposited their eggs caught in early June were sacrificed between June and August. Both groups had ovaries in a postovulatory phase with follicles 5 mm or smaller. When sacrificed in August, both groups contained follicles between 10 and 14 mm, compared with a fully mature size (18–22 mm) under natural conditions. The two groups showed the general response to stress with incomplete follicular growth.

The response to stress induced by captivity apparently varied in intensity during the gonadal phases in both males and females. Possibly, the degree of stress depends on hormonal levels and the condition of the gonads during the cycle. The response to stress in both sexes appeared to be reduced after the gonadal cycle began.

FUTURE DIRECTIONS

To date, most investigations on the reproductive physiology of the snapping turtle are limited to populations of northern latitudes and very little work has been done on those of southern latitudes. Because this species extends over a wide geographical area, extensive research is needed to gain a comprehensive information base for their entire range. Seasonal variation in the gonadal cycle among populations of the same species is common in reptiles (Licht 1982). Gonadal events such as vitellogenesis, ovulation, oviposition, luteinization, luteolysis, spermiation, and hormone dynamics must be studied in each population to gain a complete knowledge about the reproductive physiology of the species.

Circulating hormones must be monitored on freshly captured turtles to ensure values that reflect natural conditions. Most of the studies on gonadal cycles have been conducted on stressed reptiles, including turtles kept in captivity from one to several days. Some studies were based on captive reptiles shipped from biological warehouses to institutions. Data obtained from captive reptiles under stress may not accurately reflect natural gonadal events because hormone levels have been observed to change significantly after even a few hours of captivity (Licht et al. 1985a; Mendonca & Licht 1986a; Mahmoud et al. 1989). An improved understanding of their natural reproductive physiology under natural conditions is vital to assess management and conservation strategies for this species.

ACKNOWLEDGMENTS

We are grateful to John L. Plude, Department of Chemistry, University of Wisconsin Oshkosh, for his valuable comments. Special thanks to Saif Al-Barhry for his aid in formatting the figures. Preparation of this chapter was supported in part by the Fulbright Scholar Program.

7

Thermal Ecology and Feeding of the Snapping Turtle, *Chelydra serpentina*

JAMES R. SPOTILA

BARBARA A. BELL

ROWS OF TURTLES STRETCHED out on logs on a sunny day are a familiar sight. Basking is the behavior by which animals expose themselves to the warmth of the sun while resting. The classic study of Boyer (1965) made it clear that basking allows turtles to heat up and, therefore, is a form of thermoregulation. However, snapping turtles, *Chelydra serpentina,* seldom bask (Boyer 1965; Obbard & Brooks 1979; Saba 2001) and because they remain primarily in the water their body temperature (T_b) is usually near water temperature. Therefore, do snapping turtles thermoregulate? Do they actively select particular temperatures in the laboratory and field or do they avoid extremes? If they seldom bask and are usually near water temperature do they merely tolerate a wide range of temperatures or do they actually have optima for physiological and behavioral performance? How does this relate to their thermoregulatory strategy and to their feeding behavior?

Body temperature sets the stage for all biological activities in an animal. Metabolic rate is closely tied to body temperature, as are rates of chemical and enzymatic reactions. Body temperature affects the kinetic energy of the cell, so that an increase in temperature leads to an increase in reaction rates and a subsequent increase in metabolism (Clarke & Fraser 2004; Gillooly et al. 2001). Therefore, physiological performance, digestive processes, growth rate, behavior, reproduction, and survival all depend on T_b. In turn the physical environment constrains T_b by affecting biophysical aspects of heat exchange. Heat exchange in air is much more complex than heat exchange in water (Spotila & Standora 1985). Solar and thermal radiation, convection, evaporation and conduction combine to heat and cool turtles while they are exposed to air, while conduction and movement of fluid (convection) dominate heat exchange in water (Spotila & Standora 1985). Therefore, a turtle establishes its T_b

within the constraints of the physical environment. Turtles can heat up by aquatic basking in shallow water (Moll & Legler 1971; Spotila et al. 1984) and control T_b by active selection in a thermal gradient (Schuett & Gatten 1980; Williamson et al. 1989).

Unlike most freshwater turtles in the United States, North American snapping turtles are primarily carnivorous throughout their lives (Punzo 1975; Ernst et al. 1994). Snapping turtles in warmer, tropical regions tend toward herbivory in some areas (Moll 1997; Moll & Moll 2004). Since body temperature controls physiological performance (Eckert et al. 1988), including digestion (Parmenter & Avery 1990), it is probable that the carnivorous lifestyle of the snapping turtle in North America and its more herbivorous diet in tropical regions is related to its thermal biology. Temperature preference and the precision of thermoregulation vary with digestive status (Gatten 1974). Body temperature affects feeding and digestion, which, in turn, affect thermoregulation as well as growth and other life processes. In this chapter we will review the data available on thermoregulation of snapping turtles and discuss the relationship between thermoregulation, feeding ecology, and digestive physiology.

THERMOREGULATION

The terms used to characterize behavioral "choice" of temperature have been numerous and confusing (Fischer et al. 1987). This has led to a literature devoted to the terminology of thermoregulation ranging chronologically from Gunn & Cosway (1938) through Cowles & Bogert (1944) and Bligh & Johnson (1973) to Hutchison & Dupré (1992). Pough & Gans (1982) recommended that to avoid anthropomorphism the term "selected body temperature" should be used to express the temperature occupied by animals in a thermal gradient rather than "preferred body temperature." However, Hutchison & Dupré (1992) reviewed the issues related to the terminology of animal thermoregulation and concluded that there is no reason (by definition of the words or by etymology) to find "selected" less anthropomorphic than "preferred." They listed six reasons, including priority and earlier establishment in the literature (Herter 1926) to use preferred temperature to describe temperature chosen by animals in a thermal gradient in the laboratory. They also discussed the use of "eccritic temperature" to describe temperatures occupied by animals in the field (Hutchison & Hill 1976; Licht et al. 1966; Lillywhite 1971) but concluded that there was no consensus for the use of this term. We agree with these conclusions and, therefore, in this chapter we use the term preferred body temperature (PBT) to describe the temperatures selected by snapping turtles both in the laboratory and the field. The act of selection by a turtle does in fact indicate a preference within the mental capabilities of the animal. For a review of thermoregulation in turtles, see Hutchison (1979); for a discussion of their biophysical ecology and climate space,

see Spotila et al. (1990); and for a discussion of the thermoregulation of sea turtles, see Spotila & Standora (1985), Spotila (1995), and Spotila et al. (1997).

PREFERRED TEMPERATURE IN THE LABORATORY

Recently fed adult snapping turtles acclimated to 18°C have a PBT of 28.1°C (Schuett & Gatten 1980). Hatchlings acclimated to 15 and 25°C have PBTs of 24.2 to 25.0°C at 3 to 8 months of age in terrestrial temperature gradients. In an aquatic temperature gradient the 8–month-old turtles had PBT of 28.0°C, which was significantly higher than the PBTs of turtles in the terrestrial gradient (25.0°C) (Williamson et al. 1989). Thus size, rearing temperature, and age did not affect the PBTs of young snapping turtles but gradient type did.

Juvenile temperature choice is influenced by incubation temperature of eggs (O'Steen 1998). Snapping turtles exhibit temperature-dependent sex determination (Yntema 1976; Wilhoft et al. 1983; Vogt & Flores-Villela 1992), with low and high incubation temperatures producing predominantly male or female hatchlings, respectively. Hatchlings from eggs incubated at 21.5°C (a temperature that produces all male hatchlings) chose water in an aquatic gradient that was 28°C, while those incubated at 30.5°C (a temperature that produces all female hatchlings) chose water that was only 24.5°C (O'Steen 1998). Temperature during incubation can also affect the standard metabolic rate (Steyermark & Spotila 2000) and growth rate of juvenile snapping turtles (McKnight & Gutzke 1993; O'Steen 1998; Rhen & Lang 1999; but see Steyermark & Spotila 2001a).

Snapping turtles have PBT similar to that of the slider turtle, *Trachemys (Pseudemys) scripta* (29.1°C), and slightly lower than that of the box turtle, *Terrapene ornata* (29.8°C) (Gatten 1974). Snapping turtles have a higher PBT than the stinkpot turtle, *Sternotherus odoratus* (20.7°C), the painted turtle, *Chrysemys picta* (22.0°C), and the spotted turtle, *Clemmys guttata* (21.6°C). However, the last three turtle species were postabsorptive and acclimated to 15°C while the former had recently eaten (Graham & Hutchison 1979). Nevertheless, hatchling snapping turtles do not show a thermophilic response to feeding. The temperature selected by 7–month-old hatchlings in a thermal gradient was 29.8°C and did not differ with feeding (Knight et al. 1990). Bury et al. (2000) demonstrated that both hatchling and yearling *C. serpentina* and *T. scripta* clearly select T_b's through behavior in a temperature gradient and that they thermoregulate within a narrow range at or above 28°C. Steyermark & Spotila (2001b) found that environmental temperature can also affect survival behaviors in juvenile snapping turtles such as righting response (time to return to prone position when placed upside down). The authors found a negative correlation between environmental temperature and righting response times.

THERMOREGULATION IN NATURE

In nature snapping turtles typically have a T_b near water temperature but they can alter their T_b by selecting an area with appropriate water temperature. This also occurs in amphibians and fish and is perhaps best illustrated in the thermoregulatory aggregation behavior of *Rana boylii* tadpoles in a pond in Del Norte County, California during late June, where in the early morning larvae were concentrated on the bottom in the middle of a cove at 15°C, in midmorning they moved to shallow water and warmed by aquatic basking, and during the day they moved to exploit changing water temperature (Brattstrom 1962). The same principle applies to snapping turtles. By actively selecting warm or cool water they can regulate their body temperature. Snapping turtles are active at T_b's from 5 to 32.6°C (Brattstrom 1965; Punzo 1975). In Florida activity T_b's ranged from 18.7 to 32.6°C (Punzo 1975); in Philadelphia, Pennsylvania, temperatures of water, which do not necessarily reflect T_b's, that snapping turtles occupied ranged from 5.0 to 33.0°C (Saba 2001); and in Algonquin Park, Ontario, Canada, temperatures of the water occupied ranged from 7.5 to 30.0°C (Obbard & Brooks 1981a; Brown & Brooks 1991). Adult snapping turtles gain a thermal advantage when moving from warm to cold water because of their large body size. Body temperature lags behind water temperature and metabolism can generate enough heat to keep the body 0.25 to 0.5°C above water temperature (Baldwin 1925; Pell 1941; Obbard & Brooks 1981a).

Caution must be applied to the results of studies on the thermal ecology of snapping turtles in nature because typically these studies have not taken into account the immediate past thermal or behavioral history of the turtle. Most studies have reported the cloacal temperatures of turtles taken at time of capture (i.e., Punzo 1975; Obbard & Brooks 1981a, 1987) or temperatures of water occupied by turtles as measured with radio transmitters attached to the shell (i.e., Brown & Brooks 1991, 1993; Brown et al. 1994a; Saba 2001). Investigators typically record temperature of the water near where they locate a turtle and sometimes assume that this represents T_b. Manning & Grigg (1997) voiced concern about this assumption in a radio telemetry study of thermoregulation of the Brisbane River turtle, *Emydura signata*. Epaxial muscle temperatures of six turtles monitored for 112 days indicated that turtles seldom had T_b's higher than water temperature. The authors cautioned that direct measurements of body temperatures are needed in free-ranging individuals of other freshwater turtle species. This should include detailed observations on the behavior of the individual turtles.

The only studies that have reported such data were Spotila et al. (1984), who measured T_b of free-ranging slider turtles in Par Pond on the Savannah River Plant (SRP) in South Carolina with thermistors implanted near the heart; Standora (1982) who used multichannel telemetry to detail the thermoregulatory behavior and internal heat transfer of

slider turtles under natural conditions in a pond on the SRP; and Brown et al. (1990) who measured the temperatures of epaxial muscle of free-ranging *C. serpentina* in Algonquin Park. Slider turtles in Par Pond used an opportunistic strategy of thermoregulation consistent with the constraints of their climate space (Spotila et al. 1990). Par Pond was a cooling reservoir for a plutonium production reactor on the SRP. No *T. scripta* lived at the point of thermal discharge. Turtles in heated areas basked aquatically (Spotila et al. 1984). Body temperatures in spring, summer, and fall were within the PBT range measured in the laboratory by Gatten (1974). In the normal-temperature areas of the reservoir, turtles basked atmospherically on sunny days throughout the year. In summer, basking intensity had a bimodal curve (Fig. 7.1). In spring and fall, basking frequency was unimodal (Spotila et al. 1984). Slider turtles in normal-temperature areas basked on sunny, calm days in winter at air temperatures as low as 2°C (Schubauer & Parmenter 1981). Standora (1982) recorded the heat flow through the shell of basking *T. scripta* and the temperature gradients that developed within the body. He also demonstrated that in fall and winter turtles gained heat from the substrate when they were lying in the sediments at the bottom of the pond, a behavior employed by snapping turtles. His detailed measurements and behav-

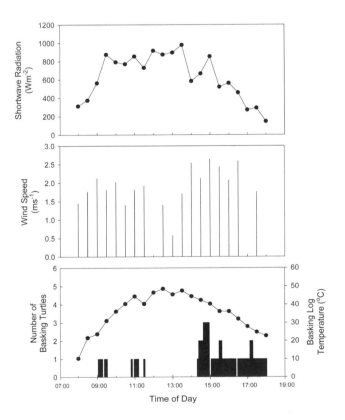

Fig. 7.1. Bimodal pattern of basking activity of *T. scripta* and microclimate conditions for a sunny July day at Susan's Swamp on Par Pond. Shortwave radiation (300–3,000 nm) shows depressions due to periodic cloud cover. Wind speeds are means for 15-min periods. A solid line connecting filled circles represents basking log temperature. Lower histogram indicates numbers of turtles basking. Data are for July 14, 1976. Adapted Spotila et al. 1984.

ioral observations presented a clear picture of the thermoregulatory behavior of this species. Brown et al. (1990) found a strong relationship between muscle temperature and water temperature in snapping turtles. Important insight can be gained into the thermal biology of snapping turtles by a detailed study of their behavior combined with multichannel, radio telemetric measurements of their body temperatures and heat transfer.

Snapping turtles are active during summer and change position daily, but as temperatures decrease, their movement decreases (Meeks & Ultsch 1990). When water temperatures drop below 15°C, they spend most of their time buried in mud rather than foraging. Mud burial is uncommon during warmer months (Ultsch & Lee 1983). Snapping turtles do not choose a final hibernation site until late autumn, when temperatures fall to 5°C (Meeks & Ultsch 1990; Obbard & Brooks 1981a; Ultsch & Lee 1983). Turtle collectors in Ohio have anecdotally reported occasional mid-winter movement during temperature increases, even when atmospheric temperatures are as low as −4 to −7°C. These collectors suggest that turtles will take advantage of brief warming periods to breathe air (Meeks & Ultsch 1990).

THERMOREGULATORY SIGNIFICANCE OF BASKING

Snapping turtles occasionally bask out of the water (Ewert 1976) and in colder portions of their range they often do so in summer (Obbard & Brooks 1979), reaching a T_b of 34°C in Algonquin Park, Ontario. In a later study Brown & Brooks (1993) found that snapping turtles at the Ontario site basked less often than in the previous study and attributed this difference to a difference in water temperature (8.6°C in the first study versus >11.5°C in the later study). Since basking has long been assumed to be of thermoregulatory significance in turtles (Boyer 1965), the fact that snapping turtles seldom bask, except in cold areas like Algonquin Park, raises the question as to the thermoregulatory significance of this behavior in this species as well as in the species where it commonly occurs. If it is important for painted turtles and slider turtles why isn't it important for snapping turtles? Manning & Grigg (1997) concluded that basking was not of thermoregulatory importance in *Emydura signata* and questioned its importance in other turtles as well. Turtles that bask do warm body temperature above air and water temperature and retain this thermal advantage for some time (hours in some cases) after returning to the water (Moll & Legler 1971; Auth 1975; Standora 1982; Spotila et al. 1984). While Manning & Grigg (1997) concluded that the period of elevated T_b is maintained for only a short portion of any 24-h period, they measured epaxial muscle temperatures and not deep body temperatures. Ernst et al. (1994) suggest that aerial basking in snapping turtles is limited by an intolerance of high temperatures.

Standora's (1982) detailed telemetry studies indicated that the deep body temperature near the heart of slider turtles stayed warmer than water temperature for long periods after basking. In 60% of readings the deep body temperature exceeded carapace and plastron surface temperatures. Heat was flowing out of the body. Outward heat flow was the most common thermal profile inside the turtles and occurred mainly in the afternoon and evening hours. Turtles spent 1.7 times as much time cooling as warming. In spring and summer turtles spent 61% of time within their PBT range of 26–32°C. After basking the turtles stayed warmer than water temperature for up to several hours. This effect was most pronounced in winter. Moll & Legler (1971) reported that cloacal temperature dropped to water temperature quite rapidly but deep body temperature remained elevated for hours. Therefore, measurements of cloacal temperature are not very useful in determining the role of basking in the physiology of a turtle because they do not reflect core T_b. We expect that basking does give an advantage to a turtle by enhancing some aspects of its physiology for a few hours because deep body temperature remains above water temperature for hours after basking.

Although additional studies are needed to determine the role of blood flow and behavior (leg position, orientation to the sun, etc.) in optimizing the effect of basking, it is clear that basking provides a thermal advantage to a turtle. But how significant is this advantage in the life of a turtle? After all, it spends most of its life at water temperature, especially in fall and winter. How does it gain an advantage by elevating its body temperature for a portion of time during a given year or season? These questions remain unanswered. Clearly emydid turtles, *Chrysemys*, *Pseudemys*, *Trachemys* and others, bask much more than snapping turtles and do so more in the spring than during the rest of the year. What is the physiological advantage to basking? Is the snapping turtle missing this advantage?

Perhaps the advantage comes in the acquiring or processing of food. Congdon (1989) hypothesized that there are four major categories of constraints on the energetics of turtles and lizards: absolute resource availability, harvest rate limitations, process rate limitations, and limitations on harvest or processing resources imposed by risk of predation. Aquatic turtles occupy relatively productive habitats with primary productivity of 2,500 $g \cdot m^{-2} \cdot yr^{-1}$ (Ricklefs 1979). However, low water temperatures may constrain the ability of temperate zone turtles to harvest and process food, especially vegetation. Slider turtles from the heated areas of Par Pond had higher growth rates and reached larger body sizes than did those from the normal-temperature areas (Gibbons 1970). They ingested more than twice as much protein as turtles from normal-temperature areas. A positive relationship exists between temperature and rate of food ingestion, rate of digestion and rate of gut clearance in slider turtles, which process very little food at temperatures of 15°C (Avery et al. 1993). Congdon (1989) predicted that the severity of the processing constraint increased with the amount of plant material in the diet and with either latitude or altitude. We predict that when water temperatures are below that for efficient absorption of food material and rapid gut clearance, turtles with full guts will bask so as to maintain T_b above water temperature whether or not they can reach PBT.

There are several studies that support these predictions. Slider turtles do bask more when fed than when nonfed (Hammond et al. 1988). Fed females bask more than fed males and both groups bask more than nonfed turtles at 10, 20, and 30°C in spring/summer. No differences occur in basking times of fed and nonfed turtles or between males or females in fall/winter, but turtles kept in 20°C water bask more than turtles kept at other temperatures. The T_b of basking turtles at 10, 20, and 30°C is 28–32°C indicating that they are thermoregulating at a temperature near the PBT measured in thermal gradients. Slider turtles have a higher PBT and greater basking rate when fed (29°C) than when nonfed (24°C) (Gatten 1974). Basking painted turtles in Pennsylvania (cloacal temperatures between 22.5 and 29.0°C; Ernst 1972) and in Minnesota (T_b of 26.3–30.2°C while basking; Brattstrom 1965) reach similar temperatures. Painted turtles become active in spring at about 8°C but do not feed until water temperatures reach 15°C (Cagle 1950; Sexton 1959). Water temperature in ponds and marshes on the E. S. George Reserve in southern Michigan typically do not exceed 15°C until May and in some areas are below 15°C in deeper water during the entire summer (Congdon 1989). Therefore, turtles can only harvest food during May to September and must restrict their activity to warmer portions of the aquatic habitat to maintain body temperatures that will allow them to process food. Basking will be advantageous in these habitats. These data support Congdon's predictions and suggest that there is a physiological advantage to basking in turtles.

Slider turtles and painted turtles have a faster turnover rate at higher temperatures (Kepenis & McManus 1974; Parmenter 1980, 1981). Turnover time at 25°C is 59 h for painted turtles and 61 h for slider turtles. At 15°C turnover time for painted turtles increases to 95 h. Diet protein digestion, sugar absorption, and energy assimilation are all faster at warmer temperatures (reviewed in Parmenter & Avery 1990). Thus when basking raises body temperature, digestive processes are enhanced. Warmer temperatures are required for digestion of plant material than the animal matter normally consumed by snapping turtles, which have a digestive turnover time about one half that of slider and painted turtles at 25°C (31 h) (Parmenter 1981). The combination of a diet of animal protein and an efficient digestive system probably obviate the need for basking in the snapping turtles for digestive reasons in most habitats. Only in cold years near the northern edge of its range does the snapping turtle gain a digestive advantage by basking.

Nevertheless, temperature constrains the digestive processes of snapping turtles as well. Hatchling snapping turtles eat little and do not grow when maintained at 15°C (Fig. 7.2).

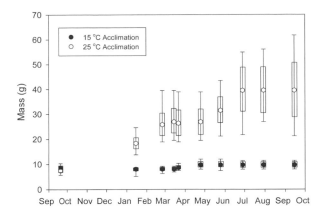

Fig. 7.2. Growth of young *C. serpentina* raised at 15 and 25°C for 1 year. Vertical lines bounded by long horizontal lines indicate range. Short horizontal lines represent means and rectangles enclose 95% confidence intervals. Sample sizes were 12 for each temperature. Adapted from Williamson et al. 1989.

They eat freely and grow when kept at 25°C (Williamson et al. 1989). Snapping turtles in a warmer, more productive habitat in southern Ontario have higher body temperatures, grow more rapidly, have a larger clutch size and a greater adjusted clutch mass than those from a cooler, less productive habitat in Algonquin Park (Brown et al. 1994a). It is in Algonquin Park that snapping turtles bask most often (Obbard & Brooks 1979; Brown & Brooks 1991).

Solar flux and heat transfer affect the basking frequency of *T. scripta* and *Pseudemys floridana* (Auth 1975). Hennemann (1979) reported that infrared radiation elicits the basking response in *Pseudemys concinna*. Operative environmental temperature (T_e), a thermal index of the microclimate of an animal (Bakken & Gates 1975; Bakken 1976), is a good predictor of basking behavior of *T. scripta* in South Carolina (Crawford et al. 1983). Turtles do not bask until T_e is 28°C or higher. Movement of the sun during the day results in spatial variation in T_e's available to turtles and influences their location and basking behavior. The T_e is also a good predictor of basking behavior of painted turtles (Schwartzkopf & Brooks 1985). In Algonquin Park painted turtles display unimodal basking patterns on cool days and bimodal patterns on warmer days like slider turtles in Par Pond (Spotila et al. 1984). Operative environmental temperature is also a good indicator of the microclimates available to *Pseudemydura umbrina*, the western swamp tortoise, in its natural habitat in Western Australia (King et al. 1998). These factors have not been thoroughly studied in snapping turtles specifically, but the same biophysical principles apply to chelydrids as to other aquatic turtle species.

Reproductive status can affect basking behavior as well. Snapping turtles in Pennsylvania begin nesting in mid to late May and continue until air temperatures reach about 30°C (unpublished data). In Michigan they nest over a similar period (Congdon et al. 1987). In Ontario they become active in early May when water temperature reaches 7.5°C and begin nesting in early June. Nesting peaks in mid June and ends by June 21–30 (Obbard & Brooks 1981a). While snapping turtles do not bask during the nesting period in Pennsylvania or Michigan they do bask in Ontario and do so more frequently when water temperature is colder (8.6°C) than warmer (>11.5°C) (Brown & Brooks 1993). Egg development is strongly related to water temperature (Obbard & Brooks 1987). The authors postulate that eggs of snapping turtles in northern regions may mature faster because the growing season is shorter. In contrast, both slider turtles and painted turtles bask more in spring than at other times of year and females bask more than males during spring (Standora 1982; Hammond et al. 1988; Krawchuk & Brooks 1998). This appears to be associated with the nesting season. There was no difference in basking duration of male and female painted turtles in Algonquin Park in July after the nesting season (Lefevre & Brooks 1995). There was no detectable relationship between date of oviposition and/or number of clutches and basking duration or frequency of basking in painted turtles in Algonquin Park (Krawchuk & Brooks 1998). It is not clear whether this was because the colder water temperatures at that site created a greater demand for basking by females than in the other studies. The exact relationship between basking behavior and egg development remains to be determined, as do the relationships between digestion of food, mobilization of lipids, development of ovarian follicles, and basking in both emydid and chelydrid turtles.

PHYSIOLOGICAL TEMPERATURE REGULATION

Turtles can control the rate of heat transfer between the body and the environment by changing peripheral blood flow. Both cutaneous and carapace blood flows are increased by heating and decreased by cooling. This effectively changes the functional insulation and, as a result, the turtles heat and cool at different rates. Snapping turtles heat 30% faster than they cool from 10 to 30°C in water. They heat and cool at the same rate in air (Weathers & White 1971). *C. picta*, *P. floridana*, and *T. scripta* heat faster than they cool in air and water (Spray & May 1972; Weathers & White 1971). In a series of theoretical and experimental studies O'Connor and Dzialowski demonstrated that changes in blood flow to the periphery and to the limbs brought about observed changes in warming and cooling rates in lizards and should do the same in other reptiles (O'Connor 1999; Dzialowski & O'Connor 1999, 2001a, b). Control of peripheral circulation can be very important in the thermoregulation of small and large reptiles (O'Connor 1999). An examination of deep body temperature, peripheral temperatures, blood flow, and behavior in snapping turtles and emydid turtles while basking aquatically and aerially would provide interesting data on this subject.

Snapping turtles have a lower tolerance to high temperature than many other species of turtles (Hutchison et al. 1966). There does not appear to be any substantial change in thermal tolerance with age in this species. The critical thermal maximum (CTM), defined as the temperature when an animal goes into spasms when heated at a rate of 1°C per minute (Hutchison 1961), of young snapping turtles was 39.1°C when acclimated to 15°C and 41.1°C when acclimated to 25°C. Adult snapping turtles had a CTM of 39.5°C when acclimated to 25°C. Aquatic chelydrid turtles (= 39.4°C) and trionychid turtles (= 40.0°C) have lower CTMs than semiaquatic emydid turtles (= 41.6°C) and terrestrial testudinid turtles (= 43.3°C). This appears to be related to habitat and not to body size. Webb & Johnson (1972) and Webb & Witten (1973) reported significant differences between head and body temperatures in the Australian long-necked turtle, *Chelodina longicollis*, during heating and cautioned that head–body temperature differences in CTM experiments could lead to spurious results. Since spasms are used for the endpoint in a CTM test, brain temperature may be most important to measure. However, the data for small (Williamson et al. 1989) and large (Hutchison et al. 1966) snapping turtles are similar, indicating that the data for large turtles are not artifacts. Thus, both selected temperature and CTM of snapping turtles appear to be related to their aquatic habitat, wherein temperatures are both more moderate and change more slowly than do temperatures on land. These aspects of their thermal biology are quite stable from hatchling to adult.

FEEDING BIOLOGY

Hatchling snapping turtles feed but do not grow at 15°C (Williamson et al. 1989), which suggests that their digestive processes slow down or stop at this temperature. In Algonquin Park snapping turtles emerge from hibernation when water temperature is about 7.5°C and begin feeding when water temperature reaches 16.0°C (Obbard & Brooks 1981a). Punzo (1975) reports that snapping turtles are omnivorous based on detailed data on the stomach contents of 59 individuals from Florida. However, an examination of his data indicates that these turtles are primarily carnivorous. They ingested annelids, crustaceans, insects, gastropods, amphibians, and reptiles (turtles and snakes), and had some plant material in their gut. Juvenile snapping turtles actively forage on vertebrates, invertebrates, carrion, and occasionally plants (Ernst et al. 1994). Budhabhatti & Moll (1990) found that of 22 adult snapping turtles examined, 21 consumed approximately equal proportions of plant and animal matter, and one spent the entire summer feeding almost entirely on duckweed.

Adult snapping turtles actively forage and also lie in ambush to capture prey such as ducks and muskrats (Lagler 1943a; Ernst et al. 1994). In Wisconsin, snapping turtles eat frog eggs in early spring. They also eat ducklings, common terns, and blackbirds (Mahmoud & Klicka 1979). Although as adults they are often "sit and wait" predators, they also are attracted to and eat carrion (Saba 2001). In our experience, it appears that North American snapping turtles are primarily carnivorous although they do ingest plant material incidentally when feeding on invertebrates and small vertebrates. However, snapping turtles in tropical regions eat more plant material than those in subtropical regions (Moll 1997; Moll & Moll 2004). Adult female snapping turtles in a eutrophic pond in southern Ontario grew more than four times as fast as adult females in an oligotrophic lake in Algonquin Park to the north (1.12% per year vs. 0.26% per year) (Brown et al. 1994). It was not possible to determine if the differences in growth rate were due to differences in diet quality, diet quantity, or both. Temperatures also differed by 3°C (19.5 and 22.6°C) although the standard deviations of the means were greater than 3°C.

The carnivorous diet of snapping turtles may explain their more rapid growth when compared with Blanding's turtles, *Emydoidea blandingii,* and painted turtles on the E. S. George Reserve (Fig. 7.3). Juvenile snapping turtles grow at a rate of 145 g yr^{-1} whereas the latter two turtles grow at a rate of 50 g yr^{-1} (Congdon 1989). Avery (1988) and Avery et al. (1993) report that the proportion of protein in the diet of slider turtles influences growth rate of juveniles. Juveniles eating a diet of 25% or 40% protein grow at a rate similar to those measured in the field. Juveniles eating a diet of 10% protein do not gain mass and their plastrons curl up. The mean crude protein content of aquatic plants in Par Pond is 13.1% (Boyd 1970). The protein content of fish carrion is about 20% (Parmenter & Avery 1990). Varying dietary protein availability accounts for differences in growth rates of slider turtles from different populations in the geographic area around the Savannah River Plant (Gibbons 1970; Parmenter 1980; Avery et al. 1993). Apparently protein content in the diet also accounts for the difference in growth rates of the turtles on the E.S. George Reserve and these studies explain why emydid turtles, in general, eat primarily a carnivorous diet as juveniles. Only after they have completed growth to adult size do they shift to a plant diet. Even then they eat animal protein when it is available (Clark & Gibbons 1969; Parmenter 1981).

Snapping turtles and emydids such as slider turtles have similar gut morphology. Some differences in folding are apparent in the duodenum and fewer differences still are seen in the colon but there do not appear to be any specialized features in the intestines of these turtles related to differences in diet (Parsons & Cameron 1977). Snapping turtles maintain the morphology and physiological capacity of their digestive system when they are not feeding (Secor & Diamond 1999). The snapping turtle, slider turtle, and the common musk turtle, *Sternotherus odoratus,* exhibit no postfeeding changes in intestinal nutrient uptake rates or in the mass of most organs measured. Small but statistically significant increases occur in metabolic rates after feeding in all three species. These increases are similar to the increases observed in other reptiles that eat frequently (Fig. 7.4) (Secor & Diamond 1999) and are much less than those observed in reptiles that eat only infrequently. The specific dynamic action (SDA), the extra energy expended for digestion, as a percentage of ingested energy was 22% for snapping turtles, 21% for slider turtles, and 17% for musk turtles (Secor & Diamond 1999). These values are similar to those for vertebrates that feed frequently and lower than those (26–37%) for snakes that feed infrequently (Secor & Diamond 1997). In contrast, the monitor lizard, *Varanus albigularis,* fasts for months in the wild and after a meal increases its metabolic rate by 900% (Secor & Phillips 1997).

There were no interspecific differences in the absorption of amino acids and D-glucose, and normalized values for uptake by fed and fasted turtles were similar to those of other ectothermic vertebrates that ate frequently (Buddington & Hilton 1987; Buddington et al. 1991). Normalized values for uptake were much higher in recently fed snakes (sidewinders, *Crotalis cerastes;* Burmese pythons, *Python molurus;* and boa constrictors, *Boa constrictor*) that ate infrequently (Secor et al. 1994; Secor & Diamond 1995). The ratios of amino acid uptake to D-glucose uptake in the small intestine were greater than 1 (Fig. 7.5), which is characteristic of carnivores and omnivores (Karasov & Diamond 1988). Thus, both carnivorous (snapping and musk turtles) and omnivorous (slider turtles) turtles are physiologically adapted to rapidly digest protein and transport amino acids across the wall of the small intestine. Apparently, adult slider turtles eat a diet primarily of vegetation when they are actually better adapted physiologically for a diet high in protein and low in carbohydrates. Perhaps this keeps them prepared to digest animal protein when it is available. It is not clear how these results relate to adult snapping turtles. Since they are "sit and wait," ambush predators, they may go for long periods between meals. In that case their digestive systems may shift from the responses measured by Secor & Diamond (1999) in 100-g

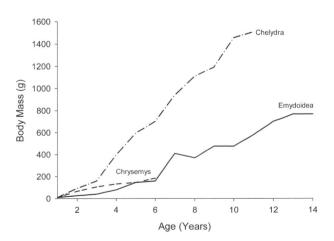

Fig. 7.3. Growth of three species of turtles on the E. S. George Reserve. Data are from hatching to maturity. Adapted from Congdon 1989.

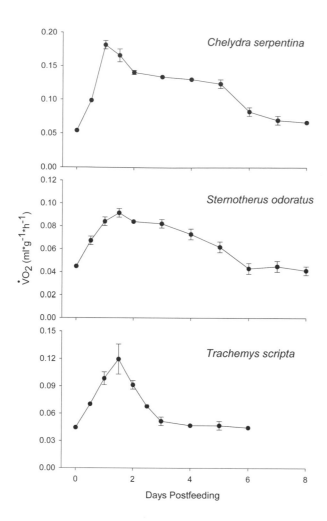

Fig. 7.4. Mean O_2 consumption rates of juvenile *C. serpentina*, adult *S. odoratus*, and subadult *T. scripta* before and up to 8 days after ingestion of meals equaling 11.3%, 5.0%, and 5.2% of body mass, respectively. Mean O_2 consumption rate at day 0 is the mean of fasting values measured over a 4-day period prior to feeding. Vertical bars represent ±1 SE. O_2 consumption peaks at 3.4, 2.1, and 2.7 times fasting rates, respectively, for these species. Adapted from Secor & Diamond 1999.

animals to those observed in reptiles that eat only infrequently. In that case we would expect greater changes in metabolic rate, morphology, and physiological processing of food in adult snapping turtles. Further experiments of this type are needed in snapping turtles.

CONCLUSIONS

Although snapping turtles seldom bask, they do thermoregulate. Snapping turtles have a lower CTM than semiaquatic emydid turtles and terrestrial testudinid turtles. Although many species of aquatic turtles bask and this gives them a thermal advantage, snapping turtles generally thermoregulate by selecting water temperatures of 24 to 28°C, as shown in the laboratory. This range of temperatures allows efficient digestion of their primarily carnivorous diet. Follicular development in spring is strongly related to water temperature. When snapping turtles cannot attain warm

temperatures in water they bask atmospherically as in Algonquin Park in Ontario, Canada. Thus, snapping turtles operate at moderate temperatures and their thermal biology appears to be related to their aquatic habits wherein temperatures are both more moderate and change more slowly than on land.

The carnivorous diet of snapping turtles is responsible for their more rapid growth than other turtles in the same habitat. Snapping turtles, slider turtles, and other emydid turtles have similar gut morphologies. They all maintain the morphology and physiological capacity of their digestive system when they are not feeding and exhibit little up-regulation in physiological performance of the digestive system when fed. Both carnivorous and omnivorous turtles are physiologically adapted to rapidly digest protein and absorb amino acids across the wall of the small intestine. Snapping turtles con-

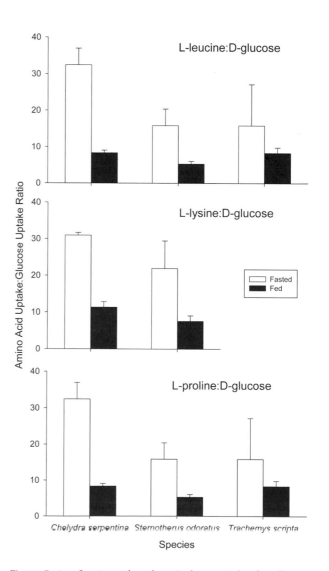

Fig. 7.5. Ratios of amino acid uptake to D-glucose uptake of small intestine for fasted and fed juvenile *C. serpentina*, adult *S. odoratus*, and subadult *T. scripta*. Vertical bars represent ±1 SE. Ratios greater than 1 are characteristic of omnivores and carnivores. Adapted from Secor & Diamond 1999.

sume mostly animal protein, maintain a moderate body temperature within a broad range ($\pm 10°C$) by selecting appropriate water temperatures, and grow more rapidly and reach larger sizes than sliders and other emydid turtles with which they share their habitat.

For a species that is generally thought to be well studied, it is surprising how much of the thermal biology and feeding/digestive biology of snapping turtles is still unknown. Considerable research is needed to better understand both aspects of the biology of snapping turtles. A detailed study of the thermal biology of snapping turtles should combine careful observations of the behavior of turtles of different sizes, over real time, in different seasons, with careful measurements of deep body, carapace and plastron temperatures, environmental temperatures and microclimate. Such a study can be accomplished using T_e models placed in different portions of the habitat and multichannel telemetry from the turtles, which will allow the investigator to determine the actual temperatures occupied and the actual behavior of individual animals. This study could be combined with measurements of reproductive performance of females to relate body temperatures, feeding, and reproductive success. Laboratory studies of digestive performance of large juvenile and adult snapping turtles will determine whether digestive performance changes from that of 100-g individuals. Finally, the metabolic rate and physiological response of liver, muscle, and gut should be determined over a range of temperature from 5 to 40°C to learn if there are optimal temperatures for physiological performance.

8

Energetics of the Snapping Turtle

ROBERT E. GATTEN, JR.

INTRODUCTION

Snapping turtles have been the subject of many studies by physiologists over the past century. This intense scrutiny has been possible in part because of the widespread distribution and relative abundance of these turtles. Furthermore, their large clutch size permits experimental designs with multiple replicates of eggs or hatchlings. This chapter reviews the factors that influence the rate of energy use by snapping turtles during embryonic development, as hatchlings, and as adults. Most research on the energetics of snapping turtles has approached the subject from an ecological rather than a biochemical perspective, and this review will necessarily reflect that emphasis.

ENERGETICS OF EGGS

When a female snapping turtle deposits an egg in a nest, the embryo is at the gastrula stage (Yntema 1968), the absolute rate of conversion of yolk or albumen into embryo is very small, and the rate of metabolism of the egg is very low. Freshly laid eggs incubated at 29°C on a relatively wet substrate (−150 kPa) completed development and hatched after about 58 days (Packard et al. 1984c; see also Morris et al. 1983). During incubation, the embryos grew from essentially zero mass to 1.2 g dry mass while the yolk shrank from 2.18 g to 0.78 g dry mass (Packard et al. 1984c; see also Morris et al. 1983). The following discussion will center on the changes in the metabolism of snapping turtle eggs during development. Other reviews that focus on the metabolism of turtle eggs are those of Packard & Packard (1984), Ewert (1985), Packard & Packard (1988a), and Vleck & Hoyt (1991).

Temporal Pattern of Metabolism during Incubation

Lynn & von Brand (1945) apparently were the first to measure the metabolic rate of eggs of snapping turtles. They found that oxygen consumption at 25°C increased in a sigmoid fashion from about 0.035 ml h^{-1} on day 10 of incubation to about 0.750 ml h^{-1} at hatching, which occurred after about 72 days. More recent studies added detail to our knowledge of the pattern of oxygen consumption during incubation. Snapping turtle eggs incubated at 29–30°C on a relatively wet substrate (–150 kPa) and measured at the same temperature had a metabolic rate that rose slowly during the first third of incubation, increased more rapidly during the second third, and then declined until hatching (Gettinger et al. 1984; Miller & Packard 1992; Birchard & Reiber 1995). The pattern of carbon dioxide production during incubation mirrored that of oxygen consumption (Miller & Packard 1992). The rise in oxygen consumption paralleled the increase in embryo mass, except for the fall in metabolism just before hatching. Rapid growth and high oxygen consumption in the last third of incubation may depend on full development of the blood vessels in the chorioallantoic membrane (Yntema 1968; Birchard & Reiber 1995). Spontaneous motility of embryos late in development may also contribute to their high rate of metabolism (Decker 1967; Ewert 1985). In all three studies in which eggs were incubated at –150 kPa and 29–30°C and oxygen consumption was measured at the incubation temperature, metabolism peaked at about day 45 at 1.2–1.3 ml O$_2$ h^{-1} (Gettinger et al. 1984; Miller & Packard 1992; Birchard & Reiber 1995).

Substrates for Metabolism during Incubation

Snapping turtle eggs at 29°C had patterns of oxygen consumption and carbon dioxide production that yielded a respiratory quotient (RQ, the ratio of the rate of production of carbon dioxide to the rate of consumption of oxygen) of 0.67–0.70 between day 10 and day 35 of incubation; RQ then rose to 0.77–0.80 by day 50 (Miller & Packard 1992). These RQ values indicated that the substrate consumed during the first two thirds of incubation was mainly lipid (Bartholomew 1982). As development continued, the use of protein as a substrate became more important.

Analyses of changes in the lipid and protein content of the yolk and embryo also indicated that both of these substrates were catabolized during development. An early study (von Brand & Lynn 1947) found that half of the organic matter consumed during incubation was protein and the other half was lipid. More recently, Packard et al. (1984a) estimated that oxidation of proteins accounted for only about 12% of the decline in egg mass, with the rest presumably due to catabolism of lipids. The results of another study (Janzen et al. 1990), however, implied just the opposite. In the latter investigation, with eggs incubated on a wet substrate, the pro-

tein content of the yolk fell by approximately 1.05 g (fig. 3 in Janzen et al. 1990) while that of the embryo (carcass) rose by about 0.35g (fig. 5 in Janzen et al. 1990); the difference in protein mass (0.70 g) was presumably due to oxidation of proteins. During the same interval, the lipid content of the yolk decreased by approximately 0.40 g (fig. 2 in Janzen et al. 1990) while the lipid content of the carcass rose by about 0.175 g (fig. 4 in Janzen et al. 1990); the difference in lipid mass (0.225 g) apparently represented catabolism of lipids. Based on the mass of protein and lipid lost during incubation, it would appear that about 76% of the decline in mass was due to oxidation of protein. Given the divergent results among these studies, additional work is needed to clarify the relative importance of catabolism of proteins and lipids during development of snapping turtles.

Effects of Substrate Moisture on Energetics of Eggs

A major theme of recent studies on turtle eggs has been the effect of the hydric condition of the incubation medium on the growth and metabolism of embryos and hatchlings. Snapping turtle eggs incubated on a wet substrate (–150 kPa) maintained their mass or lost mass more slowly than eggs kept on a dry medium (–850 kPa or –1100 kPa) (Gettinger et al. 1984; Miller & Packard 1992). Likewise, embryos within eggs incubated on the wetter environment were larger and had a higher rate of oxygen consumption and carbon dioxide production by day 40–45 than those from eggs kept on the dryer substrate (Gettinger et al. 1984; Miller & Packard 1992). Several hypotheses have been advanced to explain these effects of hydric condition during incubation on the growth and metabolism of embryos.

Intracellular Water and Egg Metabolism

In theory, turtle eggs incubated on a dry substrate might have a lower amount of intracellular water than eggs on a wet substrate, and thus a lower volume of solvent in which metabolic reactions could occur (Morris et al. 1983; Packard et al. 1983). These early studies revealed, however, that snapping turtle embryos in eggs in wet and dry environments had similar water contents (expressed as percentage of body mass represented by water) (Morris et al. 1983; Packard et al. 1983). Later studies (Packard et al. 1988; Finkler 1999) found that the hydric environment during incubation had a positive effect on the whole body water content and on the level of tissue hydration of hatchlings (based on water content adjusted for differences in body mass rather than on percentage of body mass represented by water). Therefore, hydric condition may indeed influence embryonic metabolism via intracellular water availability. At the very least, water availability does influence conversion of egg nutrients into embryonic proteins and lipids: eggs incubated on a wet substrate produced hatchlings that had converted more of the protein and fat content of the yolk into tissue than eggs kept on a dry medium (Packard et al. 1988).

Urea and Egg Metabolism

Eggs incubated on a dry substrate accumulated a higher concentration of urea than those maintained on a wet medium (Packard et al. 1984a). Because urea is a competitive inhibitor of a number of metabolic enzymes, it is plausible that the high concentration of urea in eggs kept on a dry substrate was the cause of their lower metabolic rate. However, when urea was injected into eggs so that eggs incubating on wet and dry substrates had the same urea concentration, there was no difference in the size of hatchlings or in the mass of residual yolk in the two treatment groups (Packard & Packard 1989a). This finding indicated that the high concentration of urea in eggs on dry substrates did not lead to reduced metabolism and growth.

Sex and Egg Metabolism

Packard et al. (1984b) hypothesized that the hydric environment during incubation might influence the sex ratio of hatchlings. Male and female hatchlings would potentially have different levels of gonadal hormones in their blood, which might influence their metabolic rate. However, incubation of eggs on wet and dry substrates at 29°C led to the production of 100% females; thus, the influence of the hydric environment on metabolism and growth was not due to a difference in the sex of embryos (Packard et al. 1984b).

Evaporation and Egg Metabolism

Snapping turtle eggs incubated on dry substrates lost more water by evaporation than eggs kept on wet substrates (Morris et al. 1983). Faster evaporation should result in greater cooling of eggs in dry conditions, and thus a lower egg temperature. The lower temperature, in turn, should lead to slower growth, more gradual consumption of egg nutrients, and a longer period of development. However, the magnitude of any such effect of evaporative cooling on growth and yolk depletion was too small to be measurable in snapping turtles or painted turtles (*Chrysemys picta*) (Morris et al. 1983; Packard et al. 1983).

Likewise, faster evaporation from turtle eggs incubated on a dry substrate might lead to a dry shell membrane and the inhibition of the growth of the chorioallantois, as in birds (Tazawa 1980). In turn, such slow growth of the chorioallantois might have a deleterious effect on the rate of gas exchange across the egg and on lipid utilization and growth (Ackerman 1981b). However, neither hypoxia nor incubation temperature influenced the density of vasculature in the chorioallantois of red-bellied turtles (*Pseudemys nelsoni*) or snapping turtles (Kam 1993; Birchard & Reiber 1995). Thus, it seems unlikely that hydric conditions during incubation affect metabolism through an influence on the growth of the chorioallantois.

Hypoxia and Egg Metabolism

Embryos developing under dry conditions most likely experience lowered blood volume, increased blood viscosity, decreased levels of oxygen in the circulating blood, and generalized tissue hypoxia (Packard et al. 2000; Packard and Packard 2002). Such a generalized tissue hypoxia might explain the lower metabolic rate in turtles developing under dry conditions (Gettinger et al. 1984; Miller and Packard 1992).

Energetic Cost of Development

Gettinger et al. (1984) determined that the energetic cost of development for embryos on wet and dry substrates was 14 and 8 kJ, respectively. The ratio of these costs (1.75) was similar to that of the dry masses of embryos from the wet and dry substrates on day 50 (1.65). Apparently the energetic cost of development in turtles is a function of egg mass and, presumably, original yolk mass. For eggs of 11 species of turtles kept at 29–31°C, the energetic cost of development varied with body mass according the following equation: cost of development (kJ) = 2.18 (egg mass in g)$^{0.81}$ (Vleck & Hoyt 1991).

The efficiency of embryonic growth, as measured in grams of embryo dry mass produced per volume of oxygen consumed, was indistinguishable between wet and dry treatments during incubation (Miller & Packard 1992). However, because embryos in eggs on wet substrates continued to grow when those on dry substrates had stopped growing, the former embryos were larger at hatching and consumed more oxygen, even if the patterns of metabolism over time were identical. There was no difference in the RQ between the two groups, indicating that the substrates consumed were the same (Miller & Packard 1992).

Effects of Substrate Moisture on Energetics of Hatchlings

Hydric conditions during incubation influence not only many aspects of developing embryos but also of hatchlings. Hatchlings from eggs incubated on a dry medium swam or ran more slowly, accumulated lactate faster, and accumulated more lactate than hatchlings from eggs kept under moister conditions (Miller et al. 1987). The difference in size between hatchlings emerging in August from eggs kept on wet and dry substrates, however, did not persist during the subsequent fall and winter (Finkler et al. 2002). Furthermore, hatchlings from eggs on wet substrates had lower energy reserves (body fat plus residual yolk) at the start of hibernation than turtles from dry eggs (Finkler et al. 2002). Thus, the immediate potential advantages of emerging from a well-hydrated egg may be offset by the lower amount of stored energy available during the first winter after hatching, with possible consequences for survival (Finkler et al. 2002; see also Bobyn & Brooks 1994a).

Effects of Incubation Temperature on Energetics of Eggs

Snapping turtle eggs incubated at high temperatures converted more egg nutrients into embryonic proteins than eggs kept at low temperatures, but a parallel effect on mobilization of lipids was not present (Packard et al. 1988). However, this effect of temperature on conversion of egg proteins into embryonic proteins may have been indirect, via the influence of temperature on water availability, rather than a direct action of temperature on metabolism (Packard et al. 1988).

Snapping turtle eggs incubated on a wet substrate (–150 kPa) at 30°C developed faster and had a shorter incubation time than those at 24°C (Birchard & Reiber 1995). Oxygen consumption during development peaked sooner, but at the same maximal value, at the higher temperature (Birchard & Reiber 1995). However, when the data for oxygen consumption were adjusted for the different incubation periods, the pattern and rate of oxygen consumption were indistinguishable between the two groups (Birchard & Reiber 1995). The cost of development, as estimated by the integration of oxygen consumption during development, did not differ between the two temperatures (15.0 and 14.6 kJ at 24 and 30°C, respectively) (Birchard & Reiber 1995), and was the same as that found by Gettinger et al. (1984) for eggs incubated under similar conditions.

The fact that eggs incubated and measured at 24°C had the same embryo mass and peak oxygen consumption late in incubation as those incubated and measured at 30°C suggested that the eggs completely compensated for the difference in temperature during the later portion of development (Birchard & Reiber 1995). Thermal conditions during incubation influence the metabolism not only of embryos but also of hatchlings, as will be discussed below.

Other Influences on Metabolism of Eggs

Turtle eggs in nests potentially face hypoxia because of high oxygen uptake by eggs and/or slow diffusion of oxygen from the atmosphere into the nest (Plummer 1976; Ackerman 1977, 1980). Hypoxia reduced metabolism and growth of embryos of sea turtles (Ackerman 1981b; McGehee 1990) and red-bellied turtles (Kam 1993; Kam & Lillywhite 1994), but whether this is the case for the eggs of snapping turtles is unknown.

ENERGETICS OF TURTLES

The first known reports of the metabolic rate of snapping turtles are those of Baldwin (1926a, b). Since then,

others have investigated many factors that influence their rate of energy use (Table 8.1). Previous reviews that deal with the metabolism of reptiles, or of turtles in particular, are by Bennett & Dawson (1976), Bennett (1982), Gregory (1982), Seymour (1982), Glass & Wood (1983), Gatten (1985), Waldschmidt et al. (1987), Ultsch (1989), and Jackson (2000).

Influence of Body Size on Energetics of Turtles

Because snapping turtles range in size from about 10 g (as hatchlings) to about 34 kg (Ernst et al. 1994), this species would seem to be an excellent one in which to examine the effect of body size on metabolic rate. If alligator snapping turtles (*Macroclemys temminckii*) are included, the mass range extends to 113 kg (Pawley 1987). However, only one study (Kinneary 1992) explicitly examined this relationship, and it dealt with small specimens. Kinneary (1992) kept hatchling/juvenile snapping turtles (8 to 40 g) for five weeks at 28°C in fresh water; the animals had a mass-specific rate of consumption of oxygen at 28°C that declined with mass according to the following semilogarithmic equation:

$$\log (\mu l\ O_2\ g^{-1}\ h^{-1}) = 2.24 - 0.014\ (\text{mass in g}), \quad (1)$$

n = 15, df = 13, r^2 = 0.62. The relationship between metabolism and body size did not differ between these turtles and those kept in 20% or 40% seawater (Kinneary 1992). In an earlier investigation, I measured the metabolic rate of snapping turtles ranging in mass from 1,497 to 5,630 g (mean mass = 3,487 g) that were acclimated to 25°C and studied in air at 20°C (Gatten 1978). For these animals, the relationship between total body oxygen consumption and body mass under standard conditions (postabsorptive, at rest, in the dark, during the resting phase of the daily cycle) (calculated for this chapter) is:

$$\text{ml}\ O_2\ h^{-1} = 0.09\ (\text{mass in g})^{0.777}. \quad (2)$$

n = 10, df = 8, r^2 = 0.71. The slope for this relationship does not differ from 0.75 (df = 8, t = 0.049, P > 0.05). Bennett & Dawson (1976) calculated the relationship between total body oxygen consumption and mass for 10 species of turtles (including *Chelydra serpentina*) at 20°C:

$$\text{ml}\ O_2\ h^{-1} = 0.066\ (\text{mass in g})^{0.86}. \quad (3)$$

n = 10, df = 8, r^2 = 0.61. The slopes for equations 2 and 3 do not differ (df = 16, t = 0.457, P > 0.05). If the regression line from equation 2 is extended to the size of juvenile snapping turtles, the metabolic rate predicted for a 71.98-g turtle is 2.50 ml O_2 h^{-1}, considerably above the value of 1.51 ml O_2 h^{-1} measured for animals of that mean mass at the same temperature of 20°C (Steyermark & Spotila 2000). In 1976, the existing data for reptiles in general were too limited to permit a resolution of the question of whether juveniles have a metabolic rate significantly different from that of adults (Bennett & Dawson 1976); apparently, no other analyses to resolve this question have been attempted since then.

Table 8.1

Oxygen consumption of snapping turtles

Mean mass (g)	Mass range (g)	n	T_b (°C)	Thermal history (°C)	Nutritional/ other condition	Air/water	Total $\dot{V}O_2$ (ml/h)	Pulmonary $\dot{V}O_2$ (ml/h)	Extrapulmonary $\dot{V}O_2$ (ml/h)	Reference	Notes
1700			20		freshly caught, active	air	95.54			Baldwin 1926a	
1692	1447–1920	5	20	August, Iowa	freshly caught	air	98.80			Baldwin 1926b	
1488	1322–1654	2	20	Sept–Oct in lab, Iowa	fasted 1 mo	air	29.86				
3473	1375–5705	10	10	10	standard	air	11.11			Gatten 1978	1
3473	1375–5706	10	10	10	fasted, active	air	319.17				1
3473	1375–5707	10	10	25	standard	air	26.74				1
3473	1375–5708	10	10	25	fasted, active	air	537.62				1
3473	1375–5709	10	20	10	standard	air	40.29				1
3473	1375–5710	10	20	10	fasted, active	air	622.01				1
3473	1375–5711	10	20	25	standard	air	53.14				1
3473	1375–5712	10	20	25	fasted, active	air	650.49				1
3473	1375–5713	10	30	10	standard	air	126.07				1
3473	1375–5714	10	30	10	fasted, active	air	1176.65				1
3473	1375–5715	10	30	25	standard	air	142.74				1
3473	1375–5716	10	30	25	fasted, active	air	1704.55				1
3450	1422–5788	10	10	10	standard	water		13.46		Gatten 1980	2
3450	1422–5789	10	10	25	standard	water		16.91			2
3450	1422–5790	10	20	10	standard	water		45.54			2
3450	1422–5791	10	20	25	standard	water		43.47			2
3450	1422–5792	10	30	10	standard	water		122.82			2
3450	1422–5793	10	30	25	standard	water		121.79			2
3007	1305–5290	6	4	15–20, July–Aug, MI	standard	water	16.96	15.04	1.92		
3050	1305–5290	6	20	15–20, July–Aug, MI	standard	water	116.27	110.32	5.95		
2815		1	4	15–20, July–Aug, MI	standard	water			2.31		3
2815		1	4	15–20, July–Aug, MI	standard	water			1.13		4
8519		1	4	15–20, July–Aug, MI	standard	water			3.32		3
8519		1	4	15–20, July–Aug, MI	standard	water			1.02		4
3195	2020–4520	4	25	25	fasted, resting	water	92.65	88.82	3.83	Bagatto & Henry 1999	4
101		3	30	2 days at 30°C	fasted	air	5.45			Secor & Diamond 1999	4
101		3	30	2 days at 30°C	fed 11.3% of body mass	air	18.18				5
	8–40	15	28	28	5 wk in freshwater, fasted	air	1.82			Kinneary 1992	6
	8–37	15	28	28	5 wk in 20% seawater, fasted	air	1.78				6
	12–45	14	28	28	5 wk in 40% seawater, fasted	air	1.55				6
9.3		15	25	21.5	resting, female	air	1.10			O'Steen & Janzen 1999	7
9.6		17	25	21.5	resting, male	air	1.13				7
10.6		24	25	24.5	resting, male	air	0.95				7
10.6		22	25	27.5	resting, female	air	0.82				7
10.5		23	25	27.5	resting, male	air	0.84				7
9.8		24	25	30.5	resting, female	air	0.71				7
10.7		3	31	21.5	resting, female	air	1.82				7

9.9	8	31	21.5	air	1.74	resting, male		7
11.4	12	31	24.5	air	1.50	resting, male		7
11.3	13	31	27.5	air	1.45	resting, female		7
11.1	11	31	27.5	air	1.50	resting, male		7
10.0	12	31	30.5	air	1.22	resting, female		7
9.1	8	25	24.5	air	0.93	resting, male, control	O'Steen & Janzen 1999	7,8
8.5	13	25	24.5	air	1.34	resting, male, T_3		7,9
8.9	7	25	30.5	air	0.84	resting, female, control		7,8
8.2	6	25	30.5	air	0.99	resting, female, T_3		7,9
68.28	34	15		air	1.08	standard	Steyermark & Spotila 2000	10
71.98	34	20		air	1.51	standard		10
56.92	34	25		air	3.09	standard		10
72.98	34	30		air	3.81	standard		10
63.47	138	15	26.5	air	0.91	standard	Steyermark & Spotila 2000	11
65.82	83	15	28.0	air	1.00	standard		11
66.38	25	15	30.0	air	1.05	standard		11
53.59	138	25	26.5	air	2.84	standard		11
54.90	83	25	28.0	air	3.18	standard		11
53.57	25	25	30.0	air	3.14	standard		11

1. The paper reported mass-specific oxygen consumption. Overall, 14 animals were used but only 10 were used in each group. The error introduced here by multiplying the mean mass-specific oxygen consumption of each group of 10 animals by the overall mean body mass for 14 animals is less than 7% in each case.

2. The paper reported mass-specific oxygen consumption. Overall, 11 animals were used but only 10 were used in each group. The error introduced here by multiplying the mean mass-specific oxygen consumption of each group of 10 animals by the overall mean body mass for 11 animals is less than 3% in each case. Animals were in water and oxygen consumption was calculated from the difference in oxygen content of the air flowing into and out of the chamber; the decline in oxygen content of the water, about 2%, was not used in calculations of oxygen consumption.

3. Forcibly submerged with no access to air.

4. Animals in water with access to air.

5. Peak oxygen consumption after feeding.

6. Oxygen consumption calculated from regression equation relating mass-specific oxygen consumption to mass, using mass = 20 g.

7. Thermal history is temperature of incubation of eggs that produced hatchlings. Hatchlings were still utilizing yolk in their guts so they cannot be considered postabsorptive. Data for body mass and for oxygen consumption from 1992 and 1993 are combined.

8. Eggs were treated with the control solution (ethanol) on day 33 (embryonic stage 18) at 24.5°C or on day 20 (embryonic stage 18) at 30.5°C.

9. Eggs were treated with thyroid hormone (T_3) in ethanol on day 33 (embryonic stage 18) at 24.5°C or on day 20 (embryonic stage 18) at 30.5°C.

10. All 34 turtles were studied at all four temperatures. The mass varied because turtles grew during the course of the study. Animals were evenly and randomly distributed across egg incubation temperatures of 26.5, 28.0, and 30.0°C. Turtles were kept at 25°C until measurement.

11. Thermal history is the temperature of incubation of eggs that produced hatchlings. Hatchlings were kept at 25°C until studied. All animals from each incubation temperature were measured at both 15 and 25°C.

Table 8.2

Q_{10} for oxygen consumption of snapping turtles

Measurement temperature range (°C)	$T_{accl} = 10°C$	$T_{accl} = 25°C$	Significance (P)
Standard metabolism in air			
10–20	3.6	2.0	<0.01
20–30	3.1	2.7	<0.01
Standard metabolism in water			
10–20	3.4	2.6	<0.01
20–30	2.7	2.8	NS
Activity metabolism in air			
10–20	1.9	1.2	<0.01
20–30	1.9	2.6	<0.01

Data from Gatten 1978, 1980. Statistical test for difference between two Q_{10} values from Coyer and Mangum 1973.

Influence of Temperature and Thermal History on Energetics of Turtles

The impact of an acute change in body temperature on the resting or standard metabolic rate of reptiles is well known (Bennett & Dawson 1976). The general pattern is a 2- to 3-fold rise in metabolism with a 10°C increment in temperature ($Q_{10} = 2–3$), but the rate of increase typically slows as temperature rises (Bennett & Dawson 1976). For snapping turtles acclimated to 10°C, the standard metabolic rate had a higher sensitivity to temperature from 10 to 20°C than from 20 to 30°C (Table 8.2). However, this trend was not apparent for turtles acclimated to 25°C (Table 8.2).

Hatchling and Juvenile Turtles

Snapping turtles recently emerged from eggs incubated at 21.5–30.5°C had a Q_{10} for resting metabolic rate between 25 and 31°C of 2.1–2.5 (1992 data only, O'Steen & Janzen 1999). In contrast, snapping turtles that were six months old and that had experienced temperatures during incubation of 26.5–30.0°C exhibited a Q_{10} for standard metabolic rate that varied widely over the span of measurement temperatures: $Q_{10} = 1.7$ from 15 to 20°C, $Q_{10} = 6.7$ from 20 to 25°C, and $Q_{10} =$ only 0.9 from 25 to 30°C (Steyermark & Spotila 2000). The thermal insensitivity of standard metabolism of young snapping turtles near their mean selected temperature (24.5–29.8°C) (Williamson et al. 1989; Knight et al. 1990; O'Steen 1998) is not unexpected, given the same pattern for reptiles in general (Bennett & Dawson 1976).

Acclimation of Metabolism in Adults

Snapping turtles in nature may experience body temperatures as low as 1–2°C in winter (Ernst et al. 1994) and about 22–23°C during summer in Ontario lakes (Brown et al.

1990, 1994a). However, they may raise their body temperature as high as 34°C during atmospheric basking (Obbard & Brooks 1979). One would predict that snapping turtles, like other reptiles that experience a wide range of body temperatures, would exhibit a significant effect of thermal history on metabolic rate. Snapping turtles acclimated to 10°C had a significantly lower standard metabolic rate in air than those acclimated to 25°C (Table 8.1) (Gatten 1978). However, turtles acclimated to 10 and 25°C had statistically indistinguishable standard metabolic rates when they were in water (Table 8.1) (Gatten 1980). Thus, either cold acclimation or voluntary submersion reduced standard metabolism from that seen in warm-acclimated snapping turtles studied in air, but the effect was not additive.

It seems likely that the molecular mechanisms responsible for the effect of acclimation on metabolism are different from the molecular mechanisms operating during voluntary submergence, but this topic has not been investigated. In addition, the effects noted above were for turtles acclimated to cold and warm conditions in the laboratory; how acclimatization to natural conditions influences metabolism has yet to be fully explored (but see discussion below on metabolic changes during hibernation).

Acclimation of Metabolism in Hatchlings and Juveniles

Thermal history influences metabolism not only of adult turtles but also of hatchlings. Snapping turtle eggs incubated at low temperatures produced hatchlings with a higher resting metabolic rate 2–13 days after hatching than eggs kept at higher temperatures; this result may have been due in part to the fact that the former hatchlings had higher levels of plasma thyroxine than the latter ones (Table 8.1) (O'Steen & Janzen 1999; see also Birchard & Reiber 1995, for acclimation of metabolism during incubation). Thus, the effect of immediate thermal history on resting or standard metabolism was different between hatchlings and adults: the hatchlings studied by O'Steen & Janzen (1999) showed partial positive metabolic compensation for temperature whereas the adults studied by Gatten (1978) showed inverse compensation. Partial positive compensation presumably allows embryos in cold nests and hatchlings in cold environments to compensate partially for temperature in the conversion of yolk into carcass and to reduce incubation time (O'Steen & Janzen 1999). Inverse compensation in adults reduces energy use during hibernation.

In contrast to the results above for the effect of thermal history during incubation on snapping turtles examined just after emergence from the egg, other studies revealed a different influence of thermal exposure during incubation on the metabolism of turtles at six months of age. Acclimation temperature during egg incubation had no effect on the oxygen consumption or carbon dioxide production of six-month-old turtles studied at 15°C (Table 8.1) (Steyermark & Spotila 2000). However, when measurements were carried out at 25°C, oxygen consumption and carbon dioxide production were higher for turtles derived from eggs incubated at 28.0°C than for turtles from eggs kept at either 26.5 or 30.0°C (Table 8.1) (Steyermark & Spotila 2000). Perhaps differences in the amount of residual yolk (potentially high for turtles 2–13 days after hatching and most likely zero for turtles at six months after hatching) or body water content may have contributed to differences in the results of the work of O'Steen & Janzen (1999) and Steyermark & Spotila (2000). Another difference between these two studies that may have contributed to their divergent results is that the former study used incubation temperatures of 21.5–30.5°C while the latter used a narrower range of 26.5–30.0°C. It is not clear how incubation temperature and acclimation temperature during the first six months of life interacted to generate the complex patterns of metabolism seen in these turtles.

Hypoxia, Diving, Hibernation, and Extrapulmonary Gas Exchange

Snapping turtles remain submerged for long periods. Does their metabolism during such voluntary immersion differ from that when they are at the water surface or on land, all other conditions being equal? This question has been surprisingly hard to answer, and much more complex than physiologists first envisioned. Although experiments in which snapping turtles and other species were forcibly submerged in the laboratory can lead to an understanding of their tolerance to such conditions and the limits of physiological processes (e.g., Dodge & Folk 1963; Jackson & Schmidt-Nielsen 1966; Jackson 1968; Ultsch et al. 1984; Ultsch & Wasser 1990; Reese et al. 2002, 2004), they do not necessarily reveal the types of physiological changes that occur during voluntary submersion in nature (Gatten 1981, 1984, 1985).

Hypoxia

A reduction in the oxygen concentration in the air from 21% to 2% had no effect on the oxygen consumption of snapping turtles at 25°C, at least in part because the animals breathed more rapidly and their heart rate increased (Boyer 1963). It is not clear from these experiments whether the increased ventilation and blood flow maintained tissue Po_2 at close to the normoxic level or whether the tissues relied to any extent on anaerobic metabolism. The work by Boyer (1963) is revealing because it involved manipulation of only one independent variable: the level of oxygen in the air. However, when a turtle dives, the changes are more complex: the animal does not breathe, the animal is submerged, and the animal is at near-neutral buoyancy. Which of these changes influences energy use?

Submersion

In an attempt to answer this question, I measured the pulmonary oxygen consumption of snapping turtles in both air and water at 10, 20, and 30°C after acclimation to both cold and warm conditions (Table 8.1) (Gatten 1978, 1980). For turtles acclimated to 25°C, simply placing them in shallow water (deep enough to permit complete submersion but shallow enough to allow each animal to breathe by extending its neck to the water surface while it rested on the bottom) resulted in a decline in their aerial oxygen consumption under standard conditions of 36% at 10°C, 18% at 20°C, and 14% at 30°C. Thus, it appears that mere submergence resulted in a lowering of aerobic metabolism in snapping turtles during summer conditions. However, the results were very different when similar measurements were carried out after the turtles had been acclimated to 10°C (see previous discussion on effects of thermal history on metabolism). In this case, turtles in water had the same level of pulmonary gas exchange as turtles in air at all three measurement temperatures. As noted above, it appears that either voluntary submersion or acclimation to winter conditions resulted in a decline in aerobic metabolism from that of turtles in air after warm acclimation, but the effects were not additive. The effect of submersion alone appears to reduce the metabolic rate of small snapping turtles as well as adults: hatchlings placed in nitrogen depleted their glycogen reserves and accumulated lactate faster than those submerged in anoxic water (Reese et al. 2004).

Hibernation and Extrapulmonary Gas Exchange

Snapping turtles may spend up to seven months in hibernation in water below the ice of ponds or buried in the mud (Cahn 1937; Pell 1941; Cagle 1944; Ernst et al. 1994). These and other freshwater turtles exhibit many adaptations for such prolonged periods with no access to the atmosphere (for review, see Ultsch 1989; Jackson 2000). During such intervals, opportunities for pulmonary gas exchange are limited, and snapping turtles utilize extrapulmonary gas exchange. For snapping turtles accustomed to summer temperatures and studied during voluntary diving and surfacing, extrapulmonary gas exchange accounted for only 5% of total oxygen consumption at 20°C but provided 11% of the total uptake of oxygen at a typical winter temperature (4°C) (Table 8.1) (Gatten 1980). However, when a single snapping turtle, acclimatized to winter conditions, dove and surfaced at will in a metabolic chamber at 4°C for 12 days, uptake of oxygen from the water accounted for 31% of the total oxygen consumed (Gatten 1980). Thus, it would appear that extrapulmonary routes of oxygen uptake are important for snapping turtles in winter, if not in summer, even if their capacity for aquatic oxygen uptake is considerably lower than that of soft-shelled turtles (*Trionyx spini-*

ferus) and musk turtles (*Sternotherus odoratus*) (Ultsch et al. 1984). Extrapulmonary excretion of carbon dioxide is much more important than extrapulmonary uptake of oxygen in snapping turtles in water at 25°C: 30% of the total carbon dioxide exchange occurred via nonpulmonary surfaces whereas only 4% of total oxygen uptake took place via these routes (Table 8.1) (Bagatto & Henry 1999). This pattern is likely to apply to freshwater turtles, including snapping turtles, during the lower temperatures of hibernation as well (Ultsch 1989).

Hibernation and Anaerobiosis

Reptiles have a high capacity for anaerobic metabolism (Bennett & Dawson 1976; Gatten 1985) and freshwater turtles employ anaerobic pathways, chiefly glycolysis, during hibernation (Gatten 1981; Ultsch & Jackson 1982; for review, see Ultsch 1989; Jackson 2000). During forced submergence in normoxic water at 10°C, snapping turtles relied extensively on glycolysis and exhibited a profound increase in plasma lactate level, from about 1 mM before submergence to as high as 57 mM after 21 days underwater (Ultsch et al. 1984; Ultsch & Wasser 1990). Snapping turtles kept for 150 days in normoxic water at 3°C (a more realistic hibernation temperature than 10°C) showed a relatively small (and statistically insignificant) rise in blood lactate concentration, a stable blood pH, and a decrease in blood P_{CO_2}, indicating that metabolism during this five-month interval was partly aerobic (supported by extrapulmonary uptake of oxygen and release of carbon dioxide) (Reese et al. 2002). On the other hand, snapping turtles submerged in anoxic water at 3°C did not survive as well as those in normoxic water, and the changes in blood pH, gases, ions, and lactate (up to 168 mM) indicated a heavy reliance on anaerobic metabolism (Reese et al. 2002).

Hatchling snapping turtles emerge from the nest, migrate to water, and spend their first winter underwater (Congdon et al. 1987). Such small turtles, with a higher surface-to-volume ratio and a thinner integument than adults, tolerated forcible submergence in normoxic water at 3°C for 150 days by relying on extrapulmonary gas exchange without the need to switch to glycolysis to survive (Reese et al. 2004). In contrast, hatchlings submerged in anoxic water at 3°C survived for only 30 days and relied extensively on anaerobic metabolism (Reese et al. 2004). The extent to which adult or hatchling snapping turtles rely on glycolysis during hibernation in nature has not been determined but it is likely to be very great.

Activity
Aerobic Support for Activity

Snapping turtles have been characterized as lethargic animals, swimming or crawling slowly on the bottom of lakes and streams; however, they do travel long distances during overland migrations (Cahn 1937; Klimstra 1951; Hammer 1969; Ernst et al. 1994). When forced to undergo two minutes of intense exercise in a dry metabolic chamber, the rate of oxygen consumption of snapping turtles rose above standard levels by a factor of 9–29, depending on acclimation and measurement temperatures (Table 8.1) (Gatten 1978). This ratio of aerobic metabolism during activity to that under standard conditions decreased as measurement temperature increased, and was lower in cold-acclimated than in warm-acclimated animals (Gatten 1978). In animals that ranged in mass from 1,460 to 5,609 g (mean mass = 3,469 g) and that were acclimated to 25°C and measured at 20°C, oxygen consumption during exercise scaled with body mass according to the following equation (calculated for this chapter):

$$\text{ml } O_2 \text{ h}^{-1} = 0.13 \text{ (mass in g)}^{1.038.} \quad (4)$$

n = 10, df = 8, r^2 = 0.69. The slope for this relationship does not differ from 1.0 (df = 8, t = 0.48, P > 0.05) or from that in equation 2 for standard metabolism vs. mass under the same acclimation and measurement temperatures (df = 16, t = 0.855, P > 0.05). The lack of statistical significance for the difference in slopes for metabolism during exercise (1.038) and under standard conditions (0.777) is due to small sample sizes and substantial interindividual variability. The rate of oxygen consumption of snapping turtles during exercise (Gatten 1978) is higher than that of reptiles in general, based on an allometric analysis of data from 18 to 21 species at 20 and 30°C, respectively (Bennett 1982), and this capacity is surely useful during the overland migrations of these turtles. Oxygen consumption during exercise rose with temperature with a Q_{10} of 1.2–2.6, depending on temperature interval and acclimation condition (Table 8.2), and Q_{10} during exercise was lower than when the turtles were studied under standard conditions, as is common in reptiles (Gatten 1978).

The aerobic metabolic scope for activity is the difference between the metabolic rate of an animal during exercise and that during rest, and it indicates the amount of energy from aerobic pathways that can be allocated to movement. For snapping turtles (mass range = 1,497–5,630 g, mean mass = 3,487 g during standard conditions) acclimated to 25°C and measured at 20°C (Gatten 1978), aerobic scope scaled with mass according to the following equation (calculated for this chapter):

$$\text{ml } O_2 \text{ h}^{-1} = 0.83 \text{ (mass in g)}^{1.084} \quad (5)$$

n = 10, df = 8, r^2 = 0.64. The relationship between aerobic scope and body mass at 30°C does not differ between turtles and squamates (Gatten 1978). Aerobic scope rose with measurement temperature in both warm- and cold-acclimated snapping turtles, and was higher in the former group than in the latter (Gatten 1978). The fact that summer turtles had a higher capacity for aerobic metabolism during activity than winter animals is consistent with the fact that their most vigorous and extended activity occurs during warm conditions on land (Cahn 1937; Klimstra 1951; Hammer 1969; Ernst et al. 1994).

Anaerobic Support for Activity

Reptiles have a high capacity for supporting intense locomotion with ATP generated by anaerobic pathways (Bennett & Dawson 1976; Bennett 1982). Hatchling snapping turtles forced to swim or run accumulated lactate up to concentrations of about 0.3 mg lactate g^{-1} body mass (Miller et al. 1987). For hatchlings swimming for 3 min at 29°C, the rate of lactate accumulation was 28–114 $\mu g \cdot g^{-1} \cdot min^{-1}$, whereas for hatchlings running for 2 min at 23°C, lactate accumulated at 76–86 $\mu g \cdot g^{-1} \cdot min^{-1}$ (Miller et al. 1987). In both cases, the rate of lactate accumulation was higher in turtles that hatched from eggs kept on dry substrates than in those from eggs maintained under wetter conditions (Miller et al. 1987). These rates of lactate formation are 6–25% of those for squamate reptiles exercising under similar conditions (equation 1 of Gatten 1985).

Feeding and Metabolism

After animals consume a meal, their metabolic rate rises because of the cost of processing the ingested nutrients. This phenomenon is known as the specific dynamic effect or specific dynamic action. Freshly caught snapping turtles that were presumably still digesting a meal had a rate of oxygen consumption three times that of turtles that had been fasted for a month, but some of the difference may have been due to spontaneous activity (Table 8.1) (Baldwin 1926b). A more recent study found that the rate of oxygen consumption of small snapping turtles at 30°C rose 3.4-fold 24 h after eating a meal equivalent to 11.3% of their body mass (Table 8.1) (Secor & Diamond 1999). The metabolic rate remained elevated for five days after feeding, and the total excess energy expended during these five days represented 22% of the energy ingested during the meal (Secor & Diamond 1999). These turtles are frequent feeders and experience a much smaller increase in metabolism following feeding than other reptiles that consume very large meals on an infrequent basis (Secor & Diamond 1999).

Osmotic Conditions and Metabolism

All freshwater animals face the dual challenges of osmotic gain of water from and loss of ions to an environment considerably more dilute than their body fluids, and the voiding of excess water and recovery of lost ions may carry an energetic cost. Reptiles, with their relatively impermeable skin, are less susceptible to these challenges than freshwater fish and amphibians. After hatchling snapping turtles were maintained for up to 15 weeks in freshwater, their rates of oxygen consumption were indistinguishable from those

kept in 20 or 40% seawater (Table 8.1) (Kinneary 1992). Thus, at least for these animals, the cost of osmoregulation in freshwater vs. that in a medium with an osmolality more like that of body fluids was not detectable with indirect calorimetry.

Sex and Metabolism

Relatively few studies have analyzed the possible effect of sex on metabolic rate in reptiles, and in most cases no influence of sex on metabolism was detected (for review, see Bennett & Dawson 1976; Beaupre & Duvall 1998). In snapping turtles, male hatchlings had a slightly higher resting metabolic rate than females, but the difference was primarily due to the different temperatures experienced during incubation and not sex per se (O'Steen & Janzen 1999). In this study of a species with temperature-dependent sex determination, males on average experienced lower incubation temperatures than females and, because of partial positive compensation for temperature, exhibited higher average resting metabolic rates than females (O'Steen & Janzen 1999).

Thyroid Hormones and Metabolism

Thyroid hormones can influence the metabolic rate of reptiles (for review, see O'Steen & Janzen 1999). Snapping turtles that hatched from eggs kept at 21.5°C had a higher level of plasma thyroxine and a higher resting metabolic rate than hatchlings from eggs incubated at 30.5°C (O'Steen & Janzen 1999). Furthermore, when thyroid hormone (in this case, triiodothyronine, T_3) was applied directly to the shells of snapping turtle eggs at midincubation, the resulting hatchlings had a resting metabolic rate that was, on average, 31% above that of control animals (Table 8.1) (O'Steen & Janzen 1999). Thus, it appears that thyroid hormones have a significant impact on metabolic rate and can contribute to metabolic compensation for temperature in snapping turtles and other chelonians (O'Steen & Janzen 1999).

Clutch and Metabolism

How much of the variability in metabolic rate among individual snapping turtles can be traced to the clutches from which they came, all other conditions of incubation being identical? Clutch had a measurable effect on the resting metabolic rate of snapping turtles at 25°C (1993 animals, O'Steen & Janzen 1999). Furthermore, clutch had a significant influence on the level of thyroxine in turtles incubated at different temperatures (O'Steen & Janzen 1999). In another study, clutch had a major effect on metabolic rate of snapping turtles, and clutch is likely to have a significant impact on the energy budget and thus growth and fitness of individual snapping turtles in nature (Steyermark & Spotila 2000).

Metabolism and Growth

If the amount of energy available to an animal is limited, then allocation of energy to one component of the energy budget constrains the amount available to other functions. In a test of the possible trade-off between energy allocated to maintenance and growth, Steyermark (2002) found that juvenile snapping turtles with high standard metabolic rates grew more slowly than those with lower maintenance costs. Investigations such as this should help create a fuller understanding of energy budgets in reptiles.

CONCLUSION

The two major portions of this chapter have examined the rate of energy use of snapping turtle eggs and of snapping turtle hatchlings and adults. Studies in the last few years have revealed that environmental conditions during incubation of snapping turtles eggs can have a significant effect on the physiology of hatchlings and perhaps adults, but we have only a rudimentary understanding of these relationships. Our knowledge of the causes and importance of interindividual variation in metabolism in snapping turtles and other reptiles is at an early stage, and additional study of this topic should prove rewarding. A relatively small number of studies have focused on the energetics of turtle eggs; the egg as a semi-closed metabolic system offers physiologists many opportunities for understanding cause-and-effect relationships. Most of the studies cited here have been carried out in the laboratory, and our understanding of the physiology of snapping turtles under natural conditions, especially in winter, is at a very primitive stage. I hope the coming decades will bring a clearer understanding of these topics.

ACKNOWLEDGMENTS

I am grateful to K. Miller and G. C. Packard for helpful advice on portions of an early draft of this chapter.

9

Ecology and Physiology of Overwintering

GORDON R. ULTSCH
SCOTT A. REESE

REPTILES AND AMPHIBIANS are largely a tropical assemblage. However, there is an appreciable variety of holarctic species, including several with ranges that extend to the Arctic Circle; turtles are not among these far-northern forms. In North America, 14 species of freshwater turtles have ranges that reach a latitude of about 45–50°; nine of these enter southern Canada. Snakes and lizards that occur at high latitudes are typically live-bearers, permitting females to accelerate development of their embryos by behavioral thermoregulation. In contrast, all turtles lay eggs in ground nests, with a time to hatching typically at least 60 days, longer in the northern parts of their ranges. Since embryos cannot overwinter in the egg, hatching must occur prior to hibernation, even if hibernation occurs without emergence from the nest. At the northern limits of turtle ranges, represented by painted turtles (*Chrysemys picta*) and snapping turtles (*Chelydra serpentina*), egg and hatchling survival is much lower than adult survival. For example, St. Clair & Gregory (1990) found only one live turtle in 11 nests of overwintering western painted turtle (*Chrysemys picta bellii*) hatchlings excavated in the spring in southeastern British Columbia. Galbraith et al. (1988) reported that cold weather resulted in no emergence in *Chelydra* in 7 of 11 years in Algonquin Park, Ontario, and Obbard & Brooks (1981b), studying 257 *Chelydra* nests in the same locality, found that hatchlings emerged from only 42 in the fall. Of 215 nests from which hatchlings did not emerge in the autumn, only one produced any live hatchlings the following spring, and in that clutch 16 of 27 hatchlings had died. Both studies concluded that the northern limits of the species were determined by hatching success, not by the stresses of overwintering on adults.

The hatchlings of some species overwinter in the nest, while others emerge from the nest and overwinter underwater (see below). After the first winter, all far northern aquatic species typically overwinter underwater. Among reptiles, overwintering on land would present two potential physiological challenges, freezing and desiccation, which are usually avoided among terrestrial hibernators by either finding a preformed den that is below the freeze line, or making one by burrowing. Natural freeze tolerance, as exhibited by several frog species, is not a strategy that has been demonstrated among adult aquatic turtles. Turtle hatchlings of many species do exhibit some degree of freeze tolerance, but it is questionable if freezing is common among hatchlings that overwinter in the nest, as opposed to supercooling (see below). So long as the hibernaculum does not freeze below about $-0.6°C$ (the equilibrium freezing point of turtle tissues), freezing can be avoided. Aquatic hibernation eliminates the threat of freezing, but has its own suite of physiological and behavioral challenges. For example, ion and water balance must be maintained while the turtle spends months in a hypotonic medium, so the threat of desiccation is replaced with other osmoregulatory challenges, requiring at least minimal kidney function during hibernation (Hernandez & Coulson 1957; Crawford 1991a; Warburton & Jackson 1995). In addition, some aquatic environments (e.g., marshes and shallow ponds) tend to become hypoxic or anoxic in winter, especially if covered with ice and snow (Crawford 1991b). If the turtles bury in the mud, then they will be anoxic, whatever the Po_2 of the overlying water, since mud is anoxic. Anoxia will cause a shift to anaerobic respiration, which makes fat and protein stores unavailable, and the turtle is then restricted to carbohydrate reserves (mainly glycogen) as an energy source, which in turn will induce an accumulation of lactate and a concomitant metabolic acidosis. From a behavioral viewpoint, the turtles must be able to find and identify suitable overwintering sites, which may require migration, and enter and leave them at times appropriate for other life functions, especially reproduction. In this chapter, we will briefly review overwintering in turtles, including terrestrial overwintering during the first winter by hatchlings, with an emphasis where possible on snapping turtles.

HIBERNATION DURING THE FIRST WINTER
Occurrence of Overwintering on Land

Adult aquatic turtles typically overwinter underwater; the few exceptions include mud turtles and, under some circumstances, western pond turtles (reviewd by Ultsch 2006). Hatchlings may or may not overwinter on land, depending partly on which species and partly on latitude. Hatchlings of some of the most northern species rarely overwinter terrestrially in the coldest portions of their ranges: *Apalone spinifera* (Breckenridge 1960; Costanzo et al. 1995); *Chelydra serpentina* (Hammer 1969; Petokas & Alexander 1980; Obbard & Brooks 1981b; Congdon et al. 1987; Costanzo et al. 1995, 1999; but see Toner 1940 and Parren and Rice 2004); *Clemmys insculpta* (Farrell & Graham 1991; D. M. Carroll,

personal communication; but see Parren and Rice 2004); *Sternotherus odoratus* (Risley 1933; Cagle 1944; Ernst et al. 1994); *Emydoidea blandingii* (Butler & Graham 1995; Standing et al. 1997, 1999; Congdon et al. 2000; Pappas et al. 2000). In the northern United States and southern Canada, all northern subspecies of *Chrysemys picta* typically, but not without exception for either individual nests or among hatchlings within a nest, overwinter in the nest (Breitenbach et al. 1984; Christens & Bider 1987; St. Clair & Gregory 1990; Lindeman 1991; DePari 1996; Packard 1997; Weisrock & Janzen 1999; D. M. Carroll, personal communication), as does *Trachemys scripta elegans* in Illinois (Cagle 1944; Tucker 1997, 1999; Tucker & Packard 1998). Northern common map turtle hatchlings (*Graptemys geographica*) have been reported as both emerging in the fall and in the spring (Cahn 1937; Vogt 1980, 1981; N. Paisley, M. Pappas, M. Hamernick, M. McCallum, & P. Butler, personal communications), although spring emergence may be most common (Baker et al. 2003).

Overwintering of hatchlings on land does not necessarily equate to overwintering in the nest cavity. In a study of hatchlings of several species in Nebraska, Costanzo et al. (1995) found that *Chrysemys picta bellii* overwintered in the nest cavity, while *Kinosternon flavescens* and *Terrapene ornata* burrowed deeply into the soil below the nest cavity. *Emydoidea blandingii* hatchlings may leave the nest in the fall and wander extensively in search of water without finding it before the onset of cold weather (Standing et al. 1997), but whether such sojourners can overwinter on land is unproven.

Terrestrial Overwintering, Gas Exchange, and Desiccation

On land, gas exchange by hatchlings, in particular oxygen availability, would not seem to be limiting, as most nests are in sandy soil. Gases readily permeate sandy soils so long as they are not water saturated (Ultsch & Anderson 1986; Anderson & Ultsch 1987), which may be a factor related to the usual choice of sandy soils for nesting, as these drain quickly. Furthermore, at the low temperatures associated with hibernation, and the resultant very low metabolic rates (Herbert & Jackson 1985b), it seems reasonable that hypoxia and hypercarbia would be minor. However, there are no data to verify this hypothesis for the nests of overwintering hatchlings. Two factors that could result in nest hypoxia and hypercarbia are the metabolism of the hatchlings and soil freezing. If gas exchange is significantly hindered by the freezing of soil water, then even very low metabolic rates might be sufficient to alter nest gas composition, especially for nests that are not in soil with a high sand content.

Most studies of nest gas composition to date have been on sea turtle nests. Several factors combine to produce a significant hypoxia and a striking hypercarbia in some of these nests, despite the fact they are constructed in sand. They are deep (about 0.5 m), they contain numerous eggs (about 100), and they are relatively warm (about 26–33°C)

Table 9.1

Supercooling capacity, inoculative freezing temperature, and rate of evaporative water loss for four species of turtle hatchlings from Nebraska

	Supercooling temperature (°C)	Inoculative freezing temperature (°C)	Rate of evaporative water loss (mg g^{-1} day^{-1})
Terrapene ornata	−12.0	−2.4	0.91
Kinosternon flavescens	−9.3	−3.1	2.38
Chrysemys picta bellii	−15.5	−13.6	1.90
Chelydra serpentina	−8.5	−1.5	6.28

From Costanzo et al. 2001b.

(Prange & Ackerman 1974; Ackerman 1977; Maloney et al. 1990). These studies all report a combined hypoxia / hypercarbia with the O_2 decrement approximately equal to the CO_2 increment. Minimal P_{O_2} was about 100 mm Hg, which should not be particularly stressful, but maximal P_{CO_2} was about 60 mm Hg, which should produce a severe respiratory acidosis in the developing embryo and recent hatchling, the consequences of which are unstudied. Booth (1998a) measured gas concentrations in a nest in Australia of the broad-shelled river turtle (*Chelodina expansa*), which has an incubation period exceeding 300 days. In the absence of rain, nest gases tracked soil gases, with only minor hypoxia and hypercarbia. Following heavy rains, P_{O_2} fell to near 100 mm Hg, and P_{CO_2} increased to as high as 60 mm Hg, values similar to those found in sea turtle nests, despite the much smaller number of eggs. We know of no similar studies with the hibernacula (nests) of overwintering hatchlings.

A hatchling overwintering in the nest does not drink and therefore is subject to desiccation; the degree of desiccation will depend on the nest temperature and relative humidity, the permeability of the shell and cutaneous surfaces to water, and the pulmonary ventilation rate. In a study of Nebraska hatchlings, Costanzo et al. (2001b) found that the terrestrial hibernators *Terrapene ornata* and *Kinosternon flavescens,* which burrow below the nest cavity, and *Chrysemys picta bellii,* which overwinter in the nest cavity, had a much lower rate of evaporative water loss (EWL) than *Chelydra,* which leaves the nest and overwinters underwater (Table 9.1). They suggest that risk of desiccation may be one reason why *Chelydra* hatchlings do not overwinter in the nest. The same argument could be made for softshell turtles, which also do not overwinter in the nest in the northern parts of their ranges, and which are very permeable to water relative to hardshelled species (Bentley & Schmidt-Nielsen 1970), especially in hatchlings (Robertson & Smith 1982).

Chelydra and *Chrysemys* hatchlings are roughly 80% water (our observations). The experiments of Costanzo et al. (2001b) were conducted for 10 days at 5°C in a desiccation chamber. If *Chelydra* hatchlings were to lose water at the rate reported for these conditions, they would have lost all their water in 127 days, whereas the *Chrysemys* would have lost only 30%. However, there are no data on evaporative water loss over the long term under nest conditions. The lower temperature, potentially higher humidity, crowding, and increasing osmotic content of the body fluids could in concert substantially reduce EWL as winter progresses.

Cold Tolerance of Terrestrially Overwintering Hatchlings

Gas exchange is not the first thought that comes to mind concerning hatchlings that overwinter on land—freezing is. Hatchlings can avoid freezing by burrowing downward through the floor of the nest chamber to a depth below the frost line. Mud turtle (*Kinosternon flavescens*) and western box turtle (*Terrapene ornata*) hatchlings in Nebraska dig as far down as 1 m and presumably avoid freezing (Costanzo et al. 1995). The adults of both species also hibernate on land, and also burrow deeply enough (up to 1.8 m in Wisconsin box turtles; Doroff & Keith 1990) to avoid freezing. In contrast, *Chrysemys* hatchlings in northern climates typically overwinter in the nest, and do not burrow. During winter *Chrysemys* nest temperatures often drop well below the equilibrium freezing point of the tissues for extended periods (Breitenbach et al. 1984; Storey et al. 1988, Costanzo et al. 1995; Depari 1996; Packard 1997; Packard et al. 1997); thus the turtles must either supercool or tolerate freezing. In laboratory studies, both processes occur, but both have limitations. *Chrysemys* hatchlings can survive freezing of about 50% of their body water in laboratory studies, but only down to about −4°C for periods that may not be long enough to be ecologically meaningful (Storey et al. 1988, Attaway et al. 1998; Packard et al. 1999; Packard & Packard 2001), although it is possible that naturally hibernating hatchlings may become cold-hardened and capable of more extended survival (e.g., 11 days frozen at −2.5°C; Churchill & Storey 1992). They can be supercooled to as low as −20°C, but if freezing occurs at a low subzero temperature, it is lethal. How cold a turtle becomes before it freezes depends upon the presence of ice-nucleating agents (internal and external; Costanzo et al. 2000), and it is uncertain what the role of such agents is in the field. Moreover, whether hatchlings overwintering in the nest are supercooled or frozen during potential freezing episodes is uncertain. The fact that *Chrysemys* hatchlings often survive low subzero temperatures in the nest argues for supercooling as the likely strategy throughout most of the winter, as does the poor survival of turtles frozen at moderately low temperatures (e.g., −2°C; Packard et al. 1999). Freeze tolerance might play a role during short subzero episodes in early spring before emergence from the nest.

Deaths from freezing are routine among *Chelydra* embryos that fail to hatch and hatchlings that fail to emerge (Hammer 1969; Obbard & Brooks 1981b), and also occur frequently even among *Chrysemys* hatchlings, especially during winters of little snow cover, as snow is a good insulator (Breitenbach et al. 1984; Lindeman 1991; DePari 1996; Weisrock & Janzen 1999; Nagle et al. 2000; Carroll & Ultsch 2007). The deaths could be occurring in turtles that are already frozen at high subzero temperatures and subsequently cooled to lower temperatures, or they could be occurring in deeply supercooled animals, with or without subsequent freezing.

Why Do Any Turtle Hatchlings Overwinter in the Nest?

Gibbons & Nelson (1978) used drift fences to capture hatchlings headed for water in South Carolina and found that captures in February, March, and April accounted for 158 of 164 of *Deirochelys reticularia*, 94 of 96 of *Kinosternon subrubrum*, 76 of 82 of *Trachemys scripta*, 32 of 33 of *Pseudemys floridana*, and 4 of 9 of *Sternotherus odoratus;* only one *Chelydra* was captured, and it was a fall emergent. They assumed that spring captures were of hatchlings that overwintered in the nest. They suggest overwintering in the nest allows hatchlings to enter an environment in spring when temperature is increasing and food is becoming more plentiful, whereas the reverse would be true for fall emergents. Jackson (1994) also found overwintering in the nest to occur in warmer climates (Florida) for *T. scripta, P. floridana,* and *P. concinna,* although, as with the South Carolina studies, it was not 100%. *Trachemys gaigeae* has been reported to overwinter in the nest in New Mexico (Morjan & Stuart 2001). This ecological argument is logical, but then one has to ask why it does not apply in the north, where turtle hatchlings emerging in the fall would presumably face even lower temperatures and less availability of food were they to enter the water. Some of the species listed by Gibbons & Nelson (1978) as overwintering in the nest were based on older and somewhat sketchy references and have since been found to emerge primarily in the fall (*Chelydra serpentina, Clemmys guttata, Emydoidea blandingii,* and *Sternotherus odoratus*). In fact, more northern species probably emerge in the fall than in the spring. Assuming that one mode or the other is best ecologically, then one must also assume that either those species whose hatchlings overwinter in the nest cannot overwinter underwater, or that those that overwinter underwater cannot overwinter on land.

Whether a hatchling could overwinter underwater as successfully as an adult depends on the microenvironment of the hibernaculum, in particular with regard to the P_{O_2} of the water. If the small size of hatchlings requires that they bury in mud to avoid predation, then their size is a disadvantage, as mud is anoxic, and hatchlings are not as tolerant of anoxia as adults. Survival times in anoxic water at 3°C for northern hatchlings are about 15 days for common map turtles (*Graptemys geographica*), 30 days for snapping turtles (*Chelydra serpentina*), and 40 days for western painted turtles (*Chrysemys picta bellii*), each about one-third that of adults of each species (Reese et al. 2004). Hatchlings of all three

species can survive at least 150 days of submergence in normoxic water (Reese et al. 2004). Hatchlings have much less bone than adults, largely because of their incompletely ossified shell. A hatchling *C. p. bellii*, for example, has a skeletal dry mass <2% of its wet mass (Iverson 1982; Ultsch & Reese, personal observations), while adult turtles of most species (exclusive of softshells and *Chelydra*) have skeletal dry masses that average about 25% of wet mass. Since bone, in particular the shell, is important in buffering the lactic acidosis that develops during prolonged submergence at low temperature in anoxic water (see below), as well as sequestering lactate (Jackson et al. 2001), prolonged mud burial seems precluded for hatchlings. There seems to be no physiological reason why hatchlings in the north that overwinter terrestrially (e.g., *Chrysemys* and many *Graptemys* and *Trachemys*) could not hibernate underwater, provided they could find a hibernaculum with a reliably high P_{O_2}. The fact that *Chelydra* hatchlings cannot tolerate anoxia well relative to adults means that there may be microenvironmental segregation by size during hibernation, and higher overwintering mortality among hatchlings and young turtles. Only one study, using radiotelemetry, has followed hatchling snapping turtles from their nests to their overwintering sites and into the hibernation period (G. Ultsch, M. Draud, and B. Wicklow, unpublished observations). In that study, hatchlings on Long Island, New York, initially moved to water, but later movements were both aquatic and terrestrial, and at least some left the water and hibernated in spring seeps, where they were recovered alive the following April. In contrast, in New Hampshire, all hatchlings moved directly to nearby aquatic habitats after emergence and remained submerged in shallow water among root masses throughout the winter. No adults were found hibernating with the hatchlings, but these data are too preliminary to state that segregated hibernation is the rule.

It is possible that the ecologically preferred scenario is hibernation by hatchlings on land, as suggested by Gibbons & Nelson (1978), but that certain species cannot do so, and that is why they hibernate underwater. The chief threat during terrestrial hibernation is freezing, which can be avoided by burrowing, as mentioned above for mud (*Kinosternon*) and box (*Terrapene*) turtles. It is not known if any other species that overwinter in the nest also burrow, but it is known that *Chrysemys* do not. If supercooling occurs, this still does not necessarily protect the hatchling from a metabolic acidosis, as occurs in submerged turtles in anoxia. Heart rate slows dramatically during hatchling supercooling (Birchard & Packard 1997), concomitant with a rise in whole-body lactate (Hartley et al. 2000; Costanzo et al. 2001a). Decreasing temperature from 22 to 5°C in adult *Trachemys scripta* dramatically decreases power output of the heart, heart rate, systemic stroke volume and cardiac output, and mean arterial pressure, as well as increasing vascular resistance (Hicks & Farrell 2000). All of these changes presumably can lead to a cessation of blood flow to peripheral tissues, as has been found for *Graptemys geographica* upon cooling to 5°C (Semple et al. 1970; Stitt & Semple 1971; Stitt et al. 1970, 1971), which would account for the increased lactate found in supercooled hatchlings. Moreover, not all species are equal in their supercooling abilities. *Chelydra* hatchlings, in particular, have a relatively high supercooling limit, and are also subject to inoculative freezing at a relatively high temperature (Table 9.1). Their tendency to freeze at high subzero temperatures is apparently due to their relatively high cutaneous surface area (Stone & Iverson 1999) and the lack of the cutaneous lipid layer found in *Chrysemys* hatchlings that retards nucleation by ice crystals (Willard et al. 2000). Thus it is possible that terrestrial overwintering is precluded in some species on morphological and physiological grounds, although more data are needed to justify a conclusion that terrestrial overwintering is the evolutionarily favored condition.

HIBERNATION BEYOND THE FIRST WINTER

Nothing is known about underwater hibernation in freshwater turtles during their juvenile years, which includes the first winter for many species and from the second winter onward for nearly all. Field studies primarily report on adult turtles, and laboratory studies have been either with hatchlings or with adults, but not with juveniles. Our laboratory studies indicate that both hatchlings and adults of northern species (*Chrysemys picta bellii, Graptemys geographica,* and *Chelydra serpentina*) can tolerate long periods (at least 100–150 days) of submergence in aerated water at 3°C, that adults vary in their ability to tolerate anoxic submergence, and that there is a marked reduction in survival of hatchlings compared with adults in anoxic water. Since overwintering in normoxic water does not present a serious challenge at any body size, even over extreme periods of natural overwintering at the northern limits of distribution, we will emphasize adaptations to anoxic conditions, with the caveat that there must be an ontogenetic increase in the tolerance of anoxic hibernation. We believe that the change is closely linked to progressive ossification of the shell, but do not know if the change is gradual or completed within the first year or two.

Hibernaculum Characteristics

Where turtles hibernate depends on where they are found during the nonhibernating period, the availability of hibernacula in the home range or within a distance that the turtle can migrate, the ability to tolerate hypoxia/anoxia, and probably age (Ultsch 2006). Some species typically hibernate in well-oxygenated water, which often means flowing water. *Clemmys insculpta* winters on stream bottoms, often exposed while wedged under banks, between rocks, or beneath submerged debris (Ernst 1986; Graham & Forsberg 1991; Carroll 1993); *Graptemys geographica* also winters in well-oxygenated

water and does not burrow into the substrate (Evermann & Clark 1916; Pluto & Bellis 1988; Crocker et al. 2000b). *Apalone spinifera* may bury in the mud or sand bottoms of streams and rivers, but only to a depth that will permit extension of the head into the water column (Plummer & Burnley 1997), thus permitting buccopharyngeal respiration, which is highly developed in softshell turtles (Dunson 1960; Girgis 1961; Zhao-Xian et al. 1989; Yokosuka et al. 2000a, b). *Chrysemys* have been reported to overwinter both buried in the mud (Ernst 1972; Taylor & Nol 1989) and resting on the substrate (St. Clair & Gregory 1990; Crocker et al. 2000a). The microenvironment of hibernating *Clemmys guttata* is not well-defined (in terms of water P_{O_2} or air access), but this species typically hibernates within grass tussocks, under ledges, and among roots and sphagnum moss, often migrating to these sites from terrestrial estivation forms or summer aquatic habitats (Graham 1995; Lewis & Ritzenhaler 1997; Litzgus et al. 1999).

Some hibernacula are favored over others, as evidenced by many reports of more than one turtle at a particular location. However, the same sites are not always used by the same individuals every year, and some individuals may hibernate singly in one year and in a group in another year (Brown & Brooks 1994; Lewis & Ritzenhaler 1997; Litzgus et al. 1999). This variation suggests that suitable hibernacula are not difficult to find and likely not limiting to turtle populations.

Many short reports are available on hibernation in *Chelydra,* in part because they are collected commercially. Pell (1941), Lagler (1943a, b), and Meeks & Ultsch (1990) have reported instances of *Chelydra* being caught in muskrat traps during the winter, which implies that there is at least occasional movement. Favored sites are muskrat burrows and lodges, beneath debris, and in the mouths of tributary streams (Pell 1943; Obbard & Brooks 1981a; Ultsch & Lee 1983). Brown & Brooks (1994) studied a *Chelydra* population in Algonquin Park, Ontario, for four winters. Ice coverage was typically present for five months. The turtles moved to hibernacula in late August to late September, which were outside the summertime home ranges of 35 of 59 individuals; 28 of those animals returned to their home ranges the following spring. There was incomplete site fidelity, with 13 of 18 turtles that were located for more than one winter using the same site, but not necessarily in consecutive years; ten returned to within 1 m of previous sites, and four to exactly the same site, including one that returned to the same site all four winters. Some changed sites and even bodies of water. Thus there was considerable variability in site selection, the main goal apparently being to find a suitable site, not necessarily a particular site. Three types of sites were used: buried under logs or sticks in small streams that flowed all winter (19); lakeshore sites within 5 m of shore, usually wedged under or near submerged logs or stumps and not buried in the substrate (24); and buried in deep mud or marshy areas or beneath floating vegetation beds (11).

The mean water temperature was 1.2°C, and the P_{O_2} was about 70–140 mm Hg, depending on the site. There was some wintertime movement, although usually limited (only 5% moved more than 10 m), indicating that the turtles were lethargic, but not torpid. There was no evidence of death by freezing, but in one winter there was a large mortality (65%) due to otter predation; a few uninjured turtles also died of septicemia shortly after emergence (Brooks et al. 1991).

Paisley, Wetzel, and Nelson (unpublished report of the Wisconsin Department of Natural Resources—1997–1999), studied *Chelydra* in Pool 8 on the Mississippi River between Wisconsin and Minnesota. They found that turtles moved 0.09 to 2.87 (mean, 0.53) km to hibernacula, from late August to late September. As was found by Brown & Brooks (1994), a variety of sites were used, and turtles hibernated alone or in groups.

The one characteristic in common among all the sites reported for hibernation in *Chelydra* and *Chrysemys* is the use of relatively shallow water (<-1 m). *Chelydra* often move to inlet streams or stream mouths, if they are present (Ultsch & Lee 1983; Brown & Brooks 1994), areas in which they are not typically found during the active season. When such sites are not available, the turtles move toward shore and then bury, or sequester themselves under banks, grass tussocks, vegetation, or debris. Shallow water and a place to hide seem to be the main requirements for a hibernaculum, and many types are used. These two species are the only ones studied so far that can tolerate several months of anoxia (*Emydoidea blandingii*, *Clemmys guttata*, and *C. muhlenbergii* may also be anoxia tolerant), which means that long-term burial in mud or other anoxic or severely hypoxic microenvironments is not precluded, which in turn may explain the rather catholic choice of hibernacula by these two species.

Gradations in Anoxia Tolerance

Northern adult freshwater turtles fall into two fairly distinct categories in terms of their ability to tolerate prolonged cold (e.g., 3°C) submergence in anoxic water, which would be ecologically equivalent to mud burial or hibernation in shallow, ice-covered, eutrophic bodies of water, which can become O_2 depleted during winter. One group will die in <50 days and are considered anoxia intolerant (for a turtle); included are *Graptemys geographica* (Ultsch & Jackson 1995; Reese et al. 2001), *Sternotherus odoratus* (Ultsch 1988; Ultsch & Cochran, 1994), and *Apalone spinifera* (Reese et al. 2003), in order of decreasing viability from about 45 to 14 days. A second group can survive at least 100 days; included are all northern subspecies of *Chrysemys picta* (Ultsch & Jackson 1982a; Ultsch et al. 1999; Reese et al. 2000) and *Chelydra serpentina* (Reese et al. 2002). Moreover, the greater the capability of a given species to extract O_2 from the water, the less tolerant of anoxia is that species (e.g., *Sternotherus odoratus*, *Graptemys geographica*, and *Apalone spinifera*). Thus all

northern species are capable of surviving for up to 4 to 5 months under ice if there is ample dissolved O_2, but not all are capable of tolerating anoxic submergence, a difference that has implications for habitat selection among species (see below).

Mechanisms of Anoxia Tolerance

An anoxic turtle has only one significant source of metabolic energy production: the anaerobic production of lactic acid from glycogen and glucose. Even though metabolic rate can be profoundly depressed, months of anoxic submergence lead to a striking increase in lactate. The rate of lactate accumulation has been documented for several species and subspecies, but for illustrative purposes we depict data for the most (*Chrysemys picta bellii*) and least (*Apalone spinifera*) anoxia-tolerant species studied to date and compare

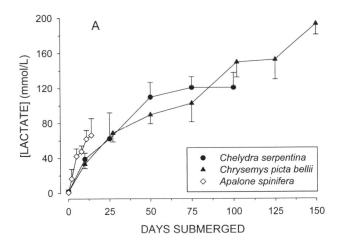

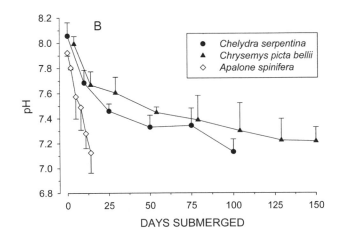

Fig. 9.1 (A) Plasma lactate accumulation in three species of temperature-acclimated turtles as a function of time submerged in anoxic water at 3°C. Final concentrations are from turtles that were near death and can therefore be considered near the maximal tolerable limits. Experiments were conducted during the normal period of hibernation. *Chelydra serpentina* and *Apalone spinifera* were from Wisconsin and Michigan; *Chrysemys picta bellii* were from Wisconsin. Data are means and 95% confidence intervals (Reese et al., 2002, 2003; Reese & Ultsch, unpublished data). (B) As in (A), but for plasma pH.

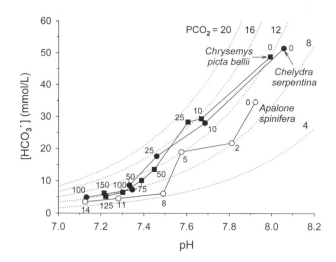

Fig. 9.2. A pH-[HCO_3^-] (Davenport) diagram, depicting the changes in plasma [HCO_3^-] and pH with time of anoxic submergence for the species shown in Fig. 9.1. Initial values are for animals in shallow water at 3°C with access to air. Numbers next to points indicate days of submergence. Data from Reese et al. 2002, 2003; Reese & Ultsch, unpublished data.

them with data for *Chelydra* (Fig. 9.1A). Although the vast majority of the protons generated by lactic acid dissociation are buffered, a lactic acidosis develops early in anoxia that ultimately reaches a lethal level (Fig. 9.1B). Approximately the same decrease in pH is lethal to the three species, although the changes in [HCO_3^-] and P_{CO_2} follow a different path in *Apalone spinifera* than they do in the other two species, with a significant fall in P_{CO_2} that produces a compensatory respiratory alkalosis (Fig. 9.2).

The greatest challenge to a turtle hibernating in anoxia is the accumulation of lactate and the accompanying acidosis. All physiological regulatory changes can be directly or indirectly related to the steadily increasing metabolic acidosis that occurs in anoxia (see reviews by Ultsch 1989; Ultsch & Jackson 1995; Jackson 2000; Jackson et al. 2001). Lactic acid is essentially completely ionized at physiological pH, and the protons must be buffered. A secondary challenge is the maintenance of electrical neutrality as lactate accumulates after protons are removed from solution. The [HCO_3^-] of turtles is among the highest of vertebrates, but it is not sufficient to buffer all the lactic acid produced by a turtle during prolonged anoxia. The characteristic feature of a turtle—its shell—is vitally important in the regulation of both pH and electroneutrality, and it does this in several ways. The shell contains calcium and magnesium carbonates, and these are mobilized during acidosis. Each CO_3^{2-} released can titrate two H^+, resulting in the formation of one H_2O and one CO_2. The CO_2 is then lost to the water by extrapulmonary diffusion. The Ca^{2+} and Mg^{2+} left behind can then electrically balance the negatively charged lactate ions, either in free solution or by complexing with them (Jackson & Heisler 1982, 1983). In addition, the shell serves as a major repository for lactate (Jackson 1997; Jackson et al. 1999, 2000a, b).

The importance of the shell is a likely explanation for the relative intolerance of anoxia of *Apalone spinifera*. The dry shell mass of *A. spinifera* is about 7.5% of wet mass, whereas it is 23.4% for *Chrysemys picta bellii*, and calcium and magnesium contents of the shell of softshells are only about 20% of those of *Chelydra* (Jackson et al. 2000b). Lack of shell ossification has also been implicated above as a cause for the reduced tolerance of anoxia among hatchlings relative to adults of both species.

Adults of the anoxia-tolerant *Chelydra* have a reduced skeletal mass relative to most other species. The skeletal dry mass of *Chelydra serpentina* averages about 10% of wet mass over a size range of 0.16–15.47 kg, with a slight increase in the percentage with increasing size (Fig. 9.3); 20–25% is typical of most species, excluding juveniles, *Chelydra*, and *Apalone* (Iverson 1984). Since *Chelydra* can tolerate anoxia much more so than *Apalone* while having skeletal masses that are closer to those of softshells than to those of most other species, the shell is clearly not the only determinant of anoxia tolerance, albeit it appears to be an important one. Another misfit in the puzzle is map turtles. They have as extensive a skeleton as *Chrysemys* (Iverson 1984), but can only

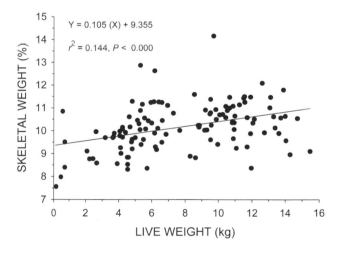

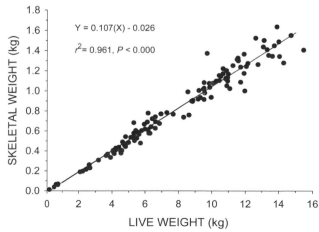

Fig. 9.3. Air-dried skeletal mass of 120 *Chelydra serpentina* as a percentage of live wet mass (*upper*) and as a function of live wet mass (*lower*).

tolerate prolonged anoxia for less than 40% of the time tolerated by either *Chelydra* or *Chrysemys*. The most likely explanation for this apparent contradiction is that map turtles do not depress their metabolic rate as much as the anoxia-tolerant species, thereby accumulating lactate and depressing pH more rapidly (Figs. 9.1 and 9.2), but there are no data on metabolic rates to support this hypothesis.

Acidosis in anoxic turtles is also ameliorated by regulatory changes in the strong ion difference (SID, the sum of the concentrations of strong cations minus that of the strong anions, all expressed in equivalents/liter) (Stewart 1981), with an increase in the SID causing an increase in the pH. In an anoxic turtle, SID is increased over the uncompensated state by increases in Ca^{2+} and Mg^{2+} (both free and complexed with lactate), an increase in $[K^+]$, and a decrease in $[Cl^-]$. These ionic responses occur in all species thus far investigated, and are not peculiar to anoxia-tolerant species. Thus it seems likely that the differences in ability to tolerate prolonged anoxia are ultimately tied to differing degrees of metabolic rate depression, at least for animals with comparable skeletal masses, and thus comparable calcium and magnesium stores.

Ecological Implications of Anoxia Tolerance

The ability to spend long periods submerged during the winter does not seem to limit the northern extent of the range of turtles. Common map turtles, western painted turtles, and spiny softshells can live and recover from at least 217 days of continuous submergence at 3°C in aerated water (our unpublished data). As discussed above, it is apparently the short duration of warm periods that limits the northern extension of turtles, primarily through limitations on successful incubation of the eggs. The result is that the northern limits of the ranges of turtles are surprisingly uniform, while the southern extents are more variable. However, the habitats that are suitable for turtles in the north do appear to depend on the ability to overwinter, specifically an ability to tolerate hypoxia or anoxia. Map turtles and softshell turtles are intolerant of submergence anoxia and thus cannot live where they might encounter severely reduced aquatic Po_2 while hibernating, even if only in an occasional winter. Therefore these species must occupy habitats with hibernacula that will have a predictably high Po_2 throughout the winter. As a consequence, both of these species are found in streams, rivers, and large lakes, and do not occur in marshes, swamps, and small shallow ponds, all of which are subject to wintertime anoxia. In contrast, *Chrysemys* and *Chelydra* are found in virtually any permanent body of water, and are often most abundant in shallow, eutrophic bodies of water (Froese & Burghardt 1975; Major 1975; Gailbraith et al. 1988), in which food is abundant during the summer, but which are subject to oxygen reduction or depletion during winter. An unanswered question, however, is how the very young survive their first winters in such habitats, consider-

ing our observations of the reduced ability of hatchlings to tolerate anoxia. It is possible that while the physiological challenges of hibernation are not a major mortality factor among adults, they may be among very young turtles. It is well known that there is a high egg-to-hatchling mortality among all species of turtles due to nest predation and that this terrestrial stage is a bottleneck in the maintenance of turtle populations; perhaps there is another bottleneck in the first winter or two due to mortality during hibernation of the very young.

Although overwintering *per se* may not kill many adult turtles, if it occurs in hypoxic/anoxic conditions, the emerging turtles will be in a seriously perturbed physiological state. Rewarming is crucial to the metabolism of lactate and the rapid readjustment of ionic and acid-base parameters (Ultsch & Jackson 1982b), as evidenced by the avid basking behavior of northern turtles recently emerged from hibernation. Even *Chelydra,* which are not routine baskers, bask more frequently near the northern limits of their range (Obbard & Brooks 1979), especially in early spring (personal observations). During the first day of emergence, turtles are lethargic and easily captured and are likely highly subject to predation by raccoons and other predators.

CONSERVATION IMPLICATIONS OF HIBERNATION IN FRESHWATER TURTLES

In many areas turtles are subject to intensive collection pressures for the pet trade (illegal and otherwise; e.g., *Clemmys* and *Glyptemys*) and for food (e.g., *Chelydra serpentina* and *Apalone spinifera*). Habitat destruction or encroachment is a serious threat for species with relatively restrictive habitat requirements and patchy distributions, such as spotted turtles and bog turtles. Road kills during autumn or spring migrations, and of females during the nesting season, are a major contributor to mortality among adult turtles (Haxton 2000). The tendency of some species to aggregate during hibernation is exploited by collectors, who either remove the turtles during the winter or as they emerge in spring, when they are lethargic and easily captured. *Clemmys guttata*, for example, often move from vernal pools that they occupy in spring and early summer to more permanent bodies of water to hibernate, and they also commonly estivate in woodlands that are outside the typical boundary zones of "protected" wetlands (Graham 1995; Milam & Melvin 2001). Map turtles frequently form hibernating congregations in specific sites in a river, moving to these sites from considerable distances in the autumn (Vogt 1980; Graham & Graham 1992; Crocker et al. 2000b; Ultsch et al. 2000).

Hibernation and estivation requirements of turtles are often not taken into account when efforts are made toward turtle conservation. Recruitment is often low in turtle populations (Congdon et al. 1994), so it is important that all aspects of a turtle's life history be considered when making conservation decisions. Since northern turtles may spend more than half of each year hibernating, it is particularly important that hibernation sites be identified and protected.

ACKNOWLEDGMENTS

We thank Ronald Brooks, Patrick Butler, David Carroll, Justin Congdon, Jon Costanzo, Carl Ernst, Terry Graham, Mark Hamernick, Peter Lindeman, Malcolm McCallum, Neal Paisley, Michael Pappas, and John Tucker for sharing unpublished information. Stephen Rogers of the Carnegie Museum of Natural History kindly weighed skeletons of *Chelydra serpentina* and provided us with both skeletal and original weights. Much of the research discussed here was supported by National Science Foundations grants IBN 96-03934 and IBN 0076592 to G.R.U.

10

Embryos and Incubation Period of the Snapping Turtle

MICHAEL A. EWERT

HELYDRA IS VENERABLE IN THE STUDY of chelonian embryos, both in vertebrate morphogenesis and in physiological zoology. This chapter briefly acknowledges the former discipline and within physiological zoology concentrates on the duration of incubation. Several exogenous and endogenous factors affect incubation time. Among exogenous factors, which include temperature, water availability, and respiratory gases, temperature is most influential (Yntema 1978). Very "dry" substrate water potentials (−850 to −950 kPa) modestly shorten incubation by 3 to 5 days in 60–70 days (Packard et al. 1987; Packard & Packard 1989b) but without clearly affecting differentiation rate (Miller & Packard 1992). Respiratory gas concentrations (low oxygen, high carbon dioxide) prolong laboratory incubation in two turtles (Ackerman 1981b; Etchberger et al. 1991, 1992), but have yet to be tested on *Chelydra*.

Two endogenous factors that affect incubation are egg size (Morris et al. 1983) and latitudinal biotype (Ewert 1979, 1985). A focus on the influences of temperature, egg size, and latitude constitutes the treatment that follows.

One goal in exploring how temperature and the embryonic stage interact is to advance an understanding of thermal compensation in amniotic ectotherms. Although avian embryos are more popular as a model for amniote development, they are constrained by narrow thermal tolerance and ontogeny of homeothermy. A practical objective for turtle research is to improve accuracy in estimating embryonic stage in thermally monitored nests (Bull 1985; Georges et al. 1994; Baumgardt 1997; Valenzuela 2001a). Although thermal responses of *Chelydra* vary geographically (Ewert 1979; 1985), the influences of egg size (Birchard & Marcellini 1996) vs. other factors toward this variation have not been evaluated, nor have these influences been associated with stage. In contrast to localities where cold or drought constrain warm seasons (Obbard & Brooks

1981), options allowing prolonged development (unrestricted egg size, leisurely response to temperature) should be prevalent in benign, mainly humid subtropical to tropical climates.

This chapter includes a presentation of some new material. Methods pertaining to these results are provided in an appendix, as well as in the figure legends and the main text.

CHELYDRA EMBRYOS AS THE "MODEL TURTLE"

Chelydra appears prominently in the work of Louis Agassiz (1857), the first work on chelonian embryology in the Americas. Modern embryology using *Chelydra* began with Chester L. Yntema (1964, 1968) at Syracuse University. The stimulus from this early work continues (e.g., Meier & Packard 1984; Burke 1989, 1991; Gilbert et al. 2001). Some advantages of using *Chelydra* have been its large clutches containing similarly aged embryos, its historical abundance, local accessibility, and its predictable nesting season. Additionally, embryos of *Chelydra* tolerate invasive procedures better than some other species (Yntema 1964, 1970a, b, 1974a, b; Burke, 1991).

GAUGING DEVELOPMENT WITH SERIAL STAGES

The classic work on *Chelydra* (Yntema 1968) combines constant temperature incubation, time interval, a concise set of descriptions, and good photographs to provide the most widely used standard series for chelonian embryos. This series, devised mainly for research on early embryos, divides postovipositional development into 27 stages: presomite (stages 0–3), somite (4–10), and stages with limbs (11–26). Beginning with stage 11, descriptions feature the limbs, but also describe the eye, eyelids, scutellation and scalation, and several other externally evident features. Yntema's series has fostered efforts toward synchronously staging other turtles (*Testudo hermanni*, Guyot et al. 1994; *Carettochelys insculpta*, Beggs et al. 2000). An important series of sea turtle stages (Miller 1985) begins with six preovipositional stages, including one at oviposition. A Miller postovipositional stage number approximately equals a Yntema stage plus six. However, there is some heterochrony among characters, and Miller slightly broadened the spacing between his postovipositional stages to omit one stage.

Observations on *Chelydra* have not included preovipositional development, when early cleavages occur. However, these stages have been described for *Glyptemys insculpta* (Agassiz 1857), *Mauremys leprosa* (Pasteels 1937), and *Caretta caretta* (Miller 1985).

First Changes after Oviposition

The stage at laying, an advanced gastrula (Yntema stage 0), is consistent in all species (six turtle families) examined to date. This stage can persist as preovipositional developmental arrest for days to weeks in *Chelydra* if oviposition is prevented (Ewert, unpublished data, 1985), but the embryo eventually deteriorates. Yntema may have known of this arrest following the observations of Cunningham (1922) on *Chrysemys picta*. However, Yntema (1964) reported only that chilling gravid females would delay development. Here, curtailment of development can be attributed to cold torpor (Yntema 1960).

An important and easily observed event, commonly coincidental with postgastrular development, is formation of a white spot on the eggshell (Yntema 1964, 1979; Ewert 1979). Usually, this white spot forms at the uppermost pole of the egg, and the embryo lies directly beneath it (Yntema 1981). Formation of the white spot occurs when the vitelline membrane of the egg yolk adheres to the inner surface of the shell membrane. This outcome is observable during excision of whitened portions of the eggshell (Yntema 1964), or simply with candling (Ewert 1985). Histology of the adhesive process is limited to just one study on a crocodilian (Webb et al. 1987). However, this adhesion seems stronger than in turtles, and thus may be different.

Histochemistry of the adhesive process needs study. Candling reveals that the initial adhesion in many *Chelydra* and perhaps all *Macrochelys* eggs is fragile and easily disrupted with slight torque. However, following disruption, the vitelline membrane appears unbroken; that is, the egg contents are not addled. As mild as the disruption in chelydrid eggs may seem, the consequence is serious. When adhesion commences and then fails in chelydrids, it does not recommence and the embryo dies within a few days. Perhaps, there is an adhesive material that can be "used" just once. Alternatively, integrity at initiation points may be prerequisite for subsequent adhesion. Whitening on viable eggs, overmature from prolonged oviductal retention, sometimes commences broadly at the sides or lower pole.

THE RANGE OF THERMAL TOLERANCE

The lowest temperature scrutinized, 10°C, prevents continuance of somite development already in progress at the time of chilling. Some embryos held at 10°C for two weeks can continue normal development after warming. At 15°C, some early development occurs but it becomes abnormal during three to four weeks of exposure (Yntema 1960). At 16.7°C, a small amount of development continues for three to four weeks. The apparent effects of 16.7°C on normal differentiation are dependent on stage. The most severe disruptions accrue from chilling presomite (stage 2–3) embryos, then, minor effects from chilling postpharyngula (stage 14) embryos, and almost no apparent effects from chilling advanced embryos (stage 22–23) (see below). At 19°C late somite–early fetal development continues through determination of embryonic sex (Baumgardt 1997). At 20°C, apparently normal development continues for about 90 days, roughly to stage 21. This temperature does not foster development to hatching, but embryos warmed by stage 21 can hatch (Yntema 1968). A nearly constant temperature averaging 21.1°C permits total development and hatching but embryonic survival, in particular during late development, is lower than at a relatively benign temperature (25.3°C). Further, the hatchlings from 21.1°C behave as if compromised neurologically. Their survival is much lower than for hatchlings from benign temperatures, at least when maintained under cool conditions (mostly ~22°C; Bobyn & Brooks 1994a, 1994b) without warmer ones (O'Steen 1998; Rhen & Lang 1999a). At 21.5°C, 95% hatching can occur, about as successful as at warmer, more benign temperatures (O'Steen 1998; O'Steen & Janzen 1999). Hatchlings from northern Illinois eggs incubated at 21.5°C express high behavioral preferences for warm temperatures and high levels of circulating thyroxine (O'Steen & Janzen 1999). A few hatchlings from Tennessee eggs also incubated at 21.5°C have been compromised neurologically, but several from Florida have seemed normal (personal observation). Therefore, 21.5°C seems close to a low threshold separating serious shortcomings from less critical phenotypic variation.

Constant temperatures of 22–30.5°C sustain normal development and yield normal hatchlings. There are temperature-dependent differences in phenotypic variation (e.g., Yntema 1976; Rhen & Lang 1995, 1999b; O'Steen 1998; O'Steen & Janzen 1999; Steyermark & Spotila 2000). At 31°C, embryonic survival declines and embryonic abnormalities become more frequent, even at adequate levels of moisture within the incubation substrate. This temperature approximates the upper threshold of tolerance for a constant temperature (Packard & Packard 1987). Embryos do not tolerate 32.5°C or 35°C beyond a week, but some have hatched following transfer to cooler temperatures (Yntema 1978). Yntema did not state the stage at which he applied hot temperatures, but apparently he always used fresh eggs (stages 0–1).

VARIATION IN THERMAL SENSITIVITY DURING DEVELOPMENT

The common reliance on the full period from oviposition to hatching (stages 0–26) to reflect standard conditions has applications (Ackerman 1994; Birchard & Marcellini 1996), but the practice overlooks some complex thermal relationships that pertain to different portions of development. The following text first recognizes the effects of temperature on blocks of serial stages and then integrates these effects into whole incubation.

It is possible to compute developmental rates at constant temperatures from the times between various stages during incubation. Comparisons among temperatures suggest that thermal sensitivity in embryonic *Chelydra* is much greater during early development than during late development

(Yntema 1968, 1978; Ewert 1985; Baumgardt 1997). Calculating the Q_{10} variable (Prosser 1973; Schmidt-Nielsen 1979), which compares rates of reaction across a 10°C difference in temperature (t), $Q_{10} = (Rate1/Rate2)10/(t_1 - t_2)$, illustrates this stage-dependent variation in thermal sensitivity (Fig. 10. 1A). In general, the early stages, ones mainly concerning establishment of organs, have high Q_{10}s (2.5–3.5) for biological processes. Stages in mid development, when major differentiation and growth of organs occurs, express more moderate Q_{10}s (2.0–3.0). Subsequently, when growth dominates but then slows, Q_{10} rapidly declines and continues at 1.0 across several advanced stages. Most likely, this late-stage thermal independence represents compensatory acclimation (Birchard & Reiber 1995; O'Steen & Janzen 1999).

According to compensatory acclimation, embryos react to divergent thermal experiences by adjusting development, usually toward a more functional, often intermediate rate (Rome et al. 1992). Q_{10} reflects this adjustment. Relevant to acclimation is whether declines in Q_{10} at constant temperatures (Fig. 10.1A) also occur in broader cases that represent variable temperatures (Huey 1992). With variable temperature, a hysteresis effect, or lag-time response common to acclimation would have to be brief or masked for Fig. 10.1A to be representative. Compensatory acclimation is demonstrated in a reciprocal shift experiment (shift warmer: shift cooler; here, 24:29°C) in which heartbeat rate in late (~stage 25) embryos adjusts to the new temperature within three days (Birchard 2000). Another observation consistent with compensation is that circulating thyroxine (T_4) varies from high in cool (21.5°C) incubated *Chelydra* hatchlings to low in warm (30.5°C) incubated hatchlings (O'Steen & Janzen 1999). Exogenous application of thyroxine (T_3) shortens incubation (O'Steen & Janzen 1999), whereas blockage of thyroid activity with thiourea prolongs incubation (Dimond 1954). Thyroid activity commences at about stage 20 (Dimond 1954), which is approximately the same stage that compensatory acclimation appears to rise to (Birchard & Reiber 1995) (Fig. 10.1A).

Late embryos may have reduced thermal dependence, as well as "translated" compensatory acclimation (e.g., Prosser 1973). Proximate thermal sensitivity of heartbeat rate declines with embryonic age (Q_{10}s of ~2.36 at ~stage 20 vs. ~1.97 at ~stage 25; Birchard 2000). Actually, heartbeat rate at 29°C gradually declines and converges toward the rate at 24°C (Birchard & Reiber 1996).

Reference to acclimation usually assumes a hysteresis. Lag-times in embryonic *Chelydra* are greater than 2–3 hours for circulating thyroxine (O'Steen & Janzen 1999) and between several hours and three days for heartbeat rate (Birchard 2000).

Whereas reciprocal shift experiments have determined lack of compensation in incubation length in one turtle (Booth 1998a), such experiments have not examined incubation length in *Chelydra*. These observations, however, are available for a second species (*Sternotherus odoratus*) with ev-

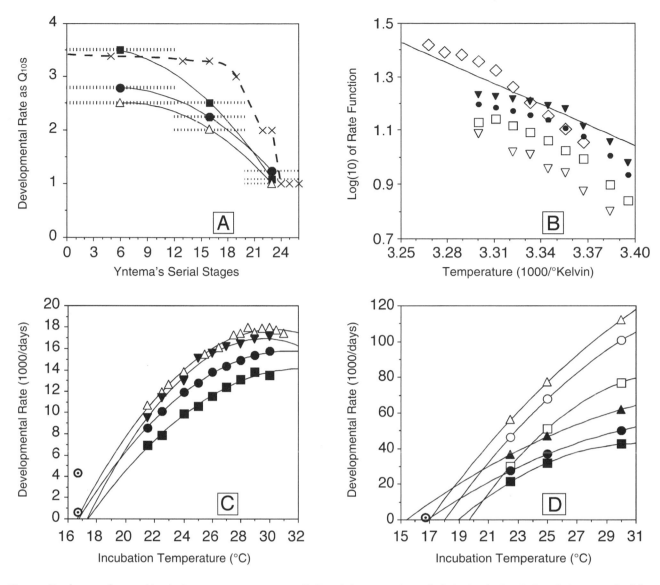

Fig. 10.1. Developmental rate and incubation at constant temperature (A, D, variation at stage intervals during incubation; B, C, variation across the full incubation period). (A) Apparent thermal sensitivity (expressed as Q_{10} values) declines in embryos from four separate origins as development advances. Horizontal dotted lines indicate the stage intervals for measure (stages 0–12, 12–20, 20–26) in three populations (filled squares, Florida; filled circles, Indiana; open triangles, northeastern Minnesota). Solid lines join midpoints of each interval for coherency. Representation of the fourth population (dashed line with × symbols, New York) is derived from Yntema (1968: table 1). (B) This Arrhenius type plot (Nobel, 1991; with "rate function" = 1,000/day) shows that relative to a Q_{10} of 2.0 (solid line) Q_{10} (slopes of lines [implied] between adjacent points) steepen with decreasing temperature (left to right) in five populations (open squares, Florida; filled circles, Indiana; filled inverted triangles, Michigan; open inverted triangles, *Macrochelys*; open diamonds, *Pelomedusa*). *Macrochelys* represents the closest living relative to *Chelydra*. *Pelomedusa* represents a species in which (unlike *Chelydra*, see panel C) developmental rate follows a linear association with temperature. The Q_{10} for *Pelomedusa* across its seven cooler rates (right-most points) is 3.2. (C) The rate of development across total incubation in four populations (open triangles, northwestern Minnesota; other symbols as in B) has a curvilinear association with temperature. Equations for the fitted lines are: MN, $y = -0.0945x^2 + 5.736 - 69.22$; MI, $y = -0.113x^2 + 6.669x - 81.80$; IN, $y = -0.0746x^2 + 4.666x - 57.157$; FL, $y = -0.0641x^2 + 4.134x - 52.52$. The bull's-eyes at the lower left represent rates estimated from incubation early in development (lower bull's-eye) and late in development (upper bull's-eye) at a cool temperature (16.7°C; see text) (northwestern Minnesota data from Baumgardt [1997] and Lang, unpublished). (D) Rates of development in three populations (symbol shape as in A): early in incubation (open symbols, stage interval 0–12; filled symbols, stage interval 12–20; bull's-eye, estimate for early development [~stage 14] at 16.7°C adjusted to stage interval 12–20).

idence for compensation (Clark 1986). Single reciprocal shifts (22.5:25 and 25:30°C) show compensation but no evidence of hysteresis (Clark 1986). Alternating shifts (2 days at 22.5°C: 1 day at 30°C) reveal no hysteresis effect. The full incubation period according to the alternating regime (56 shifts, 85 days) has the same duration as its combined counterparts (i.e., 2/3 of the period at 22.5°C + 1/3 at 30°C = 86 days) (unpublished observations). Perhaps hysteresis is

brief but not detectable. Alternatively, compensation is centripetal to the thermal shifts. Either way, Fig. 10.1A appears to convey a reasonable perspective of thermal sensitivity across stages.

High Q_{10}s of early development succeed a thermal independence expressed as preovipositional arrest (Cunningham 1922; Ewert 1985). This arrest may be prolonged in different turtles (Ewert & Wilson 1996; Kennett et al. 1993).

Slow transition out of arrest, while undocumented in *Chelydra*, could affect early-stage Q_{10} estimates tied to oviposition.

TOTAL DEVELOPMENT AND STAGE-DEPENDENT THERMAL SENSITIVITY

As reflected in total development, Q_{10} values vary across the range of thermal tolerance for *Chelydra* (Fig. 10.1B). These values are less than 2.0 above midrange temperatures for normal development and greater than 2.0 below midrange. Whereas values of 2.0 predominate in metabolic activity, in general (Prosser 1973), here, this value is the average result of complex processes. The inverse association happens for two reasons. The first pertains to thermal tolerance. Developmental rate in *Chelydra* (and in living processes, in general) declines and reaches zero at much higher temperatures (>10°C in *Chelydra*) than the rate would, for instance, following a Q_{10} fixed at 2.0. Thus fixed, an embryo that completes incubation in 50 days at 30°C would do so in 283 days at 5°C. The latter temperature is well below the developmental threshold for *Chelydra* (Yntema 1960). Ackerman (1994) proposes standardizing Q_{10} values within the thermal range that gives high viability (e.g., ~22–30°C for *Chelydra*; ~23–30°C for *Macrochelys*).

The second reason pertains to the relationship among stage-dependent thermal sensitivities, as shown in Fig. 10.1A. Here, early stages show more sensitivity than do the late ones. Across the range of favorable temperatures, the durations of early stages (0–20) vary with temperature, whereas the late stages (20–26), presumably through acclimation (Birchard & Reiber 1995), approach a fixed duration (Yntema 1968, 1978; Ewert 1985). For example, at benign temperatures (22.5–30°C) the late stages in general last 32 days for embryos from northeastern Minnesota and about 42 days for embryos from Florida. The duration of early development (to stage 20) varies inversely with temperature. Early development declines from 45 days at 22.5°C to 25 days at 30°C for embryos from northeastern Minnesota. Similarly, the duration declines from 80 days at 22.5°C to 36.5 days at 30°C for embryos from Florida. In combination, the fixed-duration phase of development constitutes a moderately large proportion (e.g., 54–56% at 30°C) of total incubation at warm temperatures and a lesser proportion (e.g., 34–42% at 22.5°C) at cool temperatures. The aggregate change in developmental rate shifts from modest changes in rate across warm temperatures to large changes at cool temperatures (Fig. 10.1C). Thus, developmental rate is nonlinear and expresses yet a further departure from a constant Q_{10} than does a linear association, as in *Pelomedusa* (Fig. 10.1B).

Thus, a nonlinear assumption rather than simple linear regression accounts better for the association of total incubation period with temperature. For the four regional samples (Fig. 10.1C), a 2° polynomial explains 99.0% (Michigan) to 99.9% (Indiana) of the variation, whereas the linear assumption explains only 90.0% (Michigan) to 95.9% (Florida).

However, curvature across the represented temperatures is sufficiently slight to give highly significant associations ($P < 0.0001$) under either assumption.

The relative thermal sensitivities among the early stages are not obvious, and high sensitivity could prevail. However, as with total development (Fig. 10.1C), two increments of early development, stages 0–12 and 12–20, both express nonlinear rates (Fig. 10.1D).

ESTIMATING DEVELOPMENTAL ZERO AT COOL TEMPERATURES

Ecological interest has motivated estimating temperatures at thresholds that are too warm or too cool to permit development (Georges et al. 1994; Shine & Harlow 1996; Valenzuela 2001a). This view of cessation applies in the sense of "stopping the clock," and thereby prolonging the time necessary to advance to a given serial stage. Extrapolation from "best fit" 2° polynomial curves for total developmental rates for *Chelydra* gives 16–18°C as "cool developmental zero" (Fig. 10.1C). Similar extrapolations from early and mid development broaden the range to ~15.5–19.7°C (Fig. 10.1D). These extrapolations, however, extend greatly beyond the actual data and risk misrepresentation. Further, it is hard to hypothesize any single temperature as being "developmental zero" from the perspective that Q_{10} changes with serial stage (Fig. 10.1A). Although embryos from a northern population do not develop at 10°C, they are able to add somites at 15°C (Yntema 1960).

I examined effects from applying the extrapolated "developmental zero" temperature (16.7°C) on *Chelydra* from southern Indiana (Fig. 10.1C and D). I inserted 21-day pulses of 16.7°C, within a background of incubation at 29°C. I applied the pulses to early embryos at stage 14 and to late embryos at approximately stage 22. Then, I estimated the fraction of development (if any) completed while at 16.7°C from the differences between incubation times at 29°C with and without a pulse.

For 16.7°C, the average rate (four clutches) of early development was 0.86 (1,000/days) (range, 0.14–1.7). By candling, these embryos advanced from stage 14 to between stages 15 and 16. Survival was high, but minor deformities, apparent as departures from the normal scute arrangement, were common. The estimated rate (four clutches) of late development averaged 4.4 (1,000/days)(3.9–5.0). Late-stage advancement, however, was too subtle to discern through candling. Survival was high and essentially without deformity.

Hatchlings from the late-chill series ($n = 33$) and the early chill series ($n = 34$) were significantly smaller than the unchilled controls ($n = 41$) (e.g., carapace length, 30.6 vs. 30.9 vs. 31.3 mm, $P = 0.04$; plastron length, 20.3 vs. 20.7 vs. 21.3 mm, $P = 0.006$, respectively). However, the average live masses of the hatchlings were similar (8.8 vs. 8.7 vs. 8.8g, $P = 0.98$). This suggests that the embryos within the different treatments had the potential to achieve similar sizes before incubation.

Apparently, while development continued at the estimated "developmental zero," embryos from the two chilled series incurred a cost of being less completely grown than the control embryos. If the chilled embryos had attained full size, completion would have required additional time and would have extended the total incubation time. Records of embryonic growth in Yntema (1968) suggest that late growth at 30°C is approximately 0.32 mm/day. Adding days to account for this growth yields an estimate of zero development at 16.7°C in the early chill series; that is, despite the observed stage advancement. During the 21-day late chill, however, a net achievement of ~3.4 days of growth (at 29°C) occurs over cost. Thus, some development occurs at 16.7°C, and this temperature, while partially equating to "developmental zero" during early stages, does not fully represent it.

Developmental rates at sublethally warm temperatures are unknown. It is not known how well a converse protocol to chilling (i.e., pulses sublethally warm temperatures inserted within benignly cool incubation) will facilitate estimation of developmental rates above the warm threshold (~ 31°C) for continuous development.

Much work has explored thermal associations of embryonic and postembryonic development across diverse taxa, including insects (various stages, Schoolfield et al. 1981; Wagner et al. 1984; Liu et al. 1995), anuran amphibians (embryos and larvae, Zweifel 1968), and reptiles (embryos, Ackerman 1994). Often, a plot of average rate against constant temperature yields a sigmoid curve, with attenuated development at sublethally cool temperatures and depressed development at sublethally warm ones. Commonly also, a linear or nearly linear association with temperature continues broadly across mid-range temperatures. This association follows well for anurans (Zweifel 1968). However, *Chelydra* does not express any linear association across mid-range temperatures (Fig. 10.1C and D) for the reasons offered.

EGG SIZE AND INCUBATION TIME

Other factors being equal, one expects the incubation period to vary directly with egg size (Morris et al. 1983; Birchard & Marcellini 1996). The data of Yntema (1968), however, suggest an alternative that similarly incubated embryos hatch at the same time, regardless of egg size. Soon after 28 days at 30°C (~stage 21), embryos in large eggs start to grow faster. By 60 days, or pipping in his population, the embryos fill large and small eggs to the same extent. The following account examines egg size in relation to overall incubation and localizes the portion of development in which egg size affects duration.

Egg size influences incubation period in *Chelydra* (Fig. 10.2). The slopes of three regression equations representing different populations of *Chelydra* bracket the slope for "all turtles" (0.144) derived by Birchard & Marcellini (1996).

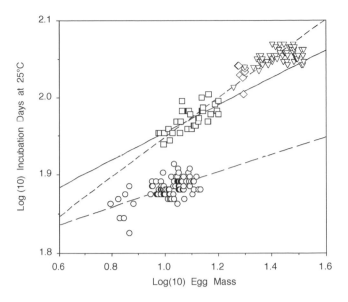

Fig. 10.2. Association of total incubation period with initial egg mass in samples from Indiana (circles, long dashes, $y = 0.11243x + 1.7689$, $r^2 = 0.343$), Florida (squares, solid line, $y = 0.17780x + 1.7769$, $r^2 = 0.532$), and Florida plus Honduras (squares and diamonds, short dashes, $y = 0.25506x + 1.6935$, $r^2 = 0.781$). Each association is significant at P <0.0001. The position of egg masses and incubation times of *Macrochelys* within this scheme is represented by inverted triangles. The format of the x and y axes was chosen to allow comparison with Birchard and Marcellini (1996).

The regression lines fitted for two distributions that involve eggs from Florida (FL alone; FL with a FL × Honduras hybrid), but not for Indiana, extrapolate to overlap data for *Macrochelys*. Some of the eggs of *Chelydra* came from the same locality as the eggs of *Macrochelys*. Viewed this way, larger size fully accounts for the longer duration of *Macrochelys*.

Because growing embryos approach filling their eggs mainly during late development (stages 20–26), this phase is likely to show an association of incubation time with egg size. Through candling, the duration of this phase and two others were determined for embryos from Indiana. The variances in phase duration were by clutch: stages 0–12, 1.27; stages 12–20, 0.20; stages 20–26, 8.23; and by individual egg: stage 0–12, 1.78; stages 12–20, 0.98; stages 20–26, 8.27. The correlation of each phase with egg size (8 clutches, 91 embryos) was, by clutch, mean: stages 0–12, $r = -0.439$, $P = 0.28$; stages 12–20, $r = 0.045$, $P = 0.91$; stages 20–26, $r = 0.818$, $P = 0.013$); and by individual egg: stages 0–12, $r = -0.29$, $P = 0.014$; stages 12–20, $r = 0.002$, $P = 0.98$; stages 20–26, $r = 0.631$, $P < 0.0001$. These determinations show that among the three phases, late development has by far the greatest variance and the strongest association with egg size.

It seems likely that support structures such as circulatory systems should take a little longer to develop in large eggs. The surface-to-volume association of the chorioallantoic system could also be limiting on larger eggs (see Birchard & Reiber 1993, 1995). The weak inverse association of stage 0–12 development with egg size at the level of the individual is puzzling. Although it could be related to preoviposi-

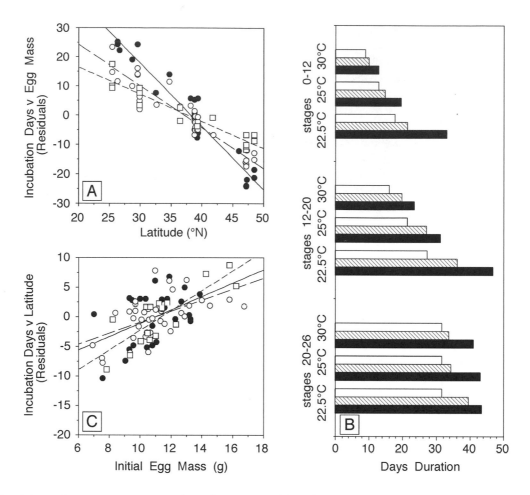

Fig. 10.3. Associations at three constant temperatures for incubation times with latitude (A, C, total incubation period; B, stage intervals during incubation). (A) Total incubation period varies with latitude after adjusting (with least squares regression) for the effects of egg size (22.5°C, filled circles, solid line, N = 25 clutch means; 25°C, open circles, long dashes, N = 38; 30°C, open squares, short dashes, N = 16; represented states: AR, FL, IN, LA, MI, MN, and TN.) (B) Three stage intervals during incubation each show an association with latitude (i.e., longest at the lowest latitude) (northeastern Minnesota, 47° N, open bars; Indiana, 39° N, hatched bars; Florida, 27–29° N, filled bars). (C) Total incubation period retains an association of egg size following least-squares regression to adjust for the effects of latitude (samples and symbols as in A).

tional arrest and thus spurious, it appears in two independent data sets.

Egg size extends late-stage incubation at 25°C by 1.4 days (0.4–2.7 = 95% confidence limits) per gram of fresh mass for Indiana eggs and a nearly identical extension of 1.4 days (0.9–1.8) per gram for Florida eggs. This association provides one reason for not expecting large eggs, such as those of tropical *Chelydra* (Ewert 1979; Iverson et al. 1997) (Fig. 10.2) or *Macrochelys* (Ewert 1976), in climates with short growing seasons.

LATITUDE AND INCUBATION TIME

Incubation time in *Chelydra* has an inverse association with latitude (Ewert 1985). However, over a short range in high latitudes, Bobyn & Brooks (1994) could not confirm this. Lott (1998) also found an inverse association but attributed it entirely to differences in initial egg mass; the eggs from his lowest sampled latitude were also his largest. Thus,

egg size can confound the association of latitude with incubation time.

Residuals analysis to adjust incubation times for initial egg mass through linear regression supports the inverse association with latitude (Fig. 10.3A). In the converse approach, residuals from regressing incubation times on latitude were associated with initial egg mass, but less strongly (Fig. 10.3C). Residuals for the third association, egg size vs. latitude, was not significant in any of these samples (P values of 0.62, 0.27, and 0.78 for 22.5, 25, and 30°C, respectively). Multiple regression yielded similar conclusions. As in Ewert (1985), the slope relating incubation time to latitude is shallowest at 30°C. Conversely, the slope relating incubation time to egg size at 30°C is steepest (Fig. 10.3A and C). This is expected because size-sensitive late development (see above) at this temperature is proportionally the largest.

Early as well as late stages contribute to the longer incubation times at low latitudes (Ewert 1985). Figure 10.3B demonstrates this relationship for three general locations (northeastern Minnesota, southern Indiana, and Florida) and three temperatures. Note that although the late stages are relatively insensitive to temperature, they, nonetheless, vary with latitude. After adjustment for egg mass (residuals analysis), the three stage intervals (early, mid, and late) show significant association with latitude at the three temperatures (22.5, 25, and 30°C) ($P < 0.005$ in each of the nine combinations). After adjustment for latitude, the late-stage intervals at 25°C and 30°C show significant association with egg size ($P < 0.05$), but the late-stage interval at 22.5°C does not ($P > 0.5$). In this last case, small sample size (nine clutches) may be a contributing factor.

Short incubation times occur in some high-latitude, natural environments, depending on exposure and weather. Eggs have completed incubation in as briefly as 67–78 days in Minnesota (47.3° N) and Wisconsin (43.4° N) (Ewert 1979; unpublished), and in 63–64 days in northwestern Illinois (Kolbe & Janzen 2002a). For Florida eggs, these times approximate or fall short of full incubation at 29–30°C, which is too warm for nests in northwestern Florida. Here (see Ewert & Jackson 1994), single measurements on each of 17 recent *Chelydra* nests (4 different years) varied from 18.6 to 24.2°C (unpublished data). Hence, at least the first stages of development would proceed gradually at much below 29–30°C.

RESPIRATION, TEMPERATURE, AND INCUBATION TIME

The pattern of oxygen consumption (Vo_2) in *Chelydra* in constant incubation environments is known for eggs from Nebraska (Gettinger et al. 1984; Miller & Packard 1992; Birchard & Reiber 1995). This pattern, also determined for samples from Minnesota and Tennessee (Fig. 10.4A), shows an exponential increase in Vo_2 then an inflection point, a peak rate and then, up to pipping, a decline to about two thirds of the peak rate. The pattern typifies many turtles and crocodilians (Thompson 1989, 1993; Whitehead & Seymour 1990; Booth 1998b). Extreme patterns vary from apparent absence of late decline (Ackerman 1981a; Vleck & Hoyt 1991) to very large declines in Vo_2 before hatching (Webb et al. 1986; Ewert 1991).

Comparison of oxygen consumption in *Chelydra* eggs at 24°C vs. 30°C in a split-clutch design revealed variation in the early, exponential phase Vo_2 (stages 0–20), whereas the late phase elevated Vo_2 remained unchanged. That is, the rise in Vo_2, beyond the inflection in rate, the height of peak Vo_2, and the overall duration of elevated Vo_2 were nearly the same at both temperatures. Birchard & Reiber (1995) attribute this finding to compensatory acclimation in *Chelydra*. This phenomenon seems atypical among embryonic reptiles in which Vo_2 has been monitored throughout incubation. Thermal dependence, or absence of compensatory acclimation, during late development is evident in three turtles (*Emydura* [*macquarrii*] *signata*, Booth 1998a, b; *Trionyx triun-*

guis, Leshem et al. 1991; *Chelonia mydas*, Booth & Astill 2001), the tuatara (Booth & Thompson 1991), and two skinks (Booth et al. 2000).

There are functional and adaptive hypotheses for why oxygen consumption declines late in development. Functionally, embryonic growth slows and sometimes stops late in development. This state follows after the chorioallantois has fully lined the inside of the eggshell (~ by stage 23; Birchard & Reiber 1995). The fully formed chorioallantois may define the maximum capability for uptake of oxygen. Quite late, the chorioallantois begins to lose contact with the eggshell, shrinks, and has its blood supply withdrawn.

An adaptive explanation for the near-term decline in respiration proposes social facilitation. Prolonged decline, perhaps physiologically unnecessary, allows nest mates that are slightly ahead or behind to hatch together. This hypothesis seems appropriate for sea turtle hatchlings, which may benefit from sharing in digging through sand to emerge (Carr & Hirth 1961). Crocodilians and precocial birds benefit because the mothers cannot wait for tardy hatchlings in broods (Cannon et al. 1986; review in Booth & Thompson 1991). With observations on late decline in respiration in two sea turtles (Thompson 1993; Booth & Astill 2001), perhaps all turtles express it. If so, however, late decline may have only a secondary connection to the social environment, as late declines are also known for turtles that produce clutches of just one or two eggs (e.g., *Kinosternon scorpioides*, Ewert 1991; *Rhinoclemmys areolata*, unpublished data).

The social facilitation hypothesis does seem applicable to *Chelydra*, given that clutches usually exceed 10 eggs. Some embryos even may await a hatchling stimulus from a companion. For instance, in partial isolation at 29–30°C, a few embryos retract their chorioallantois, sit a few days, and die without pipping. Manually pipping of clutch mates has yielded normal hatchlings (personal observation).

RESPIRATION, LATITUDE, AND INCUBATION TIME

In a comparison at 30°C between embryos from Tennessee (TN) and Minnesota (MN), embryos from Tennessee took longer to develop (64.6 ± 3.1 vs. 55.0 ± 1.0 days). Perhaps one day of this difference resulted from differences in egg mass (11.8 g, TN vs. 11.1 g, MN) but the rest is compatible with latitudinal variation. In general, the embryos from Tennessee exhibited a more gradual rise in oxygen consumption and a longer phase of high Vo_2 (Fig. 10.4A). Although embryos from Tennessee took 1.3 days (15%) longer to reach stage 12, mean Vo_2 was the same for both localities (0.066 vs. 0.063 ml/h, TN vs. MN, respectively). By stage 20, however, embryos from Tennessee had a higher average Vo_2 (0.41 ± 0.04 vs. 0.33 ± 0.02 ml/h). The difference was significant after least-squares adjustment for hatchling

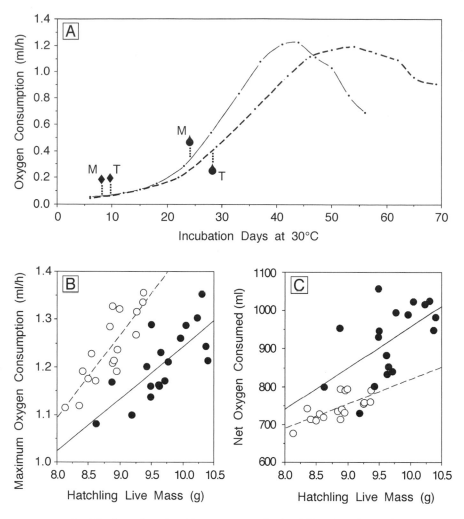

Fig. 10.4. Patterns of oxygen consumption (all values adjusted to STPD) in two samples that differ in latitude (Ericsburg, Minnesota, 48.5° N, N = 17 embryos vs. Obion County, Tennessee, 36.3° N, N = 18). (A) Changes in oxygen consumption in the samples from Minnesota (solid lines) are more abrupt than for Tennessee (dashed lines). Plotted points show ±1 SE where these values are greater than the diameters of the plot points (excepting the end points, which had tiny samples, no SE). Filled diamonds and teardrops represent the mean times of stages 12 and 20, respectively, for the two localities. (B) Minnesota embryos used oxygen more intensively relative to hatchling live mass (see text) (Minnesota, open circles, dashes; Tennessee, closed circles, solid line). (C) Minnesota embryos used less total oxygen relative to hatchling live mass (symbols as in B).

carapace length, and for body mass ($z = -4.5$, $P < 0.001$; $z = -2.46$, $P = 0.014$, respectively). Additionally, within localities, V_{O_2} was positively associated with carapace length (TN: $R = 0.525$, $P = 0.037$; MN: $R = 0.658$, $P = 0.011$). This is compatible with observations by Yntema (1968) that embryonic adjustments toward filling larger eggs may commence by stage 21. The mean peak V_{O_2} for Tennessee was significantly lower than for Minnesota (1.208 ± 0.073 vs. 1.239 ± 0.075 ml/h, respectively) after least-squares adjustments for egg mass ($P < 0.05$), for hatchling mass ($P = 0.0013$) (Fig. 10.4B), and for hatchling carapace length ($P < 0.025$). Conversely, however, embryos from Tennessee consumed more oxygen during development (923 ± 95 vs. 744 ± 34 ml STPD), which was significant after least-squares adjustment for egg mass ($P < 0.0001$), and for hatchling mass ($P < 0.05$) (Fig. 10.4C).

The average cost of development per egg according to consumed oxygen (at 19.7 kJ/liter, and a RQ of 0.71; Gordon 1972) was 18.2 kJ for Tennessee and 14.7 kJ for Minnesota. Similarly determined values give 14.6–15.0 kJ for northern Nebraska (for 30°C vs. 24°C incubation; Birchard & Reiber 1995). Costs determined by bomb calorimetry are 23.0, 14.9, and 15.5 kJ for Florida, Indiana, and Minnesota, respectively (incubation at 25°C; Clark, unpublished data, 2000). Given that the largest eggs involved in these comparisons came from Nebraska, and only the second largest from Florida, greater costs of development appear to be associated with lower latitudes, rather than with egg size. Some of the larger costs may be incurred as maintenance energy during the longer periods that low-latitude embryos spend as advanced-stage embryos (Figs. 10.3B, 10.4A). However, low-latitude embryos also hatch with more residual yolk reserves as in high-latitude embryos (1.0 g fresh mass in TN vs. 0.6 g in MN; see also Clark 2000).

CONCLUSION

This chapter proposes that response rates throughout the course of development are not associated with temperature in a uniform or linear manner. One explanation for this favors ontogeny of compensatory acclimation about midway through, with nascent thyroid activity as one mediator. Other evidence suggests that additional thermal buffering may arise during development. For the practical matter of associating developmental stage with date in nest environments, the curvilinear association renders estimation more problematic than assuming linearity (Georges et al. 1994; Valenzuela 2001a). However, rates during early development have less curvature than total development, in part, because thermal sensitivity (Q_{10}) declines less over short phases of development. Application of linearity across cool to moderately warm temperatures may crudely predict timing. Still, it should be possible first to combine soil thermal records into practically spaced categories (e.g., 0.5° intervals) and then to integrate them over a curvilinear record of development achieved per temperature. Other influences on incubation (e.g., water potential, respiratory gas concentrations) and thermal recording accuracy may cause greater unexplained variance. Egg size, however, seems unlikely to influence stage progression until late in development.

This chapter explores a geographical component to development of *Chelydra*. A latitudinal association occurs throughout development and is independent of egg size. Energy utilization, while ultimately greater at low latitudes is less constrained (lower peak V_{O_2}, more "leisurely") than at high latitudes. The cost may be only relative to the initial energy in the eggs; both eggs and hatchlings from Florida have greater energy contents and energy densities than their counterparts from Indiana, Michigan, or Minnesota (Clark 2000). The relationship is compatible with adaptation to short activity seasons at high latitudes.

ACKNOWLEDGMENTS

For assistance in obtaining eggs, I thank P. J. Clark, C. R. Etchberger, J. H. Harding, J. B. Harrel, D. R. Jackson, and the Columbus Zoo (J. M. Goode). Constant temperature incubators were provided by M. A. Watson, K. Clay, and through the Carnegie Museum of Natural History (C. J. McCoy). H. D. Prange provided access to a Gilson Differential Respirometer, and he and P. J. Clark showed me how to use it. J. W. Lang provided some incubation data on eggs from northwestern Minnesota. I thank C. E. Nelson for helping in various ways.

APPENDIX
Methods for Original Observations

Eggs came from gravid females (mostly through induced oviposition, some through captive breeding and induced nesting, and a few following dissection) and from fresh nests.

Eggs from nests were candled with a fiber-optic lamp to backestimate oviposition date, which was never more than two days. Estimates of incubation periods began with oviposition date (within 6 h of oviposition for the series involving chilling). The end of incubation was the date of pipping (within 8 h for the series involving chilling). The duration (~3 weeks) of chilling (at 16.7°C) was recorded to the minute, but included a brief thermal transition between 29°C and 16.7°C.

Eggs housed for incubation were half to fully buried in vermiculite (#2 grade, coarse) initially moistened to a 1:1 (by mass) water/dry vermiculite mixture. This gives a water potential of –55 kPa (Kam and Ackerman 1990) to –170 kPa (Steyermark & Spotila 2000). During incubation, water was added to the vermiculite several times. Overall, a slight loss in net loss in substrate water was allowed, but not enough to cause loss of turgor in eggshells. Late in incubation, random substrate measurements showed water/vermiculite ratios near 1:2 (~–250 kPa; Packard et al. 1987), which is not stressful.

The eggs and vermiculite resided in 6.8- to 7.8-liter plastic boxes with up to 30 eggs per box. Box contents obtained aeration through very small gaps in lids plus manual aeration once daily. On most occasions, upright refrigerated incubators (Percival & Precision) served to maintain controlled temperature. Temperatures were recorded from 500-ml water bottles kept adjacent to the egg boxes. The boxes were rotated within the incubators to minimize thermal variation.

Individual embryos provided sequential developmental stages through candling (Ewert 1985: Beggs et al. 1999). Stage declaration emphasized easily discernable stages. For *Chelydra*, two stages with wide spacing are stage 12 (dark retinal pigmentation) and stage 20 (an even, early darkening of the skin over most of the embryo). Candling also allowed close timing for stage 14 (the expanding allantois nearly equals the embryo in size) and approximate timing for stage 22–23 (high embryonic motility; Decker 1967).

Measurements of oxygen consumption employed up to 10 channels on a 20-channel Gilson (manometric) respirometer. For each session, the protocol allowed two hours equilibration for the equipment and an additional two hours for the eggs within their chambers. There were four readings (each across intervals of 15, 20, or 30 minutes, depending on demand) per egg per session. Between readings, the system was reequilibrated to ambient pressure (with a new ambient barometer reading) to freshen the atmosphere within the system. Granules of barium hydroxide served as a carbon dioxide absorbent. Only about half of each experimental group of eggs (from Ericsburg, Minnesota, latitude 48.5° N vs. Obion Co., Tennessee, latitude 36.3° N) was in the Gilson at any one time. Oxygen consumption of each egg was tracked through 11–13 sessions (readings every 3–5 days).

The estimates of oxygen consumption (standard temperature and pressure, dry) were plotted on pages from a single stock of 1-mm graph paper. Then, adjacent points were connected free-hand. A planimeter was used to integrate V_{O_2} in computation of oxygen "cost" of development for each embryo, with stage 8 (from candling) to pipping as the standard period of development.

Individual embryos monitored by this protocol showed predominantly smooth patterns of oxygen consumption (e.g., Ewert 1991) and yielded expected costs of development (Vleck & Hoyt 1991).

11

Growth Patterns of Snapping Turtles, *Chelydra serpentina*

ANTHONY C. STEYERMARK

GROWTH IS BIOLOGIC SYNTHESIS, or the production of new biochemical and cellular units. Broadly considered, growth is a facet of development concerned with the increase (or decrease) of biological substance, and includes cell multiplication (hyperplasia), cell enlargement (hypertrophy), and incorporation of exogenous material (Brody 1945). It is this last type of growth, incorporation of exogenous material into the individual to form somatic tissue, which this chapter will treat. Specifically, this chapter will describe general patterns of growth in various snapping turtle life stages, discuss various factors affecting growth, and discuss the relationship between growth and fitness.

Growth is typically measured as change in either linear dimension or in mass, per unit of time. Note, however, that other measures of growth can be used, such as developmental changes, that are not accompanied by changes in mass over time. For our purposes, we will consider growth in the body size, rather than the developmental, fashion. Although when measured over short periods growth appears linear, growth through development to maturity is characteristically sigmoid in many reptiles (Andrews 1982), accelerating during a short initial period, declining as maturity approaches, and then approaching zero. Several functions have been used to describe the relationship between change in mass (or linear dimension) and time. Most functions define dW/dt, or the rate of growth, as a function of weight at the time rate of growth was measured.

Growth can be described as a power function, $G = aW^b$, where,

G equals rate of growth,
W equals adult body mass, and

a equals a group specific constant
b equals a power function (≈0.72)
(Andrews 1982).

Common growth functions include the logistic equation, the Gompertz function, and the Bertalanffy function (Blaxter 1989). Relative growth rate, $(1/W)(dW/dt)$, has often been used to describe the relative decrease in mass gained with time (Andrews 1982; Blaxter 1989; Brody 1945). Accurate measurement of growth can be affected by many variables, including change in body composition (fat, water, glycogen), the presence of food in the digestive tract, or by measurement error.

Why Does Growth Matter?

Growth is an important adaptive trait that allows for a pattern of life such that reproducing individuals can invest a relatively small amount of energy into offspring production, and the offspring can then incorporate exogenous resources into somatic tissue to reach a sexually mature size. Variation in growth patterns between species allows for variable life history strategies. Typical of low-energy-flow vertebrates such as snapping turtles, variable growth rates allow individuals to persist at times or places of low resources as an energy conservation strategy (Christiansen & Burken 1979; Pough 1980; Wikelski & Thom 2000).

EMBRYONIC GROWTH
General

A major challenge in vertebrate evolution was the movement from the aquatic environment to a terrestrial one. Among the many physiological and morphological adaptations that allowed this major lifestyle shift was the evolution of an egg that allowed the aquatic habitat to be brought to the embryo, rather than having the embryo brought to the aquatic habitat (Romer 1957; White 1991). Two major adaptations allowed terrestrial vertebrate eggs to move out of the aquatic habitat. First, the amnion envelops the embryo and contains amniotic fluid that allows the embryo to continue development in a liquid environment. Second, shelled, or cleidoic, eggs allow gas and water exchange between the environment and the embryo while retaining a moist environment for the embryo while being in a terrestrial setting. However, all nutrients necessary for the embryonic maintenance and growth must be contained within the egg. Although these two adaptations do not completely isolate the embryo from its environment, they allow eggs to develop in relatively dry habitats.

Rate of growth changes as a function of life stage. Embryonic growth tends to be logistic in pattern (Birchard & Reiber 1996; Gettinger et al. 1984; Morris et al. 1983; Vleck & Hoyt 1991), but relative, or mass specific, growth slows at hatching (Brody 1945). Changes in the rate of growth and

growth curve at hatching reflect changes in energy source, energy acquisition, and energy allocation (Andrews 1982): excepting growth, developing embryos limit energetic expenditures to maintenance, and yolk provides an energetic source with low processing costs (Andrews 1982). The yolk sac provides an overlap between embryo and neonatal growth: it provides the only source of energy for developing embryos, and it also provides posthatching energy for some reptiles that have either high energy demands or are unable to process exogenous energy shortly after hatching (Congdon et al. 1983b; Fischer et al. 1991; Kraemer & Bennett 1981; Troyer 1983; Wilhoft 1986).

Intrinsic Factors Affecting Embryonic Growth
Maternal Effects

The course of an embryo's development will be in large part affected by its immediate surroundings, e.g., the egg contents, and small differences in egg production between females can lead to large differences in embryonic and hatchling phenotype. Such sources of phenotypic variation, such as egg size, lipid composition, and presence of hormones, are considered maternal effects, which occur when the phenotype of an individual is determined not only by its own genotype and the environmental conditions it experiences during development, but also by the phenotype or environment of its mother (and/or father) (Bernardo 1996a; Falconer & Mackay 1996).

Perhaps one of the most widely recognized maternal effects is that of egg size (Bernardo 1996b). In general, because larger eggs contain more material that can be used by the embryo for somatic tissue synthesis than do smaller eggs, they tend to result in larger offspring (incubation conditions being equal). Larger snapping turtle eggs contain both more solids (Congdon & Gibbons 1985; Congdon et al. 1983b; Finkler & Claussen 1997; Steyermark & Spotila 2001a; Wilhoft 1986) and water (Finkler & Claussen 1997; Steyermark & Spotila 2001a) than do smaller eggs. Accordingly, larger eggs tend to produce larger hatchlings relative to smaller eggs (Bobyn & Brooks 1994b; Brooks et al. 1991; Finkler 1999; Janzen 1993a, 1995; O'Steen 1998; Packard & Packard 1993; Rhen & Lang 1995; Steyermark & Spotila 2001a, b). In addition to size at hatching, egg size also correlates positively with incubation duration (Packard et al. 1987; Steyermark & Spotila 2001a), although Bobyn & Brooks (1994b) reported no such correlation, and even reported a negative correlation between egg size and incubation duration in two clutches.

Increases in solid and water components with increasing egg size do not appear to be proportional, but larger eggs contain proportionally greater amounts of water than smaller eggs (Finkler & Claussen 1997; Steyermark & Spotila 2001a). It is not clear why solid and water content do not scale proportionally with each other, but egg size/egg number tradeoffs may constrain energy allocation per egg.

In addition to variation in egg size is variation in egg quality (Congdon & Gibbons 1990). Specifically, egg yolk contains both polar and nonpolar lipids. Polar lipids tend to be used in cell membranes and other structural components, whereas nonpolar lipids are used for lipid reserves. Thus female decisions regarding the proportion of polar to nonpolar lipids can affect embryonic and posthatching growth (through inclusion of more polar lipids relative to nonpolar lipids) and posthatching survival (through inclusion of more nonpolar lipids relative to polar lipids). Snapping turtles use about 50% of polar lipids during embryonic development, which leaves a significant amount of endogenous structural lipids that may be used for posthatching growth (Congdon et al. 1983b; Wilhoft 1986). Effects of egg water, lipid, and protein content and quality on hatchling performance and survivorship in the field are not well understood (but see Finkler and Kolbe, Chapter 15).

Maternal effects are not limited to egg size and composition. Variation in hormone levels, such as thyroxine T_4, may affect embryonic growth. In snapping turtles, there were significant differences in yolk estradiol-17β (E_2) among clutches from different females (Elf et al. 2002a). E_2 yolk levels varied from a clutch mean of 1.38 to 4.55 ng/g. In addition, Janzen et al. (1998) measured among-clutch differences in yolk testosterone (T) levels. Both E_2 and T can affect sex differentiation in reptiles, and may also affect posthatching growth and behavior (Elf et al. 2002a). Finally, note that because nesting site selection is a maternal effect (Messina 1998; Resetarits 1996; Roosenburg 1996), factors such as the nest's hydric and thermal environment should also be considered as maternal effects on embryonic and posthatching development. In this way, temperature and water availability can affect both the developing embryo (see Ackerman et al., Chapter 13, and Ewert, Chapter 10) as well as posthatching growth (see below).

Genetic Effects

Genetic effects can exert a strong influence over embryonic snapping turtle growth. Many studies report "clutch effects" on parameters such as incubation time (Bobyn & Brooks 1994b; Packard & Packard 1993; Steyermark & Spotila 2001a) and hatchling size (Bobyn & Brooks 1994a, b; Brooks et al. 1991; O'Steen 1998; Packard & Packard 1993; Rhen & Lang 1999a; Steyermark & Spotila 2001a), two parameters associated with embryonic growth rate. These effects have typically been designated as "clutch" effects rather than as "genetic" effects because it is difficult to separate genetic effects from maternal effects. In some cases, however, significant clutch effects on incubation duration and mass at hatching have been detected after controlling for the main maternal effect, egg mass (Packard & Packard 1993; Steyermark & Spotila 2001a), which suggests that the embryo's genotype may affect embryonic growth patterns. Alternatively, other maternal effects may significantly affect

snapping turtle embryonic growth, such as nest site selection and yolk hormones (see above). Although logistically difficult, controlled studies involving breeding designs, and analysis of eggs for a greater suite of maternal effects, would indicate the extent to which genetic effects act on embryonic growth. Partitioning phenotypic variation between genetic and maternal effects in embryonic growth (and posthatching) growth parameters is important in understanding the extent to which natural selection (or genetic drift) can affect important traits.

Extrinsic Factors Affecting Embryonic Growth
Egg Incubation Temperature

Egg incubation temperature affects both incubation duration and mass at hatching. Temperatures within single natural nests in Nebraska ranged from about 16 to 32°C within a 60- to 70-day incubation period (Packard et al. 1985). In general, higher incubation temperatures tend to decrease incubation time (Birchard & Reiber 1996; Bobyn & Brooks 1994a, b; Packard et al. 1987; Steyermark & Spotila 2001a), by increasing the rate of enzymatic reactions, and thereby increasing the rate of the growth process. However, apparently incubation times vary greatly between studies. For example, Birchard & Reiber (1996) reported incubation times of 71 days at 24°C and 56 days at 29°C for eggs from Nebraska, whereas Steyermark and Spotila (2001a) reported incubation times of 71 days at 26.5°C, 70 days at 28.0°C, and 68 days at 30.0°C for eggs from Pennsylvania. Such population differences in incubation duration may be due to genotypic differences, differences in egg composition, or differences in other incubation environment features.

Incubation temperature affects body size of hatchlings: embryos that develop at extreme incubation temperatures (<22°C, >30°C) tend to be smaller at hatching than embryos incubated at more intermediate temperatures (24–28°C) (Bobyn & Brooks 1994b; Brooks et al. 1991; O'Steen 1998; Packard et al. 1987, 1988; Steyermark & Spotila 2001a). However, some studies have shown no effect of egg incubation temperature on body size (mass, plastron length, and carapace length) at hatching (Bobyn & Brooks 1994a; Janzen 1995; Rhen & Lang 1999a). In addition to body size effects, egg incubation temperature also affects body composition of snapping turtle hatchlings: high incubation temperature results in more residual yolk and less body fat (Packard et al. 1988; Rhen & Lang 1999a). Ecological effects of such developmentally induced variation are wholly unknown.

Water Potential

Water potential is the tendency of movement of water molecules, and can be used to indicate how readily a substrate, such as soil, would give up its water. Because water is necessary for embryonic development, and water vapor can pass through snapping turtle egg shells (see Ackerman et al., Chapter 13), the water potential of the substrate in which snapping turtle eggs incubate affects embryonic growth. Water potentials in natural nests tend to be stable around −30kPa, but can reach dry levels of about −150 kPa (Ackerman et al, Chapter 13). Thus, in general, the soil surrounding natural nests tends to be on the "wet" side, with infrequent periods of drying (Ackerman et al., Chapter 13). In general, "wetter" incubation substrates tend to produce larger hatchlings (both body mass and carapace length) than "drier" incubation substrates (Bobyn & Brooks 1994b; Finkler 1999, 2006; Miller & Packard 1992; Morris et al. 1983; Packard & Packard 1988, 1989a, b, 1993; Packard et al. 1981, 1987, 1988, 2000; but see Rimkus et al. 2002). Embryos in wetter substrates consume yolk energy reserves faster than do those in drier substrates, resulting in heavier hatchlings (wet mass and dry carcass mass) with smaller yolk reserves (Morris et al. 1983; Packard & Packard 1988, 1989a; Packard et al. 1988). In addition, those hatchlings from wet incubation substrates have more body fat than do hatchlings from drier substrates (Packard et al. 1988). Thus hatchlings from wetter substrates are heavier, contain more lean body mass, more fat, but less residual yolk, than hatchlings from drier substrates.

Substrate water potential affects mass of some body organs (Packard et al. 2000). When adjusted for body size, hatchlings from dry substrates had heavier hearts than turtles incubated in wet substrates. Masses of other organs (liver, stomach, lungs, kidneys, small intestine) of hatchlings from dry and wet substrates were similar. When hatchlings from dry substrates were placed in water, body mass, and liver, stomach, lung, kidney, and small intestine mass all increased, but heart mass did not. Packard et al. (2000) suggested that turtles incubated in dry substrates may be hypovolemic as compared with turtles incubated in wet substrates, and once turtles are allowed to fully hydrate, circulatory volume is restored, and turtles are released from a constraint on growth. In addition, hypovolemia may make turtles more susceptible to desiccation during overland movement (see Finkler and Kolbe, Chapter 15).

In addition to effects on body size and body composition, substrate water potential affects incubation duration. Wet incubation substrates tend to increase the incubation period relative to dry incubation substrates (Bobyn & Brooks 1994b; Miller & Packard 1992; Packard & Packard 1988, 1989b, 1993; Packard et al. 1987). Relatively long incubation times may expose the embryo to a greater probability of mortality (from predation, or an unfavorable hydric or thermal environment, see below), whereas the relatively heavier hatchling (resulting from wetter substrate) may have a greater probability of survivorship (see below). This potential trade-off between time spent in the embryonic stage versus size at hatching has not been well explored in turtles.

Field studies have confirmed the general patterns observed in the laboratory. In one study, Packard et al. (1993) divided eggs from three snapping turtles between two nests. Eggs from one nest tended to absorb water during development, and those eggs resulted in larger hatchlings with less residual yolk than hatchlings from the second nest, whose eggs tended to lose water during development. In another study, Packard et al. (1999) incubated eggs in natural nests for the first 8 weeks of development, thus exposing them to varying water potentials, and then brought the eggs into the laboratory to finish incubation. Embryos from eggs that absorbed water during the 8 weeks in the field were larger, consumed more yolk, and contained more water, than did embryos from eggs that lost water in the field (Packard et al. 1999). But Rimkus et al. (2002) suggest that disturbance of the nest during excavation may have affected the water potential of the substrate in contact with the eggs. This may have led to an unnatural condition, causing eggs from one nest to be in a drier substrate than they would be in an undisturbed nest. Rimkus et al. (2002) suggest that snapping turtle eggs in the field are always in a positive water balance and that they would gain sufficient water such that embryonic growth and hatchling mass would be unaffected by water uptake. Finkler (2006), however, observed differences in hatchling mass and carapace length corresponding to differences in soil water content within the range of natural variation in soil hydration.

Results in the literature illustrate that process of water uptake by the egg, and its effects on embryonic growth, embryonic energy use, and hatchling mass, are not completely understood. Some laboratory data suggest that patterns of embryonic growth depend on substrate water potential: snapping turtle embryos incubated on wetter substrates are larger at hatching than embryos incubated on drier substrates partly due to greater accumulation of water and thus greater relative hydration and partly due to greater conversion of yolk to tissue than embryos incubated on drier substrates. But laboratory conditions may not reflect natural field nest conditions (Kam & Ackerman 1990; Rimkus et al. 2002; see Ackerman et al., Chapter 13). Further, it is not clear precisely how substrate water potential mediates its effects on the conversion of yolk to tissue, which results in greater dry mass of hatchlings from wetter eggs compared with hatchlings from drier eggs. One hypothesis had been that dry substrates cause an increase in blood urea concentration (Packard et al. 1984a). Although embryos incubated on relatively dry substrates do have higher blood urea levels than do embryos incubated on wetter substrates, urea injection experiments indicated that urea concentrations, irrespective of incubation conditions, were not correlated with mass at hatching (Packard & Packard 1989a).

Substrate water potential may also affect embryonic growth by influencing rate of calcium and phosphorus absorption. Approximately 56% of necessary calcium for embryonic development is obtained from the eggshell, and the remaining 44% is obtained from the yolk (Packard et al. 1984b). The rate of calcium withdrawal from the yolk is similar for embryos incubated on either wet or dry substrates, and almost all available calcium in the yolk is depleted by hatching. However, embryos in wet substrates obtain more calcium from the eggshell than do embryos incubated in dry substrates. Embryos obtain all phosphorus used during development from the yolk, but embryos incubated on wet substrates mobilize phosphorus more rapidly than embryos incubated on dry substrates (Packard & Packard 1989b).

Last, although it is not clear whether variation in body composition and energy reserves affects hatchling survivorship or individual fitness, results from desiccation studies on snapping turtle hatchlings suggest that low water potentials, perhaps through effects of hypovolemia, can detrimentally affect overland movements (Finkler 1999, 2001; Finkler et al. 2000).

Gas Exchange

Gas tensions in turtle nests can affect embryonic development and survival, although whether gas tensions affect development in natural snapping turtle nests is unclear. Fresh *Chelodina rugosa* eggs can be submerged for up to 12 weeks and then undergo successful development. This Australian species can lay eggs underwater in flooded areas, and embryonic development resumes once floodwaters recede (Kennett et al. 1993). However, 19-day-old Florida red-bellied turtle (*Pseudemys nelsoni*) eggs submerged for six days to simulate flooding showed 100% mortality (Kam 1994). Laboratory exposure to chronic hypoxia (10% O_2) in red-bellied turtle eggs slowed growth, depressed metabolism, and reduced hatchling mass compared with eggs exposed to normoxia, although the incubation period was comparable (Kam 1993), but mass at hatching and incubation period were similar for red-eared slider eggs (*Trachemys scripta*) exposed to 15%, 21%, and 30% O_2 (Etchberger et al. 1991). Thus exposure to hypoxia differentially affects embryonic development of freshwater turtle species; some species can clearly stand significant submerging, while others are less tolerant. It is likely that snapping turtle eggs can withstand some submerging (probably less than 5 days), though those experiments need to be performed.

In the field, Ackerman (1977) showed that oxygen tension of sea turtle nests are typically about 14–19%, but can fall to as low as 5% near the time of hatching, and laboratory experiments demonstrated that hatching success and incubation period are affected by oxygen tensions in the nest (Ackerman 1981). Thus gas exchange by sea turtle eggs may constrain nest structure, clutch size, and incubation period in sea turtles (Ackerman 1977, 1980, 1981), and may affect embryonic development in freshwater turtles as well. However, it is unclear whether snapping turtle embryonic growth is constrained by gas exchange in natural nests.

Population/Geographic Variation

It is difficult to assess whether population/geographic differences exist in embryonic growth. Most studies have used eggs from only one population or geographic area. Because slight variations in incubation substrate, or differences in variation around mean incubation temperature or water potential (see Shine 1995) can affect embryonic growth, the comparison of studies may be difficult. However, Bobyn and Brooks (1994a, b) examined growth of embryos from different populations of snapping turtles in Canada, and found that population of origin significantly affected mass at hatching. In addition, Finkler et al. (2004) reported significant variation in egg composition across a longitudinal gradient, and thus population differences. When accounting for egg mass, proportion of water increased and proportion of solids decreased from east to west. The authors suggest that differential needs based on climate of both water availability to the embryo and posthatch energy reserves may explain the variation. This variation may translate into variation in embryonic and posthatching growth. More studies of this type are needed to determine the extent to which geographic variation occurs in embryonic growth.

POSTHATCHING GROWTH
General

Posthatching growth is not as well studied as embryonic growth, perhaps because it is easier to follow the growth trajectory of an embryo for a few months than it is a posthatching turtle for many years. Moreover, much of what is known about snapping turtle posthatching growth is from laboratory experiments, most of which unfortunately do not represent natural field conditions.

In the laboratory, growth tends to be slow or absent for the first few weeks, as hatchlings do not feed and subsist on their yolk reserves (M. Finkler, unpublished data; pers. observ.). In nature, after nest emergence in late August through September, hatchlings most likely find their way to a body of water, and then "hibernate." During the fall and winter, young snapping turtles likely subsist on yolk and fat body reserves and do not feed. During this time, little or no growth takes place, because growth does not occur unless feeding takes place (Wieser 1994). Turtles emerge in the spring to begin feeding, and then experience their first bout of growth.

Several factors may affect snapping turtle posthatching growth, including embryonic developmental conditions, maternal effects, climate space, and nutrient availability. Although turtles are commonly thought to have indeterminant growth (Patnaik 1994), many turtles may reach an asymptotic size after which growth is very slow or negligible (Andrews 1982; Cagle 1946; Christiansen & Burken 1979; Congdon et al. 2001). Posthatching snapping turtle growth appears to best fit the von Bertalanffy model (Andrews 1982). Nothing is known about senescence in snapping turtles, but a study of the Blanding's turtle (*Emydoidea blandingii*) indicated that older female Blanding's turtles increased their reproductive output through increases in clutch size, reproductive frequency, and adult survivorship relative to younger females (Congdon et al. 2001).

Intrinsic Factors Affecting Posthatching Growth

The two main intrinsic factors affecting posthatching growth in snapping turtles are maternal effects and genotype.

Some investigators have reported clutch effects on posthatching growth in snapping turtles (Bobyn & Brooks 1994a, b; Brooks et al. 1991; McKnight & Gutzke 1993; Rhen & Lang 1995; Steyermark & Spotila 2001b), whereas other studies reported no clutch effects on growth (O'Steen 1998; Rhen & Lang 1999b). The discrepancy between results may be due to low statistical power: studies that reported no significant clutch effects had low sample sizes (2–4 clutches), whereas those studies that reported significant clutch effects had larger sample sizes (>6–24 clutches). Thus, it is likely that with sample sizes greater than five clutches, investigators would detect clutch effects on posthatching growth.

The term clutch effects encompasses both genetic and maternal effects. Once effects of egg mass, the most commonly studied maternal effect, are removed through covariate analysis, clutch still remains as a significant effect on posthatching growth, suggesting either a genetic basis or other maternal effects. Clutch-based variation in growth has been reported for several taxa, including anurans (Newman 1994; Travis 1980), crocodilians (Hutton 1987), and lizards (Sinervo 1990b; Sinervo & Adolph 1989). Both maternity and paternity affected growth and metamorphic mass of individual *Bufo woodhousei* tadpoles produced from breeding crosses in the laboratory and field (Mitchell 1990). Travis (1980) reported among clutch variation in specific growth rates for larval *Hyla gratiosa* raised under similar conditions that controlled for different thermal environments and eliminated competition. He concluded that genetic composition affected larval growth characteristics, and that the divergent growth curves were due to "genetic variation at the many loci." Thus, while it remains unclear whether genotype affects posthatching growth in snapping turtles, removal of some maternal effects through covariate analysis, and results from breeding designs from other vertebrate species, suggest that genotype differences between individuals may contribute to variation in growth. Such genetic effects on growth rate may be realized at one or both of two levels: energy acquisition or energy expenditure. Genetic variation in energy acquisition may be due to differences in quantity of food eaten, gut length, digestive efficiency, or processing time, whereas genetic effects at the level of energy expenditure may be due to differences in mitochondria number, mitochondria density, organ mass, enzyme kinetics, or hormone levels.

Maternal effects on posthatching growth include residual energy stores in the form of fat and residual yolk. Although incubation effects (such as egg incubation temperature) are a result of oviposition site, and thus could be considered as maternal effects, they will be discussed below as extrinsic factors. Snapping turtle neonates commonly hatch in the laboratory with some residual yolk, ranging from about 0 to 2.2 g wet mass (Finkler 1999; Packard et al. 1987, 1988, 2000; Wilhoft 1986), which can contribute to between 1 and 20% of total neonatal body mass (Finkler 1999; Packard et al. 1987, 1988, 2000; Wilhoft 1986). Amount of residual yolk may depend on incubation conditions (Finkler 1999; Packard et al. 1987, 1988, 2000). Although snapping turtle neonates derive energy from their yolk sacs for the first few weeks of life (M. Finkler, unpublished data; pers. commun.), it is not known whether variation in residual yolk sac between individuals affects posthatching growth or survival. Other maternal effects that can affect posthatching growth include variation in hormone levels (see below).

Finally, maternal effects can persist long into ontogeny: Steyermark and Spotila (2001b) reported that turtles that were larger than their conspecifics at hatching tended to be larger at six months of age.

Extrinsic Factors Affecting Posthatching Growth
Incubation Effects

Some egg incubation environment factors appear to affect posthatching growth of snapping turtles. Among the most commonly studied incubation effects are substrate water potential and egg incubation temperature. Substrate water potential does not appear to affect posthatching growth (Bobyn & Brooks 1994b; Brooks et al. 1991; McKnight & Gutzke 1993; Miller 1993).

In contrast to the hydric environment, egg incubation temperature does appear to affect growth. In general, eggs incubated at cooler temperatures tend to produce hatchlings that exhibit faster posthatching growth than do eggs incubated at warmer temperatures (Brooks et al. 1991; O'Steen 1998; Rhen & Lang 1995, 1999b). However, two studies found the reverse: eggs incubated at cooler temperatures produced hatchlings that exhibit slower posthatching growth than did eggs incubated at warmer temperatures (Bobyn & Brooks 1994a, b), and one study found that egg incubation temperature did not affect posthatching growth (Steyermark & Spotila 2001b).

Methodological differences between studies that have examined the effects of egg incubation temperature on posthatching growth make it difficult to compare results. However, some trends have recently emerged. Brooks et al. (1991), O'Steen (1998), and Rhen and Lang (1995) housed juvenile snapping turtles in common environments with access to basking sites. Turtles could thermoregulate and compete for basking sites as well as for food. Thus, growth differences between individuals may emerge based on thermoregulation and feeding, rather than on intrinsic effects of egg incubation temperature. In addition, O'Steen (1998) reported that egg incubation temperature affected the preferred body temperature of juvenile snapping turtles: turtles hatched from cooler incubation temperatures chose warmer water when given a choice. The temperature choices made by the juvenile turtles were repeatable after a six-month hibernation period at $7°C$. Growth of turtles is typically positively correlated with body temperature (Avery et al. 1993; Rhen & Lang 1999a; Williamson et al. 1989). This finding may explain the results of Steyermark and Spotila (2001b), who maintained juvenile snapping turtles from hatching until 6 months of age in individual plastic containers, such that all turtles experienced similar feeding and temperature conditions. They reported that egg incubation temperature did not affect posthatching growth, which suggests that egg incubation temperature itself does not affect posthatching growth. Thus, it appears that egg incubation temperature affects preferred body temperature, which in turn affects growth rate. Rhen & Lang (1999a) also found that egg incubation temperature affected posthatching temperature choice. However, they then placed six-month-old turtles that had been given the opportunity to thermoregulate in constant-temperature pools. They found that the turtles' growth rate was still inversely correlated with egg incubation temperature. Thus, egg incubation temperature can have indirect effects on growth, by affecting the preferred body temperature of juveniles, though it is not clear whether egg incubation temperature directly affects growth.

The mechanisms by which egg incubation temperature may affect growth are unknown. Sex appears to have no effects on posthatching growth (O'Steen 1998; Rhen & Lang 1995, 1999a; see Janzen, Chapter 14, for discussion of temperature-dependent sex determination). Egg incubation temperature could directly or indirectly affect posthatching growth in several ways. First, egg incubation temperature negatively correlated with snapping turtle hatchling plasma thyroxine T_4 levels (O'Steen & Janzen 1999), and thyroxine T_4 positively correlated with juvenile growth in slider turtles (Denver & Licht 1991; Stamper et al. 1990). Thus, turtles from colder egg incubation temperatures may have higher levels of thyroxine T_4, which enhances growth, relative to turtles from higher egg incubation temperature. Differences in thyroxine T_4 levels due to egg incubation temperature might also affect selected body temperature in turtles through behavioral thermoregulation, as it does in western fence lizards (Sinervo & Dunlap 1995). In addition to thyroxine, experimental manipulation of growth hormone levels in snapping turtles affected growth, food consumption, and liver protein, water, lipid, and glycogen levels (Brown et al. 1974).

Second, dietary fat composition affects cell membrane lipid composition in desert iguanas, which influences membrane fluidity. Iguanas compensate for changes in membrane

fluidity by selecting different body temperatures (Simandle et al. 2001). Perhaps egg incubation temperature might affect cell membrane lipid composition, and thus membrane fluidity, thereby leading to differences in selected body temperature.

Geographic, Environmental, and Genetic Effects on Growth

Geographic variation may affect snapping turtle growth patterns in multiple ways. In general, differences in temperature and productivity may lead to differences in individuals' growth rates between populations.

Very little is known about geographic or population variation in snapping turtle growth, except that variation in growth does occur. Galbraith et al. (1989) reported that female snapping turtles from Algonquin Park, a low-productivity area in northern Ontario, Canada, grew more slowly than those from more productive areas in Michigan and Iowa. Also, the smallest (24 cm) and youngest (17 years old) sexually mature females at Algonquin Park were larger or among the largest, and older, than the smallest (Michigan, 20 cm [Congdon et al. 1987]; South Dakota, 26 cm [Hammer 1969]; Tennessee, 20 cm [White & Murphy 1973]), and youngest (Iowa, 9–10 years [Christiansen & Burken 1979]; Michigan: 12 years [Congdon et al. 1987]; South Dakota, 9 years [Hammer 1969]) sexually mature female snapping turtles in other populations, suggesting that turtles from low-productivity areas may take longer to mature than those from more productive areas. Brown et al. (1994a) reported that Algonquin Park female turtles had a lower growth and reproductive output (smaller clutches) than those from southern Ontario populations. Given that both diet quality (Avery et al. 1993; Gibbons 1967; Moll 1976; Parmenter 1980) and body temperature (Avery et al. 1993; Cagle 1946; Parmenter 1980; Williamson et al. 1989) affect turtle growth rate, low primary productivity and water temperatures of Algonquin Park (Brown et al. 1994a; Galbraith et al. 1988) relative to more southern populations may explain, in part, geographic variation in snapping turtle growth and age at sexual maturity.

Temperature affects the rate of chemical reactions, and as such, body temperature affects many body processes. Temperature may affect turtles both directly, through enhancing digestive parameters and extending the growing season, and indirectly, by enhancing environmental productivity. Snapping turtles typically show very little growth during the winter months. In Iowa, shell growth of juveniles and adults begins during the last two weeks of May, is most rapid in June, July, and August, and takes place during an estimated feeding and growing period of about 145 days, while the estimated activity period is about 200 days (Christiansen & Burken 1979). Almost no growth was observed between mid September and mid May. Growth rate was similar for males and females for the first two to three years, but then males grew more rapidly than females. Some males continued to grow for more than 35 years, whereas most females showed little growth past 20 years. In the laboratory Williamson et al. (1989) found that juvenile snapping turtles kept at 25°C ate freely and grew rapidly, whereas those kept at 15°C fed and grew very little.

In addition to environmental factors such as productivity and body temperature that affect posthatching growth in turtles, geographic variation in growth also appears to have a genetic basis. Bobyn & Brooks (1994b) incubated eggs in the laboratory from four populations of snapping turtles in Ontario, Canada. Juveniles that were raised in the laboratory for one year from a northern population (Algonquin Park) grew more slowly than juveniles from three more southern Ontario populations. These differences in growth between populations may represent a genetic adaptation to a habitat with lower primary productivity and a shorter growing season than the more southern Ontario populations (Bobyn & Brooks 1994b). Whereas it appears that growth may also have a genetic component, it is not clear whether this is an adaptation to areas with lower productivity (Bobyn & Brooks 1994b), or whether the genetic differentiation is simply a matter of genetic drift, and is nonadaptive.

Social Interactions

Social interactions among individuals in laboratory settings can affect growth and body composition of juvenile snapping turtles. Competition for food appears to be substantial when animals are grouped at high population densities, affecting growth (Mayeaux et al. 1996; McKnight & Gutzke 1993) and dominance hierarchies (Froese & Burghardt 1974). Snapping turtles housed individually in general had higher masses after 14 weeks of age than those housed in groups, even though the density of individuals was similar for both groups (McKnight & Gutzke 1993). Effects of egg incubation temperature on individual mass were different between individuals housed in solitary and those housed in groups (McKnight & Gutzke 1993). Two-month-old snapping turtles stocked at high density (58 animals/m²) had lower growth rates, higher mortality, higher feed consumption, less efficient feed conversion, lower liposomatic index, and lower productive protein value than turtles stocked at low density (29 animals/m²; Mayeaux et al. 1996). Mayeaux et al. (1996) point out the need for digestibility and basic nutrition studies in snapping turtles to better understand their nutritional requirements. Competition between individuals can affect resource acquisition: in laboratory experiments, larger juvenile turtles outcompeted smaller turtles for food (Froese & Burghardt 1974). However, whether competition for resources between snapping turtles occurs in nature, whether body size affects competitive outcome, and whether competitive outcome affects survivorship, reproductive output, or fitness, is largely unknown. Future research should investigate individual and population variations in energet-

ics to understand how genes and the environment affect patterns of growth in snapping turtles. Snapping turtles may make an especially useful model in this regard because they extend over a wide geographic area, encompassing areas of ranging productivities and climates.

GROWTH AND FITNESS

At the earliest stages of development, growth has the potential to affect survival, and thus fitness. Egg size, incubation temperature, and substrate water potential all affect incubation duration. Eggs with relatively long incubation durations may be at greater risk to predation and unfavorable hydric and temperature regimes than eggs with short incubation durations; for example, embryonic development ceases below 20°C (Yntema 1978). In addition, although speculative, eggs with very fast incubation times may be at a disadvantage because hatchlings would be exposed to very high temperatures during the overland migration. Thus, egg incubation time may affect fitness.

At the posthatching stage, the relationship between mass at hatching, growth, and juvenile survival in snapping turtles and other freshwater turtles is unclear despite several experiments designed to investigate these relationships. Janzen (1993) released 112 three-week-old juveniles at a nesting site near the Mississippi River in Illinois, and recaptured 66 with a drift fence. Janzen found that larger turtles had a higher probability of recapture than did smaller turtles, suggesting that larger juveniles exhibited a higher survivorship than did smaller ones during the nest-to-water migration. In a subsequent study, Janzen (1995) released 121 two- to three-month-old juveniles into an artificial pond in Michigan in November and recaptured 16 turtles in May. Body size was not correlated with probability of recapture, suggesting that overwintering survivorship was independent of body mass. In a replicated field study at the E.S. George Reserve in Michigan, Congdon et al. (1999) released 62 5- to 15-day-old turtles in 1995, and recaptured 44 in drift fences. Neither body mass nor carapace length was correlated with probability of recapture. In 1996, they released 147 hatchlings from experimental nests; 86 were recaptured at drift fences. Body mass was not correlated with probability of recapture, but carapace length was. Fitness function analysis suggested stabilizing selection on both body mass and carapace length, suggesting that nest-to-water migration survivorship favored neither large nor small turtles. Congdon et al. (1999) also analyzed survivorship data collected for 675 hatchlings released over seven years. Sixty-one of those hatchlings were recaptured. For the seven years combined, probability of recapture was not related to carapace length or body mass. Data from Congdon et al. (1999) suggest that selection on body size in young snapping turtles is infrequent during the nest-to-water migration, and when selection does occur, it follows a stabilizing pattern. Finally, Kolbe & Janzen (2001) released 463 3- to 14-day-old hatchlings from eight release points within an enclosed area on a slope near the Mississippi River, and recaptured 291 turtles. They reported no effect of body size on probability of survival. Comparisons between Janzen (1993, 1995), Congdon et al. (1999), and Kolbe & Janzen (2001) are difficult because of methodological differences between the studies. Further, it is likely that temporal and population variation exists in selection pressures on neonatal body size (e.g., Ferguson et al. 1982; Sinervo 1990; Sinervo et al. 1992). However, available data do not support the "bigger is better" hypothesis in snapping turtles.

No data exist on the relationship between growth and survivorship in snapping turtles older than one year, mainly because very few long-term demographic studies on snapping turtles are available (but see Congdon et al. 1994). Habitat use tends to differ between smaller juvenile snapping turtles and larger ones at the E.S. George Reserve, with smaller turtles being found in more shallow areas than larger turtles (Congdon et al. 1992), although it is not clear whether ontogenetic changes in habitat use were due to the distribution of food resources, predator avoidance, or differences in swimming abilities of juveniles, foraging mode, or thermoregulatory behavior (Congdon et al. 1992).

Reproductive output appears to be positively correlated with female body size, although apparently significant population variation exists in the relationship between body size, egg size, and clutch size. Egg size was positively correlated with female body size in Nebraska (Iverson et al. 1997) and Michigan (Congdon et al. 1987), but not in Pennsylvania (Steyermark & Spotila 2001a). Further, mean egg mass was not correlated with mean female carapace length from 14 snapping turtle populations (Iverson et al. 1997). However, clutch size was positively correlated with female body size in Nebraska (Iverson et al. 1997), Michigan (Congdon et al. 1987), and Pennsylvania (Steyermark & Spotila 2001a), and mean clutch size was correlated with mean female carapace length from 22 snapping turtle populations (Iverson et al. 1997). Thus, larger female snapping turtles tend to have a higher reproductive output relative to smaller females by increasing the number of eggs in a clutch. However, studies that examine lifetime reproductive output, reproductive success, and fitness in terms of body size in snapping turtles are entirely lacking.

FUTURE DIRECTIONS

Despite the scores of studies that have examined embryonic and posthatching growth in snapping turtles, many questions have been left unanswered. Three future areas of study should include: the ecological relevance of laboratory experiments; the evolutionary significance of growth; and underlying physiological mechanisms explaining between-individual variation in growth.

Although abundant laboratory studies have detected the effects of various factors on growth (egg incubation temperature, water potential), one is left wondering whether such laboratory results are at all relevant to snapping turtles in the field. Although it is certainly important to isolate and study individual effects in the laboratory, one must be careful to ensure that laboratory conditions are ecologically relevant. A better understanding of conditions that snapping turtle embryos and neonates inhabit in the natural field will yield better experimental designs in the laboratory.

Of equal importance to ecological relevance is evolutionary significance. That is, is the trait in question important to survivorship or fitness? Body size, long assumed to be a determinant of neonatal survivorship, may be of questionable consequence. Does it matter whether one has a higher growth rate than a conspecific? The answer to that question is unknown. Recent breakthroughs in technology may soon allow individual neonates to be permanently tagged, and then recaptured months to years later. Only after repeated, replicated studies are conducted, in various geographical locations, will we begin to understand whether embryonic and neonatal growth are important.

Despite many studies that have detected significant effects of various factors on growth, very few studies have examined underlying physiological mechanisms. How does egg incubation temperature affect thermal preference? Why do some females produce neonates with relatively fast growth rates, while some females produce neonates with relatively slow growth rates? Are differences among offspring growth attributable to maternal effects or to heritable genotype differences?

Much remains to be understood in vertebrate ectotherm growth patterns. Snapping turtle studies to date have provided an excellent launching point and have shown that this species makes a robust model to understand the evolutionary significance and mechanistic basis of variation in growth.

PART 3 LIFE HISTORY AND ECOLOGY

12

Reproductive and Nesting Ecology of Female Snapping Turtles

JUSTIN D. CONGDON
JUDITH L. GREENE
RONALD J. BROOKS

Snapping turtles evoke impressions of a primitive nature and time, because of their appearance and because they shared the earth with the dinosaurs; they are "entitled to regard the brontosaur and mastodon as brief zoological fads." (Gilbert 1993)

THE GOAL OF THIS CHAPTER is to summarize what is known about the reproductive traits and nesting behaviors of snapping turtles (*Chelydra serpentina*) in light of their historical and present success. We will use a summary of geographic, local, and temporal patterns of variation to explore how variation in traits may influence individual fitness. Because traits of living organisms represent solutions to evolutionary problems, we will also examine the traits of female snapping turtles to see if and how their reproductive tactics might be interpreted as attempts at problem solving (Popper 1999). Where comparisons can be made, the chapter draws on published and unpublished long-term data from the Savannah River Site near Aiken, South Carolina, the Edwin S. George Reserve near Hell, Michigan, and the Algonquin Provincial Park near Dwight, Ontario, Canada.

A life history is a suite of coevolved, age-specific traits such as survivorship and reproductive output, levels and types of parental investment, and age at maturity. Differences in lifetime reproductive success result from the overall variation in life history trait values among individuals. The process of evolution is translated through differential reproduction of individuals with survival and death both mediated through their effects on births (i.e., the evolutionary currency of death is births [Williams 1966]).

Construction of a successful neonate is the base from which other reproductive traits must follow. Suites of reproductive traits provide potential opportunities for a female to maximize the number of surviving offspring she produces at a given time: (1) egg size, representing parental investment that is a major component of offspring size and quality; (2) clutch size or reproductive output that is determined by levels of parental investment and resources allocated to a reproductive bout; and (3) clutch frequency, or how often to reproduce. Females can influence their reproductive success by varying the timing, frequency, and relative allocations to the components of reproduction: (1) parental investment, (2) reproductive output, and (3) nesting activities such as pre-nesting migrations, nesting migrations, nesting site selection, and nest construction.

The combination of viewpoints taken by Ewert (1979, 1985), Packard & Packard (1988b), Congdon & Gibbons (1990), and Booth & Thompson (1991) provide an excellent, broad overview of eggs and embryo development in turtles. Parental investment in turtles consists entirely of allocation to individual eggs, and patterns of parental investment reflect the lack of postovulatory, trophic parental care. In contrast to organisms that provide postovulatory parental care, female turtles cannot make differential allocations among offspring based on embryo genotypes or phenotypes. In addition, eggs of all turtle species are essentially closed, zero-sum energy systems that provide embryos with all of their energy needs from the time the eggs are shelled until the hatchling leaves the nest and consumes its first meal (Kraemer & Bennett 1981; Wilhoft 1986; Congdon & Gibbons 1985, 1990; Rowe et al. 1995; Nagle et al. 1998, 2003). Female turtles make preovulatory parental investments in embryogenesis and embryo growth (PIE) and parental care of hatchlings (PIC) after they leave the egg (Congdon & Gibbons 1990).

For all turtle species examined, all trophic parental care takes the form of hatchling yolk sacs and fat bodies that represent a substantial portion of total parental investment. Among the turtles examined, the proportion of egg lipids that remain in the hatchling at emergence from the egg ranges from approximately 40% in Kinosternidae (Nagle et al. 1998) to >70% in *Apalone mutica* (Nagle et al. 2003) and the sea turtle, *Natator depressus* (Suhashini & Parmenter 2002). Resources in yolk sacs and fat bodies fuel standard and activity maintenance costs, early development, and possibly some growth of hatchlings (Kraemer & Bennett 1981; Congdon et al. 1983b; Congdon & Gibbons 1990; Finkler et al. 2002). Because female turtles make large investments that provide hatchlings with yolk reserves, parental investment and optimal egg size are more tightly coupled than they are in organisms with postovulatory trophic parental care.

Nesting is the last important step for females to complete reproduction (Fig. 12.1). For almost all turtles, nesting requires a terrestrial migration and construction of a terrestrial nest. The following aspects vary among females: (1) reproductive investment, (2) levels of parental investment, (3) timing of nesting, (4) selection of a nesting area, and (5) selection of a specific nest site; all these combine to influence the fitness of the parent. In combination these factors alter the success of the present reproductive event, the probability of future reproduction, and the phenotypes of neonates (including the sex of hatchlings; see Janzen, Chapter 14).

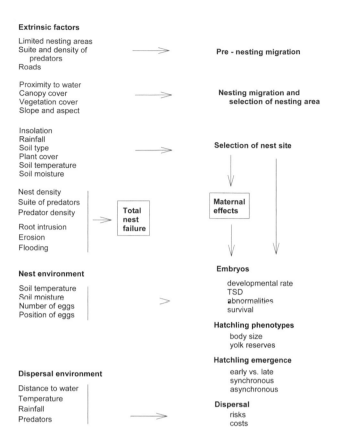

Fig. 12.1. Components of nesting and nest environments

Differences in nesting activities among females may indicate variation in attempts to solve a combination of biotic and biophysical problems that influence their reproductive success (Fig. 12.1), but just how much control females have over subsequent incubation environments that influence hatchling phenotypes is unknown and difficult to document.

Overview of the Snapping Turtle

In North America, snapping turtles range from the southern portion of eastern Canada south into the eastern two-thirds of the United States (Gibbons et al. 1988). Snapping turtle biology is relatively well known compared to other turtle species. The demography (Brooks et al. 1988; Congdon et al. 1994; Cunnington & Brooks 1996), life history and reproduction (Congdon et al. 1987; Iverson et al. 1997), nesting ecology, characteristics of nesting areas and nest site selection (Ewert 1976; Loncke & Obbard 1977; Petokas & Alexander 1980; Congdon et al. 1987; Obbard & Brooks 1987; Hotaling 1990; Kolbe & Janzen 2002a), and developmental dynamics (Ewert 1976; Ewert et al. 2005; Brooks et al. 1991a) have been studied across their broad geographic range. Snapping turtles are long lived, and because they are large bodied, adults experience additional mortality because they are killed (commercially and by individuals) for meat. The combination of being long lived and commercially harvested has resulted in the collapse of some alligator snapping

Table 12.1
Geographic variation in reproductive characteristics of female snapping turtles

Location	N	Latitude/ longitude	SizMat CL (mm)	AgeMat (yr)	CL (mm)	Body mass (g)	Clutch size (min–max)	Egg Diameter (mm)	Egg Mass (g)	Ref.
ON	223	45.5/78.5	246	17	285	5262	36.0 (12–69)	27.6	11.9	1
NY	28	43.0/75.5			259	4370	30.9	–	–	2
SD	291	43.0/101.5			319	7364	49.0 (31–87)	–	–	3
MI	149	42.5/84.0	175	11	249	3790	25.6 (5–44)	28.0	9.2	4
NE	77	42.0/102.5			325	7878	46.8 (20–73)	27.2	11.4	5
NJ	314	40.5/74.5			261	4220	32.3 (15–52)	25.7	9.7	6
PA	44	40.5/76.5			295		30	26.1	10.1	7
VA	22	38.5/77.0			260	3963	30	27.6	11.1	8
SC	43	33.3/81.8	220	7	262	4493	29.9 (3–55)	25.8	9.6	9

Data are means for all columns except for the minimum size and age at maturity. Minimum and maximum clutch sizes are presented in parentheses.

1. Ontario, Brooks, unpublished data (see also Obbard 1983, Galbraith et al. 1989, Galbraith & Brooks 1987; Loncke & Obbard 1977; Brown et al. 1994a).

2. New York, Petokas & Alexander 1980.

3. South Dakota, Hammer 1969.

4. Michigan, Congdon et al. 1987; Congdon, unpublished data.

5. Nebraska, Iverson et al. 1997.

6. New Jersey, Hotaling 1990.

7. Pennsylvania, Carl Ernst, personal communication (cited in Iverson et al. 1997).

8. Virginia, Wilgenbusch, personal communication (cited in Iverson 1997).

9. South Carolina, Congdon & Gibbons 1985; Gibbons unpublished data.

turtle (*Macrochelys temminckii*) populations (Pritchard 1989). The same is probably true for some other snapping turtle populations, and there are many anecdotal accounts of local declines or extirpations.

COMPONENTS OF REPRODUCTION
Follicle Development

Females allocate resources to the development of follicles during the activity season in the year before ovulation and nesting. In Tennessee follicle enlargement began in late July (6.8 mm diameter) through August (10.8 mm) and by November follicles averaged 23.3 mm (White & Murphy 1973). In some cases additional follicle enlargement may occur in the following spring since warmer springs result in earlier initiation of nesting activity (Congdon et al. 1987; Obbard & Brooks 1987).

Eggs

Eggs of most snapping turtles are round, but in the smallest females they are more spheroid. Freshly laid eggs are 26–28 mm in diameter, weigh from 9 to 12 g (Table 12.1), and are composed of 68–72% water (Congdon & Gibbons 1985; Wilhoft 1986). Eggshells are 28–30% of the total dry mass of the egg (Wilhoft 1986; Rowe et al. 1995), and consist of approximately 40% inorganic ash by dry weight (Lamb & Congdon 1985). Although snapping turtle eggs feel rigid, their shell components are more similar to species

with flexible-shelled eggs than they are to those with rigid-shelled eggs (Lamb & Congdon 1985). The shells of freshly laid snapping turtle eggs make clicking sounds as they dry when exposed to dry air, but the flexible eggshells of Midland painted turtles (*Chrysemys picta marginata*) and Blanding's turtles (*Emydoidea blandingii*) do not (Congdon, pers. observ.). As a component of total yolk dry mass, snapping turtle egg lipids were 13.2% in Nebraska (Rowe et al. 1995), 21.6% in Michigan (corrected to remove shell mass, Congdon et al. 1983b), and 33.8% in New Jersey (Wilhoft 1986).

Hatchlings

Neonate snapping turtles ($n = 2{,}227$) on the E. S. George Reserve (ESGR) in Michigan were 29.5 mm in carapace length (CL) and weighed 9.1 g (76% of mean egg wet weights). One week after hatching, neonates from Michigan that had been incubated at constant 25°C on wet and dry substrates weighed 7.7 g and 6.6 g, respectively (Finkler et al. 2002). Even though the average egg weights were similar (11.7 and 11.6 g) to average egg mass on the ESGR (12.0 g), hatchlings from the wet and dry incubation substrates were 1.4 g (15.4%) and 2.5 g (27.5%) lighter, respectively, than hatchlings from natural nests on the ESGR. In addition, the percent lipids in hatchlings incubated on wet (6.2%) and dry (8.5%) substrates (Finkler et al. 2002) were lower than those from hatchlings from natural nests in Michigan (12.9%, Congdon et al. 1983b) and New Jersey (19.7%, Wilhoft 1986), but similar to hatchlings from Nebraska (6.5%, Rowe et al. 1995). Incubation at con-

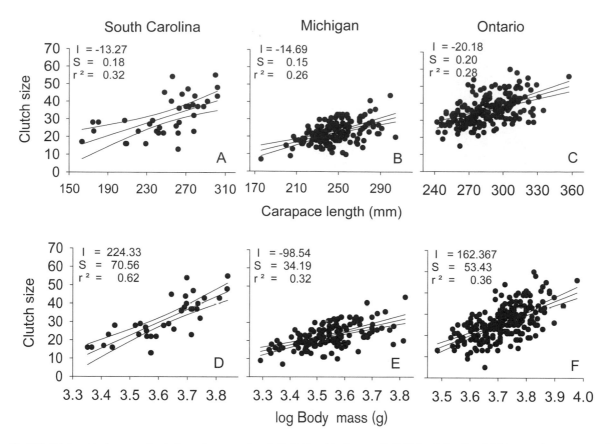

Fig. 12.2. Relationships between clutch size and carapace length (A, B, C) and body mass (D, E, F) South Carolina, Michigan, and Ontario, Canada, respectively

stant temperatures, even if similar to average temperatures in natural nests, may result in both lower hatchling mass and lower amount of lipid reserves than hatchlings from natural nests. The PIC index (hatchling lipids/egg lipids) was 53%, 60%, and 33% in snapping turtles from Michigan, New Jersey, and Nebraska, respectively.

Parental Investment

Because resources are allocated to eggs for embryo development and hatchling reserves, hatchling size is probably not the only major component of hatchling fitness (Congdon 1989; Congdon & Gibbons 1990). For example, hatchling snapping turtles of intermediate size can have greater yolk reserves, grow faster, and survive longer in captivity than do smaller or larger clutch mates (Bobyn & Brooks 1994b). Hatchling yolk reserves fuel all activities during emergence from nests, dispersal, and movement to water. However, because most snapping turtle nests are not far from water, most yolk reserves certainly remain to fuel hatchlings after they reach wetlands.

In Michigan, the proportion of lipids in bodies of hatching snapping turtles ranges from 7.5% (Finkler et al. 2002, the mean of wet and dry incubation treatments) to 19.1% (Congdon et al. 1983a, corrected to remove eggshell mass). Hatchlings from Nebraska were composed of 13.2% (Rowe

et al. 1995) and 17.8% (Congdon & Gibbons 1985) lipids in South Carolina. In New Jersey, hatchling percent lipids by dry weight were the highest reported (bodies and yolk sacs were 37.3% and 19.7%; Wilhoft 1986) and came from eggs with substantially higher proportion of lipids (33.8%; Wilhoft 1986) than eggs from other sites (Nebraska, 16.8 % [Rowe et al. 1995], and Michigan, 7.5–12.9% [Congdon et al. 1983b; Finkler et al. 2002]). However, even though snapping turtle hatchlings from New Jersey and Michigan have very different lipid levels in eggs, they have similar PIC ratios (hatchling lipids/egg lipids) of 52% (Wilhoft 1986) and 53% (Congdon et al. 1983b), respectively.

Phenotypes such as body size, yolk reserves, metabolic traits, and growth rates of snapping turtle hatchlings are primarily influenced by the amount of resources allocated to each egg (Hotaling 1990), and secondarily by selection of nest sites. Effects of differences in maternal investment among females are strong enough to remain detectable across a variety of experimental treatments (Hotaling 1990; Brooks et al. 1991a; Rhen & Lang 1995; Osteen 1998; Osteen & Janzen 1999; Steyermark & Spotila 2001a)

Clutch Size

The number of eggs produced at one time (clutch size) is a primary component of reproduction. Compared to other turtles, common snapping turtles produce relatively large clutches of small eggs (Congdon & Gibbons 1985). Across their geographic range, average clutch sizes range from 26 to 55 eggs (Table 12.1), but more than 100 eggs have been reported for clutches from Nebraska (Miller et al. 1989; Packard et al. 1990).

A positive relationship between clutch size and body size has often been reported within populations at widespread geographic locations (Ash 1951; Congdon et al. 1987; Hotaling 1990; Obbard 1983; Pell 1941; White & Murphy 1973; Wilhoft 1986; Yntema 1970c; Iverson et al. 1997). Among these studies, plus data from South Carolina, Michigan, and Ontario (Greene, Congdon & Brooks, unpublished data), the amount of clutch size variation explained by female body size varied from 23 to 77%. Body mass (log transformed) explains almost twice the variation in clutch size in South Carolina (Fig. 12.2d) compared to Michigan (Fig. 12.2e) or Ontario (Fig. 12.2f).

In contrast to clutch size, the relationship between egg size and body size within populations has been documented less frequently. Egg size did not increase with body size in a population in Pennsylvania (Steyermark & Spotila 2001a), but it did in Michigan (Congdon et al. 1987), New Jersey (Hotaling 1990), and Nebraska (Iverson et al. 1997). In the New Jersey population only large females increased egg mass substantially, and the relationship between body size and egg size apparently contributed to the significant positive relationship between hatchling size and female body size (Hotaling 1990). In the Nebraska population, the rate of increase in egg size with body size was greater in smaller- than in larger-bodied females (Iverson et al. 1997). Egg diameters increased with body size in well-studied populations in Ontario and Michigan (Fig. 12.3a, b). However, long-term data from the ESGR in Michigan indicate that among-population comparisons of relationships between body size and reproductive traits based on small sample sizes taken during one year should be made with caution. In a series of samples of the same females over 28 years, the relationships between clutch size and egg size with body size were not significant in eight (28.6%) and seven years (25%), respectively, and in five years (17.8%) neither relationship was significant (Congdon, unpublished data). Similar variations were evident in long-term data from Ontario (Brooks, unpublished data).

Geographic comparisons of relationships among reproductive traits and body size do not show clear trends. Mean clutch sizes were 25.6 (3–44) in Michigan, 29.2 (3–55) in South Carolina, and 36.1 (12–69) in Ontario, Canada (Greene, Congdon & Brooks, unpublished data) and the largest average clutches were from populations in Nebraska and North Dakota (Table 12.1). The positive relationship between reproductive traits and latitude reported by Iverson et al. (1997) confounds effects of longitude with latitude because two areas with the highest clutch sizes (South Dakota and Nebraska) are also the westernmost areas. When small sample sizes and data from the South Dakota and Nebraska pop-

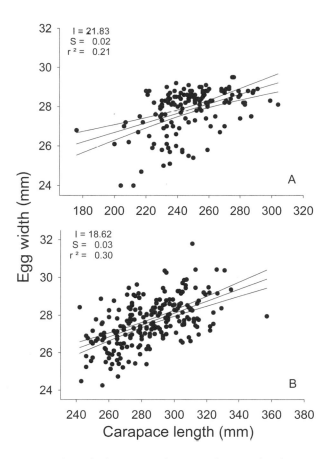

Fig. 12.3. Relationship between egg diameter and carapace length in snapping turtles from Michigan (A) and Ontario, Canada (B)

ulations were excluded there were no significant relationships between body size (Fig. 12.4a), clutch size (Fig. 12.4b), or egg diameter (Fig. 12.4c) and latitude.

Body Size and Age at First Reproduction

The minimum size of primiparous females (reproducing for the first time in their lives) from Michigan is substantially smaller than those from Quebec (200 mm CL; Mosimann & Bider 1960), Algonquin Park, Canada (Table 12.1), Tennessee (145 mm plastron length = approximately 200 mm CL; White & Murphy 1973), or South Carolina (Table 12.1). Minimum ages at maturity increased from 7 to 15 years with increasing latitude (Table 12.1). The among-population variation in size and ages of primiparous females suggests that the length of growing season is not the only factor influencing the characteristics of females at first reproduction.

Clutch Frequency

In many species of turtles, more variation in annual reproductive output is due to variation in clutch frequency than clutch size (Gibbons 1982; Congdon et al. 2003). The same may be true even in turtles that produce a maximum

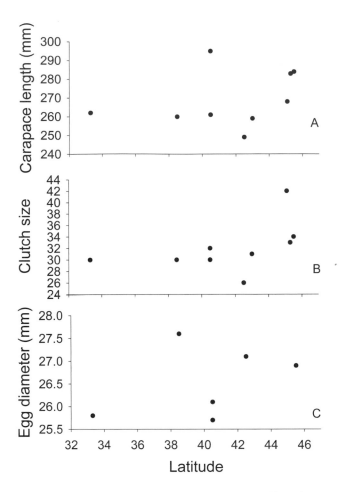

Fig. 12.4. Relationships between reproductive parameters and latitude among populations of snapping turtles

than one per season (but see Iverson et al. 1997). The actual proportion of female snapping turtles that skip reproduction is difficult to document, but at least 83% of the adult females in Ontario, Canada (Obbard 1983), and 60% on the ESGR in Michigan (Congdon et al. 1987) reproduce in a given year and the number of females that skip reproduction in some years is variable.

As data accumulate from intensely studied populations of several species, it appears that low frequency of reproduction (Bull & Shine 1979) is more common than previously expected (Gibbons & Greene 1978; Landers et al. 1980; Tinkle et al. 1981; Congdon et al. 1983a, 1987). Reports of females skipping annual reproduction come from various taxa, including species that produce more than one clutch annually, *Deirochelys reticularia* (Gibbons & Greene 1978), *Chrysemys picta* (Tinkle et al. 1981; Congdon et al. 2003), and those that produce a maximum of one clutch per year, *Gopherus polyphemus* (Landers et al. 1980), *Emydoidea blandingii* (Congdon et al. 1983a, 2001); *Chelydra serpentina* (Obbard 1983; Congdon et al. 1987).

Indeterminate Growth and Reproductive Traits

Increases in annual fecundity in individuals can be due to increases in clutch size or clutch frequency with body size or age. Increased lifetime reproductive output can also be the result of extending reproductive life span (with or without increases in annual fecundity). Because a positive relationship between clutch size and body size exists in turtles, increased fecundity with age can occur if larger body size of adult females is associated with increased age (indeterminate growth). However, even in species with indeterminate growth, the assumption that larger turtles in a population are older may not be correct (Carr 1952; Halliday & Verrell 1988; Congdon et al. 2001). Differences in growth rates of juveniles and differences in ages at maturity can cause a substantial amount of the variation in body sizes of adults in a population, and in some turtles such as *Emydoidea blandingii,* older adults may not grow for as long as 20 years (Pappas et al. 2000; Congdon et al. 2001).

Juvenile snapping turtles from a river near Otsego, Michigan, grew about 32 mm per year from ages 1 to 6 years (Gibbons 1968), and annual growth rates of juveniles between 1 and 10 years of age averaged ~19 mm on the ESGR in southeastern Michigan. In the ESGR population, growth rate of adults between ages 11 and 25 years was about 3.5 mm per year, and then slowed to 0.4 mm per year (ages 30–50 yr). Variation in juvenile growth rates and ages at maturity (11–17 years; Congdon et al. 1987) results in the average and range of primiparous snapping turtle females representing 88% and 59% of the average and range, respectively, of carapace lengths (CLs) of all adult females. Thus, the positive relationship between body size and age of snapping turtles is primarily a juvenile trait and secondarily a weak adult trait (i.e., reproductive benefits associated with body size in-

of one clutch annually (Congdon et al. 2001). Evidence for multiple clutches within a season can be gained by dissecting gravid females or from intensive field observations. The simultaneous presence of enlarged follicles or corpora lutea and oviductal eggs, or corpora lutea and ovulatory-sized follicles, indicates production of second clutches. However, dissection data cannot establish the frequency of second-clutch production even if all females in a population are sampled (i.e., a dissected female may have subsequently produced a second clutch). Intensive field observations coupled with radiotelemetry can provide evidence for the occurrence of reproductive frequency. However it remains difficult to reliably document the lack of reproduction of an individual without controlling ingress and egress of reproductive females from a wetland. Even under the best of conditions, clutch frequency within a population remains one of the most difficult reproductive traits to document.

Early suggestions that common snapping turtles might produce second clutches of eggs in a reproductive season (Ernst & Barbour 1972; Pritchard 1979) have not been substantiated, and intensive studies in North America (Hammer 1969; Obbard 1983; Congdon et al. 1987; Iverson et al. 1997) indicate that clutch frequency in many populations is less creases due to adult growth are small and would be discounted by mortality rates). As a result, substantial increases in reproductive output derived through the reduced rate of indeterminate growth of an adult will take years to manifest themselves. However, two snapping turtles marked by Henry Wilbur on the ESGR (1968–1973) were alive after 2000 and were a minimum of 53 and 55 years old. Although increases in reproductive output associated with increased body size obtained through indeterminate growth should correctly be viewed as a mechanism that enhances natural selection for longevity (a mechanism for increasing the proportion of late versus early births of individual females), it does not appear to be a major factor for snapping turtles.

COMPONENTS OF NESTING
Seasonal Phenology

Snapping turtles nest in spring or early summer and nesting seasons are earlier at lower latitudes (Table 12.2; Iverson et al. 1997). Over 25 years in Algonquin Park, the onset of nesting ranged from May 26 to June 18 and the end of nesting from June 14 to July 7, and over 30 years in southeastern Michigan, the onset of nesting ranged from May 16 to June 8 and the end of nesting from May 30 to July 9. Over eleven nesting seasons in New Jersey, the duration of nesting averaged 17 (13–23) days, and the onset and cessation of nesting varied by 16 and 19 days, respectively (Table 10 in Hotaling 1990). In South Carolina (1977–1999), the earliest nesting activity was observed on April 16, the latest on June 4.

Warmer springs result in more rapid completion of follicle development, ovulation, shelling of eggs, and earlier initiation of nesting activity (Congdon et al. 1987; Obbard & Brooks 1987; Hotaling 1990; Iverson et al. 1997). In general, shorter nesting seasons occur when rainfall is frequent and sufficient to keep soil moisture high and when relatively high nighttime temperatures allow females to nest earlier and later in the day. Longer seasons occur when extended cool weather occurs and during extremely dry periods when soil moisture is so low that construction of a nest cavity is difficult and many nesting attempts are aborted. Cessation

of nesting may be influenced by the number of summer days remaining with soil temperatures above the minimum required for embryo development and hatchling emergence (i.e., some minimum amount of time is required for hatchlings to disperse, reach wetlands, and locate overwintering sites).

Migrations

Movements made by gravid females can be divided into two categories. First, nesting activity may begin with a prenesting migration (females move from resident wetlands to wetlands adjacent to nesting areas) before or during nesting seasons and return to their resident wetlands after nesting. Prenesting migrations can be relatively short (<1km; Congdon et al. 1987) or relatively long (>11 km) and may have an overland component (0.5 km; Obbard & Brooks 1980). They often take more than one day (1–15 days, Obbard & Brooks 1980; 1–9 days, Congdon et al. 1987), and some females take similar routes in successive years (Obbard & Brooks 1980; Congdon et al. 1987).

Second, females make nesting migrations (movements from wetlands to nest sites and then return to wetlands) that are generally shorter than prenesting migrations and are almost always completed within one day (but see de Solla & Fernie 2004), usually within a few hours. Over 30 years of study at the E. S. George Reserve and Algonquin Park, no nesting migrations lasted longer than one day (Congdon and Brooks, unpublished data). Because snapping turtles have substantially higher water-loss rates (0.64 g/h) than many other aquatic turtles (Ernst 1968; Finkler 2001), they may become dehydrated and stressed by staying on land for long periods.

Levels of risk to reproductive females during migration presumably increase with the density and diversity of predators, duration and distance of overland travel, and the number, proximity to, and traffic volume on roads that females may cross (Fig. 12.1). Risks from humans include accidental and deliberate injuries and death associated with roads (Stoner, 1925; Gibbs & Shriver 2002; Steen & Gibbs 2004;

Table 12.2.

Geographic variation in seasonal nesting phenology of snapping turtles

Location	Earliest	Latest	Mean duration (days)	References
Ontario	May 26	June 18	19 (12–34)*	Loncke & Obbard 1977; Brooks, unpublished data
Quebec	June 16			Robinson & Bider 1988
South Dakota	June 5	June 30	25	Hammer 1969
Michigan	May 16	June 28	19 (8–32)	Congdon et al. 1987; Congdon, unpublished data
Nebraska	June 1			Iverson et al. 1997
New York	May 28			Petokas & Alexander 1980
New Jersey	May 17	June 21	17 (13–23)	Hotaling 1990
Virginia	May 17			Mitchell 1994
South Carolina	April 16	June 4		Gibbons et al., unpublished data

* Minimum and maximum durations are presented in parentheses.

Table 12.3
Daily nesting activity of snapping turtles

| Location | Begin/end | | % morning | Reference |
	Morning	Evening		
Algonquin	5:00–9:00 A.M.	8:00–11:30 P.M.	35	Brooks, unpublished data
Michigan	5:00–11:00 A.M.	8:00–11:30 P.M.	65	Congdon et al. 1987
New York			44	Petokas & Alexander 1980
South Dakota	5:00–9:00 A.M.	5:00–9:00 P.M.	94	Hammer 1969
Nebraska	5:30–6:30 A.M.	9:30–10:30 P.M.	36	Iverson et al. 1997
New Jersey		3:00–10:00 P.M.	63	Hotaling 1990

Gibbs & Steen 2005; Aresco 2005), and the taking of females for food during nesting migrations (even in states that have laws against it). Although natural risks to large-bodied female snapping turtles appear to be low during nesting activity, females have been captured that had had their hind limbs and tails mutilated (Saumure 2001). The hind limbs and tails of two female snapping turtles on the ESGR were severely injured (apparently by coyotes) during the past 10 years; one turtle was never seen again, and the other was found dead some years later (Congdon, unpublished data).

Because snapping turtles are long-lived, have relatively low fecundity, and high nest mortality, even small or moderate chronic increases in mortality rates of adult females can result in severe population declines (Brooks et al. 1991b; Congdon et al. 1994).

Daily Phenology

In almost all geographic regions, nesting activity of snapping turtles is bimodal with morning and evening peaks (Table 12.3), but Cahn (1937) suggested that almost all females nested in the morning in Illinois. Of 613 snapping turtles observed nesting on the ESGR in southeastern Michigan 64.9% were completed in the morning (5:30–11:59 A.M.), 15.3% in the afternoon (12:00–5:59 P.M.), 15.6% in the evening (6:00–9:59 P.M.), and 4.1% at night (10:00 P.M.–5:30 A.M.). Nesting during the afternoon was almost always associated with rainfall. The actual number of nests constructed at night on the ESGR is underrepresented because completion times were not recorded after 1:30 A.M. on all days of the nesting season. In contrast, most snapping turtles in Algonquin Park in Ontario, Canada, nested in the evening (Table 12.3).

Nest Site Selection and Nesting

Snapping turtle nests are located in open canopy areas with minimal vegetation and sandy soils. Natural nest sites include open areas caused by tree falls, muskrat mounds, beaver lodges and dams, beaches, sandbars, and islands. Disturbed nest sites include agricultural areas, flowerbeds, lawns, sawdust piles, old fields, gravel pits, railroad grades, road-

sides, driveways, and firelanes. Although not common, some females on the ESGR successfully constructed nests in active ant mounds. Whether hatchlings could have emerged from the nests is not known, because predators subsequently destroyed all of them (Burke et al. 1993).

In Michigan, when females are in an area suitable for nesting they often wander erratically while dragging their chins on the substrate, whereas in Algonquin Park they pause, arch their necks, and push their snouts and chins into the ground. Some females stop and immediately begin to dig a nest hole, whereas others sample a number of areas by repeatedly stopping and scraping dirt with their front feet, before beginning to excavate a nest hole. The following historical account is a colorful (and inspirational for those conducting field studies of nesting) description of Professor Jenks' observations of a snapping turtle's nesting activity (initially presented in Babcock 1919 and quoted in Cahn 1937).

Leaving my horse unhitched, as if he, too, understood, I slipped eagerly into my covert for a look at the pond. As I did so, a large pickerel ploughed a furrow out through the spatterdocks, and in his wake rose the head of an enormous turtle. Swinging slowly around, the creature headed straight for the shore, and without a pause scrambled out on to the sand. She was about the size of a big scoop-shovel; but that was not what excited me, so much as her manner, and the gait at which she moved; for there was method in it and fixed purpose. On she came, shuffling over the sand toward the higher open fields, with a hurried, determined seesaw that was taking her somewhere in particular, and that was bound to get her there on time. I held my breath. Had she been a dinosaurian making Mesozoic footprints, I could not have been more fearful. For footprints on the Mesozoic mud, or in the sands of time, were as nothing to me when compared with fresh turtles eggs on the sands of this pond. But over the strip of sand, without a stop, she paddled and up a narrow cow-path on all fours, just like another turtle, I paddled, and into the high, wet grass along the fence. I kept well within sound of her, for she moved recklessly, leaving a trail of flattened grass a foot and a half wide. I wanted to stand up—and I don't believe I could have turned her back with a rail—but I was afraid if she saw me that she might return indefinitely to the pond; so on I went, flat on the ground, squeezing through the lower rails of the fence, as if the field beyond were a melon-patch. It was nothing of the kind, only a wild, uncomfortable pasture, full of dewberry vines, and very discouraging. They were excessively wet vines and briary. I pulled my coat-sleeves as far over my fists as I could get them, and with the tin pail of sand swinging from between my teeth to avoid noise, I stumped fiercely, but silently on after the turtle. She was laying her course, I thought straight down the length of this dreadful pasture, when, not far from the fence, she suddenly hove to, warped herself short about, and came back, barely clearing me, at a clip that was thrilling. I warped about, too, and in her wake bore down across the corner of the pasture, across the powdery public road, and on to a fence along a field of young corn. I was somewhat wet by this time, but not so wet as I had been before wallowing through the deep, dry dust of the road. Hurrying up behind a large tree by the fence, I peered down the corn-rows and saw the turtle stop and begin to paw about in the loose soft soil. She was going to lay. I held on to the tree and watched, as she tried this place, and that place, and the other place—the eternally feminine. But the place, evidently, was hard to find. What could a female turtle do with a whole field of possible nests to choose from? Then at last she found it, and whirling about she backed quickly at it, and, tail first, began to bury herself before my staring eyes.

During the nesting migration described above, the female traveled approximately 47 m and made three nesting attempts before reaching the area where she finally nested. Because movements were circuitous, the straight-line distance from nest to water was probably less than 30 m.

Distances of snapping turtle nests from nearest water averaged 27.4 m (1–89 m) (Petokas and Alexander 1980), 37.1 m (1–183 m) (Congdon et al. 1987), 25 m (maximum, 100 m) (Iverson et al. 1997), and from 1 to more than 60 m at sites in Wisconsin, Minnesota, and Michigan (Ewert 1976). Of 465 nests on the ESGR, 92% were within 70 m of water (Fig. 12.5a). Average distances from water for nests that survived were similar to those destroyed by predators (Congdon et al. 1987). More than 90% of nests in Algonquin Park are placed less than 10 m from water, presumably because forest canopy restricts open sites to areas close to shorelines (Brooks, unpublished data).

Nest Construction

Female snapping turtles maintain the position of their bodies in relation to the nest hole by fixing the position of their front feet during the duration of nest construction. A nest chamber is excavated by alternate use of the hind feet. Nest chambers are flask shaped with tops and bottoms approximately 100–110 mm and 120–180 mm from the soil surface, respectively (Congdon et al. 1987). In Nebraska females are large-bodied and the tops and bottoms of nest chambers average 174 mm and 257 mm below the soil surface, respectively (Iverson et al. 1997). The depth of nest ceilings and floors are correlated (Hotaling 1990). Eggs are po-

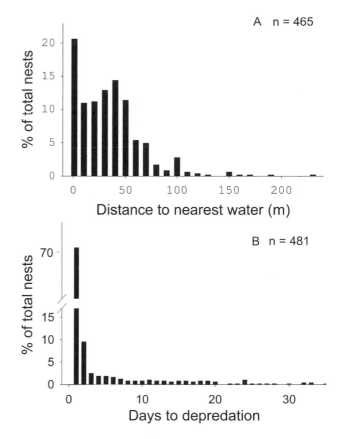

Fig. 12.5. (A) Distance of nests from water. (B) Days to predation of snapping turtle nests on the E. S. George Reserve in Michigan.

sitioned within the nest with the hind feet, and all nest covering is done with alternating use of the hind feet.

After digging was initiated, construction (digging, egg deposition, and covering) of 39 nests on the ESGR averaged 130.1 (70–250) minutes. Four (10.3%) nests took more than 2 hours to complete (Congdon, unpublished data). Females that had nest excavation impeded by roots or rocks abandoned the site after extended attempts to dislodge the obstructions.

Compared to the well-hidden nests of many other turtle species, completed nests of snapping turtles are visually very distinctive. After the female leaves the nest, two mounds of dirt with an imprint of the tail in the center remain, forming an "arrow" that points to the eggs in the nest chamber (Fig. 12.6). It has been suggested that the two mounds of dirt are beyond the reach of the female's hind legs (Iverson et al. 1997). However, it appears to us that because the female placed the dirt there with her hind feet, the mounds are well within reach. It may be that because the large number of eggs in a clutch requires a large nest cavity, there is just too much dirt to attempt to disperse and camouflage it. Alternatively, because snapping turtles have a strong musky odor that may be left at the nest site, visual camouflage may not be as effective at deterring predators as it is in other species.

Nesting areas and nest sites are warmer at lower latitudes

Fig. 12.6. (*left*) A female snapping turtle nesting on the Edwin S. George Reserve in southeastern Michigan. Photo by O. M. Kinney. (*right*) A female snapping turtle depositing eggs in a nest at Algonquin Park in Ontario. Photo by R. J. Brooks.

(Congdon & Gibbons 1990 fig. 8.2) and remain warmer for a longer period during the summer. At higher latitudes, only nests exposed to at least 3 hours of sunlight per day are warm enough for sufficient embryo development to allow hatchlings to emerge in the fall. Although rare, some nests on the ESGR were located under closed canopies, none of which produced hatchlings in fall, and all embryos were killed over winter. At high latitudes, survival of snapping turtle hatchlings overwintering in nests appears to be very rare (Obbard & Brooks 1981b; Congdon et al. 1987). At lower latitudes nests can be placed in areas that are exposed to fewer hours of direct sunlight and the probability of hatchlings successfully overwintering in nests may be higher than at high latitudes.

Nest Predators

Most embryo mortality of freshwater turtles results from destruction of nests by predators (Legler 1954; Hammer 1969; Plummer 1976; Tinkle et al. 1981; Congdon et al. 1983a, 1987, 2000; Iverson 1997; Burke et al. 1998; Spencer & Thompson 2003). In North America, raccoons destroy the majority of snapping turtle nests, with foxes, skunks, mink, coyotes, and burrowing mammals destroying a smaller portion (Hammer 1969; Christens & Bider 1987; Congdon et al. 1987). In three instances, crows destroyed uncovered nests on the ESGR (Congdon, unpublished observations). In the southern areas, nest predation by introduced fire ants has been documented for some species of turtles (Buhlmann & Coffman 2001), and they may prevent any snapping turtle hatchlings from overwintering in nests (Conners 1998).

Nests are at highest risk within the first 24 hours after construction (Congdon et al. 1987, Robinson & Bider 1988; Ernst et al. 1994) (Fig. 12.5b) and some eggs are taken while females are still ovipositing. On the ESGR, detection and destruction of both snapping turtle and Blanding's turtle nests after 6 days is usually associated with rainfall (Congdon et al. 1983a, 1987, 2000).

Nest predation rates varied from 46 to 63% in South Dakota (Hammer 1969), and from 30 to 100% (mean = 70%) in southeastern Michigan (Congdon et al. 1987). Nests destroyed by foxes were farther from water and older than those destroyed by raccoons (Congdon et al. 1987); perhaps raccoons more frequently patrol nesting areas closer to water.

Studies examining the relationship between nest density and predation rates have yielded mixed results. In a one-year study at one location, areas with high-density nests had lower predation rates than did sites with lower nest densities (Robinson & Bider 1988). In another study, higher predation rates on experimental nests occurred in areas with higher density (Marchand & Litvaitis 2004), and a multiyear study found no relationship between nest density and predation rates (Burke et al. 1998). Mixed results suggest that the relationship between nest densities and predation rates is complex, and at minimum related to species of turtles, site, predator assemblage, and whether predators attack just nests, or females and nests (see Spencer & Thompson 2003). We agree with Burke et al. (1998) that, as a single factor, density is not necessarily a good predictor of the probability of predation.

Studies of predation rates on experimental nest arrays show promise of identifying some patterns of nest risk and female behaviors (Wilhoft et al. 1979; Spencer, 2001; Spencer & Thompson 2003; Marchand & Litvaitis 2004), but risks to natural nests and experimental nests may be substantially different, because many predators trail females to natural nests.

Other Causes of Embryo Mortality

At high latitudes, nests that escape predators can fail if they receive insufficient insolation because of closed canopies. In addition, some nests fail because of flooding, erosion, and intrusion of roots into the nest cavity. Accidental destruction of nests by humans results from agricultural practices, disking firelanes, and grading dirt roads. In northern portions of the snapping turtle's range (Ontario, Canada), low nest temperatures result in embryo developmental rates that are not sufficient to allow hatchling emergence in fall except in the warmest summers. As a result almost all embryos die during winter (Obbard & Brooks 1981b; Brooks, unpublished data).

Incubation and Hatchling Emergence

Below 20°C, *Chelydra serpentina* embryos failed to develop (Brooks et al. 1991a). The thermal maximum above which turtle embryos cease development and die is about 33°C (Yntema 1960; Ewert 1979). Short periods above and below these thermal limits are not lethal. At high latitudes, nest temperatures are sometimes below the lower extremes early in nesting seasons; but, because snapping turtle nests are deeper than those of many other freshwater turtles, it is unlikely that embryos are exposed to temperatures above the upper lethal limits even at lower latitudes.

Successful incubation and emergence of hatchling snapping turtles takes from about 75 to 100 days (Obbard & Brooks 1981b; Congdon et al. 1987). On the ESGR the time from nest construction to hatchling emergence for 42 nests averaged 89.9 days with emergence occurring from late August to early October. Emergence from 61.9% of surviving nests was synchronous (occurred on one day) and 31.9% were asynchronous (occurred over 2 or 3 days). Synchronous and asynchronous nests produce 22.3 and 25.3 hatchlings per nest, respectively. Embryo survivorship of all nests averaged 22% (Congdon et al. 1987).

Hatchlings apparently remain in the nest until all or almost all individuals have internalized their yolk sacs and then dig their way out. The route to the surface is usually through soils that were previously softened by the female when she excavated the nest. Almost all hatchling emergence from snapping turtle nests takes place in the fall following egg laying, but there are persistent reports of a few snapping turtle hatchlings overwintering in the nest or on land (Toner 1940; Bleakney 1957; Ernst 1966; Obbard & Brooks 1981b; Congdon et al. 1987).

Incubation Conditions and Hatchling Characteristics

Variation in the developmental dynamics of embryos can contribute to variation in hatchling characteristics (Rhen & Lang 2004), and the variation in turn may influence survivorship during neonatal dispersal (Kolbe & Janzen 2001) and beyond. Snapping turtle eggs are porous and exchange water with the surrounding soil (Packard & Packard 1988b, 1993). In addition to variation in temperature of natural nests, variation in the amount of available moisture results in differences in incubation times and hatchling phenotypes. For example, compared with hatchlings from dry nests, em-

bryos from wetter nests have higher rates of growth and metabolism (Morris et al 1983; Gettinger et al 1984; Miller & Packard 1992); as a result hatchlings are larger and have fewer yolk reserves (Brooks et al. 1991a; Packard & Packard 1993; Bobyn & Brooks 1994b). In addition, hatchlings from wetter nests have higher water content and are able to lose more water before locomotor performance is affected (Finkler 1999). However, temperature is clearly the most important factor impacting developmental rates, hatchling phenotypes, and survivorship of snapping turtles (Brooks et al. 1991a).

Dispersal from Nests

After emerging from the nest, a hatchling snapping turtle must solve problems related to dispersing from the nest site, orientation, and reaching water. How hatchlings find water remains poorly understood at best, but visual and olfactory cues and geotaxis have been suggested as mechanisms (Noble & Breslau 1938; Ernst & Hamilton 1969; Ehrenfeld 1979). Most evidence from marine and freshwater turtles indicates that visual cues are the most important for hatchling orientation and dispersal from the nest (Ehrenfeld 1979). We do know that under a variety of conditions with respect to closeness of water and exposure to dark or light horizons, the orientation and dispersal of snapping turtle hatchlings is not random and they tend to move toward open, more intensely illuminated areas (Noble & Breslau 1938; Pappas, Congdon & Brecke, unpublished data).

As hatchlings disperse from nests to wetland habitats hatchling snapping turtles are exposed to risks from predators (Janzen et al. 2000b), high temperatures, and desiccation (Finkler 2001). However, whether risk of death during migration is high in general is an open question. For example, survivorships of juvenile snapping turtles were similar for those released as hatchlings at natural nest sites or directly into marshes (Congdon et al. 1999). Levels of risk may be increased in proportion to the duration of time in the terrestrial habitat. As expected, distance from wetlands is related to the time some hatchlings spend on land; however, body size was not correlated with time spent moving from nest to a fence near a wetland (Congdon et al. 1999). In another study, hatchling survivorship increased when nest locations and nest-site areas had less ground cover, and the time required for dispersal from nest to water increased as dispersal distances increased, when nests were farther from water and on flatter terrain (Kolbe & Janzen 2001).

The widespread assumption that larger neonates perform and survive better than smaller individuals (Williams 1966; Smith & Fretwell 1974; Brockelman 1975) was formalized as the "Bigger is Better" hypothesis (Packard & Packard 1988a). However, studies of relationships between survival and body size of neonate turtles during dispersal and movement to water (Janzen 1993a; Congdon et al. 1999; Janzen et al. 2000a, b; Tucker 2000; Kolbe & Janzen 2001; Valenzuela 2001b) have produced conflicting support for the hypothesis. Results of four tests of the "Bigger is Better" hypothesis have found that bigger was better (Janzen 1993a) and not better (Congdon et al. 1999; Janzen et al. 2000b; Kolbe and Janzen 2001). Perhaps some of the conflicting results are due to differences among methodologies and vegetation at release sites as suggested by the results from Kolbe and Janzen (2001). Certainly acceptance of the concept of "Bigger is Better" without qualification is not justified (Gutzke & Packard 1987; Packard & Packard 1988a; Congdon et al. 1999).

Evidence of Problem Solving: Allocation to Reproduction

The relationship between clutch size and egg size has been discussed as an evolutionary solution for maximizing lifetime fitness (Smith & Fretwell 1974; Brockelman 1975). For example, female snapping turtles in the northeast produce fewer and larger eggs than do those from Nebraska. Examining variation in reproductive traits and wetland characteristics (past or present) from various regions might shed light on whether level of parental investment is a response to differences in wetland productivity, suites of hatchling predators, or other selective pressures.

Although wetlands are relatively resource-rich environments, snapping turtle females on the ESGR skip reproduction (Congdon et al. 1987) frequently enough to have substantial impact on lifetime reproductive output (Congdon, unpublished data). If skipping reproduction represents a response to reduced resources, it is probably related to minimizing the impact on survival of females with a potentially long future reproductive life. It would be interesting to know the physiological-resource state of females that forgo reproduction, and what level of costs they would bear if they reproduced.

Evidence of Problem Solving: Nesting

Whereas the behaviors related to nest construction appear to be stereotyped and conservative, other activities associated with nesting are plastic and broad patterns of variation may provide evidence of problem solving. One consideration is how females balance ecological factors (characteristics of nesting areas and nest sites), risks (probability of and susceptibility to injury and death from predators or thermal stress), and reproductive success of the current nesting attempt.

Variation in prenesting migrations or in selection of nesting areas may represent responses of females to environmental cues. In the broad sense, prenesting migrations are solutions to being able to reside in the best aquatic habitats where adequate nesting areas are not available in adjacent terrestrial areas. However, even though most females that reside in the southwest wetlands of the ESGR have always nested in adjacent terrestrial areas, a few often travel about 1 km to the southeastern part of the ESGR, nest, and then return. That both patterns of activity occur in the same nesting season suggests that in some cases prenesting migrations are more of an intrinsic than an environmental response.

Is reproductive success of females related to whether females nest early or late in a season, or in the morning or evening? Variation in such traits could be random or represent responses to environmental conditions. Although hatchling phenotypes and survival can be influenced by variation in the seasonal and daily phenology of nesting, the conditions relating hatchling traits to female reproductive success are probably highly variable, difficult for a female to predict, and very difficult to document.

Although the claws and strong hind legs of females allow them to dig in hard-packed soils such as the tracks of dirt roads, it is sometimes difficult or impossible for hatchlings to dig their way out of such nests. Female snapping turtles select open areas with well-drained soils that support minimal vegetation. They exhibit "chin rubbing" or "sand sniffing" behavior and will often dig at many areas before settling on a nest site. Although "sand sniffing" has not been reported to occur in males or in females outside the nesting season, it is not known whether or how the behavior provides information on the suitability of nest sites. It may be that some of the prenesting activities help stimulate hormone balances related to initiating nest construction.

Nesting activity is usually greatest during and after warm rains when females appear to take advantage of water-softened soils to more easily and rapidly construct nests. Another benefit may be that signs of nesting and female scent are washed away by rain. On warm, humid nights, some females atypically begin to nest after 11:00 A.M. and finish as late as 2:30 A.M. the next morning (Congdon, unpublished data).

In combination, the observed within- and among-individual variations in allocation to reproduction and behaviors related to nesting appear to be sufficient to allow snapping turtle populations to respond to some climatic and environmental changes. Such variation can be interpreted as "problem solving" and it probably is. However, clear documentation that it is adaptive remains extremely difficult.

13

Water Relations of Snapping Turtle Eggs

RALPH A. ACKERMAN

TODD A. RIMKUS

DAVID B. LOTT

FEMALE SNAPPING TURTLES (*Chelydra serpentina*) produce clutches containing numerous eggs that are typically buried in soil where they incubate unattended by the female for several months. Like other environmentally incubated reptile eggs snapping turtle eggs contain the embryo and the resources (produced by the female) necessary to support embryonic growth and development. These resources include the yolk and albumen and also the eggshell and eggshell membranes. The embryo and its resources are contained by the eggshell and associated membranes (Packard & Seymour 1997) which form a boundary separating the egg contents from the external environment. The egg contents are coupled to the environment by any exchanges (principally heat, water, oxygen, and carbon dioxide) that occur across the eggshell and membranes. Eggshell characteristics vary among turtle (and reptile) species (Packard & DeMarco 1991) but as yet no unequivocal explanation for this variation is available. We can speculate that eggshell variation may be related in some way to environmental exchange but as yet few data support this speculation.

Snapping turtles produce 10–50+ spherical eggs with a hard but expansible eggshell (Packard & DeMarco 1991). The initial mass of the eggs can range from less than 10 g to 25 g and they are deposited in a nest chamber excavated in soil to a depth of about 5–20 cm. After deposition, the nest chamber and eggs are covered with the excavated medium and the eggs are left to be incubated by the soil environment for roughly 2–3 months. At the end of incubation, the eggs hatch and the hatchlings dig out of the nest and make their way to water.

Embryonic development is fueled by the metabolic conversion of yolk and albumen into embryonic tissues. The metabolic process is aerobic, consuming O_2 while producing CO_2, water, and heat as by-products. Exchange between the eggs and soil surrounding the clutch supplies O_2 and dissipates CO_2 and heat. After some period of development, the inner surface of the shell membrane is invested with the extra embryonic, chorioallantoic membrane whose capillary beds expose embryonic blood directly to gas exchange across the eggshell. The heat generated by metabolism is carried to the eggshell by conduction and by convection (the embryonic circulation) and then exchanged with the surrounding soil by conduction (Ackerman et al. 1985b). Water is incorporated into new tissue as it is formed, presumably hydrating the tissue to an appropriate osmotic pressure. However, neither the maintenance of tissue osmotic pressure nor, for that matter, fluid balance of embryonic tissue is well understood. Three potential sources of water for the growth are: (1) metabolic water, (2) water stored in the albumen and other egg contents, and (3) water acquired via exchange with the environment. Unfortunately, neither the relative nor the quantitative importance of any of these water sources is understood for C. serpentina (or any reptile embryo).

Water exchange with the environment can occur because the eggshell and membranes must be open for the exchange of O_2 and CO_2 between the embryo and its environment. Water exchange depends on the relative concentration of water in the egg and in the soil surrounding the clutch. The influence of environmental water exchange on embryonic development has been the subject of lively and extensive discussion (Packard et al. 1977, 1981, 1999, 2000; Packard & Packard 1984, 1988b, 1993; Ackerman et al. 1985a; Kam & Ackerman 1990; Ackerman 1991; Packard 1991, 1999). Environmental water exchange and its influence on embryonic development appear to vary among species. The embryos of snapping turtles apparently can tolerate either net water gain or loss (Morris et al. 1983; Packard & Packard 1988b; Packard 1999; Packard et al. 1999, 2000). However, it is clear that a number of developmental processes are influenced by egg water exchange (Miller et al. 1987; Packard & Packard 1988b; Packard et al. 1999, 2000; Janzen et al. 1990; Miller 1993) over some range of net egg water exchange. What is not especially clear, as yet, is the extent to which the range of egg water exchanges used in the laboratory represents the range of egg water exchanges found in natural nests.

The evolutionary significance of the exchange between the reptile egg and its environment and the role of various egg structures in the exchange have also been subjects of considerable interest (Packard et al. 1977; Packard & Seymour 1997). A strong argument has been presented that the eggshell and shell membranes and especially the chorioallantoic membranes are implicated in the evolutionary transition of vertebrate eggs from fully aquatic to fully terrestrial incubation (Packard & Seymour 1997). Skulan (2000) provides a contrary perspective on this issue.

We will focus in this chapter on the water exchange of C. serpentina eggs. Numerous studies of water exchange by snapping turtle eggs have been performed in laboratory environments (Packard et al. 1980, 1981, 1984, 1988, 1999, 2000; Packard & Packard 1984, 1988a, b,1993; Packard 1991, 1999) and a few have examined the effects of water exchange in the field (Packard et al. 1985, 1993a, 1999). C. serpentina eggs may be, with the possible exception of Chrysemys picta eggs, the most intensively studied eggs of any reptile species. It is probable that what we think that we understand about the role and importance of egg and embryonic water exchange for reptile eggs rests largely on the study of the eggs of C. serpentina.

We have organized our discussion into three topics. In the first section, we briefly examine the influence of egg water exchange on embryonic development. In the second section, we examine the process of egg water exchange and the factors that are most likely to influence this process, namely the water potential of the soil, the water potential of the egg, and the water vapor conductance of the eggshell. In the third section, we examine how soil and boundary layer climates influence the water content (and water potential) of the soil in which clutches of eggs are buried and thus the water exchange of eggs in the clutch. Among other things, we will suggest that C. serpentina eggs are incubated in environments that are characterized by relatively constant and predictable water potentials and that tend to be more humid than the environment used in the laboratory. This has important implications for our understanding of the role of egg water exchange to the developmental biology of C. serpentina.

WATER EXCHANGE AND DEVELOPMENT

C. serpentina inhabits a wide geographic range, including most of the central and eastern United States, ranging from southern Canada to Florida and from western Nebraska to the Atlantic coast (Zug 1993; Ernst, Chapter 1). Rainfall is frequent over this humid region and the variability from year to year is low relative to the rest of North America (Trewartha & Horn 1980). Egg water exchange and embryonic development tend to occur in moist and humid circumstances when compared with regions outside the range of C. serpentina. The influence of the hydric environment has been extensively documented in the laboratory using partially buried C. serpentina eggs (Miller et al. 1987; Packard et al. 1980, 1981, 1987, 1988; Janzen et al. 1990; Miller & Packard 1992; Packard & Packard 1988b, 1993; Packard 1999; Morris et al. 1983). Treatment effects for a wide range of hatchling and developmental variables have been repeatedly reported (Packard 1999). These include hatchling and yolk size and water content, duration of incubation, lipid and protein mobilization, egg metabolism, and hatching success. Although the treatment effects are unequivocal, the interpretation of these effects is not. There are several reasons for this. Principal among them is that the treatment effects are typically analyzed with analysis of variance (ANOVA) (and

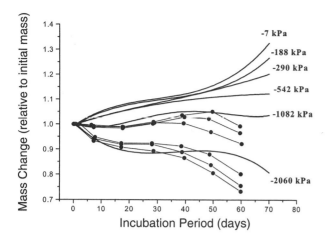

Fig. 13.1. Data from Rimkus et al. (2002) compared with data from Packard (1999) for eggs incubated partially buried in vermiculite. Both sets of data are normalized to initial egg mass. The data from Packard (1999) were scanned and digitized using an X-Y plotter and depicted as filled circles.

sometimes analysis of covariance [ANCOVA]). The difficulty with interpreting ANOVA results is well described (Lewontin 1974). The ANOVA model relates the variation produced by the experimental treatment to itself. It is thus tautological and yields no insight into cause. In this context, water potential (ψ) is used as a treatment class. It is a proxy, describing in some undefined way the relative moisture level of a given treatment. Thus, it is inferred that there is a treatment effect associated with treatment moisture level. There seems to be an implicit (and at times explicit) notion present in these experimental designs that it is egg water exchange that is causing the treatment effects. However, egg water exchange is typically not reported nor is its correlation with phenotypic variation assessed. This is a problem because the interpretation of ANOVA results is limited to the range of treatments used. If the range is not well defined or is addressed by proxy then it is very difficult to make sense of any response, especially in regard to what might be happening in the field. C. serpentina eggs partially buried in vermiculite appear (by qualitative visual estimation) to experience water exchanges in the range of −20% of initial egg mass to about +10% of initial egg mass (Fig. 13.1). This range does not encompass the full range of water exchanges that can occur in the laboratory. It is apparent that eggs can take up a good deal more water than +10% of initial egg mass (Packard et al. 1980; Kam & Ackerman 1990; Packard 1999), especially when fully buried in vermiculite (Kam & Ackerman 1990; Packard 1999). What is needed is a description of the functional response of embryonic phenotype to variation in egg water balance, but this information is only just now becoming available (Rimkus et al. 2002) and is far from extensive. The lack of functional information makes it difficult to compare and interpret experimental results from different laboratories. The extent to which clutches vary at a location

or among different locations and laboratory results can only be extrapolated to the field (Packard et al. 1993a, 1999) with difficulty. It would appear that, in large measure, the impact of environmental and genetic contributions to C. serpentina embryonic and hatchling phenotype as a function of egg water exchange remains to be described.

We have, however, examined some functional relationships for a few C. serpentina hatchling variables (Rimkus et al. 2002) and these include functional relationships (Ackerman & Lott 1999; Lott 1998) for eggs incubated in field nests (Fig. 13.2). These data indicate that there may be a much wider tolerance range for egg water exchange than implied by previous laboratory studies. It is especially interesting that many developmental variables appear to be independent of egg water exchange over some part of the range of egg water exchanges. Hatchling mass and size appear to be independent of egg water exchange for all water uptakes greater than +10% of initial mass. The effects of mass gains and below this threshold (and of water loss) are qualitatively consistent with earlier work (Packard 1999). We do not as yet have functional relationships for populations of C. serpentina and so the possibility of geographic variation cannot be assessed. Our limited results suggest that, if females can put their eggs into a suitable environment, developmental processes are independent of egg water exchange. How this environment might look, especially for C. serpentina eggs, is suggested later in this chapter. These functional relationships can also be interpreted to suggest that the egg and embryo are, in some

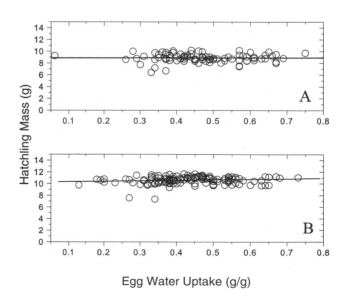

Fig. 13.2. Functional relationships for hatchling C. serpentina wet mass as a function of egg water balance at the end of incubation. The eggs were incubated in field nests (Lott 1998; Ackerman & Lott 1999). The average water potential around those nests is shown in Table 2. The slope of the linear regression line is not significantly different from 0. The same results were obtained for hatchling water content and incubation time.

way, regulating the internal osmotic environment, freeing the embryo from egg water uptake. Rimkus et al. (2002) found that eggs could take up half again their initial mass in water with no effect on the water content of the embryo. Clearly, the embryo must be isolated from this water exchange. This implies a regulatory mechanism that is at present unknown. Altogether, a great deal remains to be learned about the functional implications of egg water exchange for *C. serpentina* embryonic development.

WATER EXCHANGE BY EGGS

The snapping turtle egg is surrounded by a calcareous shell deposited on a shell membrane as the egg is produced (Packard & DeMarco 1991; Stewart 1997). Once deposited in the soil, O_2 and CO_2 exchange occur across the shell and membranes between the external environment and the chorioallantoic membranes underlying the shell. This process occurs most intensively later in development (incubation) when the metabolism (Gettinger et al. 1984) is most intense. However, water exchange occurs through the same structures from the earliest point of shell deposition. This exchange can occur as liquid or as vapor but must occur as liquid when the egg is assembled and invested with albumen in the oviduct prior to oviposition (Packard & DeMarco 1991). After oviposition, the egg is no longer surrounded by a liquid environment but by soil (a porous medium) comprising granular, solid particles, with a pore space between the particles. The pore space can contain a gas mixture (N_2, O_2, CO_2, and water vapor) and liquid water, which is held by surface forces in the pores. When the pore space is full of liquid water, the soil is considered to be saturated. The properties of soils are complex, nonlinear, and generally well described (Koorevaar et al. 1983; Campbell 1985; Brady 1990; Jury et al. 1991). Respiratory gases and water vapor move though gas-filled pores principally by diffusion, though there can be a convective component (Campbell 1985; Packard & Seymour 1997). The rate of movement is reduced exponentially as liquid water progressively displaces gas from the pores. When an egg or egg clutch is surrounded by a soil, the exchange between the gas-filled soil pores and the egg interior occurs across the eggshell and its associated shell membranes (Packard & Seymour 1997). It is essential that the eggshell and shell membranes have a significant gas-filled pore space in communication with the gas-filled soil pore space to ensure the O_2 exchange necessary for development. A fluid-filled shell precludes adequate O_2 exchange because of the very low diffusivity of O_2 in water (Deeming & Thompson 1991; Packard & Seymour 1997). If the eggshell were flooded with fluid, as it must be at oviposition (Deeming & Thompson 1991), it can be calculated that the partial pressure of O_2 (Po_2) inside the shell would fall to intolerable levels quite early in incubation (Rimkus 1996), using the measured O_2 consumption ($\dot{M}o_2$) of *C. serpentina* eggs (Get-

tinger et al. 1984). Thus, the eggshell and shell membranes must dry out to some extent after oviposition (Packard & Seymour 1997), opening a gas space in the shell for diffusion. That this occurs is indicated by the increasing area of the chalky white spot on many reptile eggs (Thompson 1985) and by measurements of shell drying (Rimkus 1996). The presence of a trapped and stable gas pore space in the shell and shell membranes means that the egg must also exchange water vapor with its environment through the shell. An intriguing problem remains: how can the shell and shell membranes, which are flooded with fluid at egg oviposition, dry out when the egg is deposited in an environment more humid than the egg and with a liquid water connection between the egg interior and the water in the soil? Seymour & Piiper (1988) address this issue for avian eggs where no liquid water connection across the shell is possible, but it is not clear how their hypothesis can be applied to buried reptile eggs where such a connection has been shown to exist (Thompson 1987; Rimkus 1996). The problem is that most reptile eggs are incubated in soils that are more humid (<–800 kPa) than they are themselves (>–800 kPa). This means that the eggshell can not dry by losing water to the more humid surrounding soil. The alternative is that water is withdrawn from the shell inward into some internal compartment. This can only result in shell drying if water is withdrawn from the shell at a rate exceeding the capacity of the surrounding soil to replace it. The observed hydraulic conductivities of the soil are so large (Ackerman 1991) that this is not likely to occur. Nonetheless, eggshells dry out! How they do so remains a fascinating question, the answer to which may provide great insight into the function and evolution of the reptilian eggshell.

Egg water exchange can be modeled most simply by analyzing the mass of materials (Ackerman et al. 1985b; Packard & Packard 1988a) entering and leaving the egg across its boundaries, the eggshell:

$$D_{Mass}/dt = \text{(mass of water in - mass of water out)}/\text{time.} \tag{1}$$

At a respiratory quotient (RQ) of 0.7, which is likely for an egg metabolizing lipids, the mass of O_2 entering the egg balances the mass of CO_2 leaving the egg. Any change in mass, therefore, is due to water entering or leaving the egg. Metabolic transformation of water does not enter into this balance because mass does not change (no system boundary is crossed). Weighing the eggs periodically provides data suitable for such a balance if the RQ is appropriate (about 0.7). Even when it isn't, the error is relatively small. When water vapor transport is the principal mode of exchange, the mass balance can be described by:

$$dM/dt = K_a \cdot A(P_a - P_e) \tag{2}$$

where K_a ($mg \cdot cm^{-2} \cdot kPa^{-1}$) is the eggshell permeability, A (cm^2) is eggshell area, and P_a (kPa) and P_e (kPa) are the water vapor pressures of the interior of the egg and the environment, respectively. The product, $K_a \cdot A$, is commonly called the eggshell water vapor conductance ($G_{H2O} = K_a \cdot A$). This model has been used extensively to describe the gas exchange across the avian eggshell (Paganelli 1991) and assumes, as its chief hypothesis, that water is transported as vapor by diffusing through the gas spaces in the porous eggshell and shell membranes. G_{H2O} appears to be fixed for most avian eggs (Rahn & Ar 1974; Rahn et al. 1979), but this cannot be true for flexible shelled eggs such as those of *C. serpentina* that change size (and presumably shell area and thickness) as they exchange water. In this case, we might expect to find both K_a and A to be functions of egg mass and therefore time.

It also has been argued that reptile eggs exchange liquid water with their environments (Tracy et al. 1978; Thompson 1987). Using the simplest model, liquid water flows through the shell in proportion to the water potential difference between the liquid water inside the egg and the liquid water in the environment:

$$dM/dt = K_l \cdot A(\psi_a - \psi_e), \qquad (3)$$

where K_l ($mg \cdot cm^{-2} \cdot kPa^{-1}$) is the eggshell hydraulic (liquid) permeability, A (cm^2) is eggshell area, and ψ_a (kPa) and ψ_e (kPa) are the water potentials of the interior of the egg and the environmental water, respectively. It should be noted that partial pressure (P) and water potential (ψ) have the same units (Pa, energy per unit mass) but very different values. One pascal of liquid pressure is equivalent to $2.3 \cdot 10^{-5}$ Pa of vapor pressure, principally due to their difference in density (Ackerman 1991). Among other things, this means that vapor permeability is 5 orders of magnitude smaller than liquid permeability at the same driving pressure; this means, among other things, that if water transport can be limited to vapor transport, ingress and egress are reduced by a factor of about 100,000. The possibility of liquid water transport across the eggshell is problematic for several reasons, both empirical and theoretical. No one has provided, as yet, a model of liquid water exchange that predicts the observed patterns of egg water exchange for any reptile egg. In particular, it does not appear possible that liquid water transport models can account for the thermal effects (Kam & Ackerman 1990) that have been hypothesized (Packard et al. 1980; Ackerman et al. 1985b; Kam & Ackerman 1990) to influence egg water exchange. So far, the measured values for K_l are too large (by orders of magnitude) to account for egg water exchange. Kutchai & Steen (1971) report that the K_l for chicken eggshell membranes is about 14,600 $g \cdot m^{-2} \cdot kPa^{-1} \cdot day^{-1}$, which is likely to be a reasonable upper limit for an uncalcified membrane. Ar and Ackerman (unpublished) measured a K_l for the 5-g eggs of the lizard *Agama stellio* and 9-g eggs of the snake *Natrix tessellata* to be about 1,390 and 2,511 $g \cdot m^{-2} \cdot kPa^{-1} \cdot day^{-1}$. Robinson (unpublished data) has measured a K_l of about 2,800 $g \cdot m^{-2} \cdot kPa^{-1} \cdot day^{-1}$ for *C. serpentina* eggshell by using the same techniques. These values are all within the range that might be expected for vertebrate membranes (Dick 1966). Clearly, for reptile eggs, in general, and snapping turtle eggs,

in particular, any egg with a K_l of this magnitude would experience water exchanges many orders of magnitude larger than ever observed. One difficulty is understanding how vapor transport and liquid transport can operate simultaneously through the same pore space, if only because liquid in pores form menisci occluding the pore. There could be a range of pore sizes with the smallest pores occluded and the largest pores open, but this would require a pressure difference across the shell to open the pores in the first instance and then to keep the pores open. No such pressure difference has been reported.

A predictive model has been constructed using water vapor exchange (Ackerman et al. 1985b). The G_{H2O} of *C. serpentina* eggs (Packard et al. 1979; Ackerman 1991) and other reptile eggs (Packard et al. 1979; Plummer & Snell 1988) have been measured. The values are many orders of magnitude smaller than K_l values and more in line with the observed water exchanges of reptile eggs. Packard et al. (1979) reported the first G_{H2O} for *C. serpentina* eggs by using a drying technique developed for avian eggs (Ar et al. 1974). Subsequently, Ackerman (1991) reported a larger value for *C. serpentina* G_{H2O}, also by using a drying technique but in a much more humid environment (Ackerman et al. 1985a).

Equations 2 and 3 both imply that egg water exchange should be a linear function of environmental water concentration. In some situations this appears to be the case for *C. serpentina* eggs (Packard et al. 1980; Kam & Ackerman 1990; Packard 1991) (Fig. 13.2) when they are fully buried. However, numerous descriptions of *C. serpentina* egg water exchange are available (Packard et al. 1977, 1980, 1981; Morris et al. 1983; Packard 1999) in which the patterns of egg water exchange are convexly curvilinear (Fig. 13.1). Packard et al. (1980) suggested that this shape was attributable to the impact of egg heat production and heat exchange on water transport. Ackerman et al. (1985b) were able to support this conjecture theoretically by adding a heat balance to the mass balance and demonstrated that the patterns of water exchange were similar to those reported for half-buried eggs incubated in the laboratory. To construct the appropriate heat and mass balances, they assumed that water vapor exchange was the principal mode of water exchange. The temperature effect occurs because the vapor pressure of the egg increases as egg temperature increases, in principle, because of the progressive increase in egg heat production. This is due in turn to the progressive increase in embryonic heat production. Heat production increases throughout incubation and peaks near the end of incubation (Gettinger et al. 1984). This pattern correlates well with the pattern of water exchange in thermally insulating media. Kam & Ackerman (1990) documented the effect of heat exchange for fully buried *C. serpentina* eggs and pointed out that the presence of such a thermal effect was inconsistent with the operation of the liquid water transport model (equation 3). These observations all point to the fact that the thermal properties of the laboratory incubation medium can influence the heat exchange and, therefore, the water exchange of reptile eggs. A larger question is, to what extent? In general, only two kinds of incubation medium have been used in laboratory studies, vermiculite (Packard et al. 1977, 1980, 1981; Morris et al. 1983) and sand (Packard et al. 1987; Kam & Ackerman 1990). Although different experimental designs were used (either partially [Packard et al. 1987] or fully [Kam & Ackerman 1990] buried eggs) both reported statistically significant but small differences between eggs incubated with vermiculite and sand. It is possible that the media used by Packard et al. (1987) were at different water potentials due to differences in their hydric properties. Sand dries out much more quickly than vermiculite and if water were replaced in the drying surface layer at the same frequency, then the sand would tend to have a lower water potential (Kam & Ackerman 1990). Subsequent analysis with the models of Ackerman et al. (1985b) and with experimental work using eggs *fully buried* in sand and vermiculite at similar water potentials (unpublished) indicates that, for fully buried *C. serpentina* eggs, the differences in water exchange due to heat exchange will tend to be small and limited to the last half to third of incubation for water potentials less negative than −100 kPa. The convex-curvilinear patterns (Packard et al. 1977, 1980, 1981; Morris et al. 1983; Packard 1999) disappear when the eggs are fully buried (Packard 1991; Rimkus et al. 2002).

Experimental design is always an important issue and ultimately depends on the intent of the investigator. It is clear that the hydric and thermal environments experienced by fully and partially buried eggs are different and that these differences should be carefully considered when considering experimental design. Fully buried eggs are exposed uniformly over their entire surface to the local hydric and thermal environment. If water replacement at the surface is carefully controlled, then the water and solute content and thermal conductivity of the medium around the egg can be controlled with some precision (Rimkus et al. 2002). Partially buried eggs are exposed, in part, to the medium and, in part, to the atmosphere over the medium. Assessing the extent of each part can be an issue. Each part of the egg will also experience a different thermal environment because of the different mixes of solid, liquid, and gas. This difference in composition also means that each part will experience a different hydric environment as well, with a fraction of the egg exposed only to vapor and the remaining fraction exposed to liquid and vapor. Moreover, the water vapor composition above the medium will not be the same as the water vapor composition in the medium because a loss of water takes place from the surface of the medium to the atmosphere over it. This necessitates a periodic replacement of the lost water (Packard et al. 1977, 1980, 1981; Morris et al. 1983; Packard 1999) and must mean that the atmosphere over the medium is dryer than the medium. It must also mean that the surface layer of the medium is, on average, dryer than the more deeply buried medium. The assumption that the medium and atmosphere over the medium are in vapor pressure equilibrium (Packard et al. 1980) cannot be true, but the magnitude of the error introduced by this assumption is unknown. The establishment and control of a water vapor pressure environment around incubating eggs requires careful consideration; fully burying eggs is one solution (see Kam & Ackerman 1990; Rimkus et al. 2002), but the vapor pressure must be controlled in some way. Because the differences in thermal and hydric properties of sand and vermiculite are likely to be larger than those found among all types of natural soils, thermal effects may be considerably reduced by using natural soil media (Al-Nakshabandi & Kohnke 1965). Rosenberg et al. (1983) point out that soil thermal conductivity may be independent of soil type when the soils are at the same water potential (they will have different water contents, however).

C. serpentina eggs in natural nests are in a more interesting and a more complicated environment than is found in the laboratory. In natural nests, the eggs are fully buried in a clutch at some distance from the soil surface. They may contact each other, soil, or clutch atmosphere (or all three) depending on position in the clutch. Quantifying contact is difficult. In some cases, each egg may be surrounded entirely by soil (Hotaling et al. 1985), as for example, when the soil loses cohesiveness (see Spangler & Handy 1982) due to drying and falls among the eggs. In any event, the atmosphere in the clutch is clearly not analogous to the gas phase over partially buried eggs in the laboratory. The pressure of water vapor, O_2, and CO_2 in the clutch atmosphere is established by exchange with the eggs and the soil surrounding the clutch and not with the general atmosphere (e.g., an incubator) as in the laboratory. Moreover, the volume of the

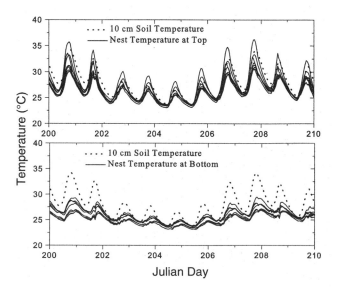

Fig. 13.3. Temperatures at the top (*upper*) and bottom (*lower*) of *C. serpentina* nests in the field. The dashed line represents temperature at a depth of 10 cm away from nests. Note that depth acts as a filter reducing the peak temperatures experienced at the top of the nest relative to the bottom of the nest but not influencing the minimum temperature. Thus the lower temperatures are cooler because the temperature peak has been reduced not because the minimum temperature has been lowered.

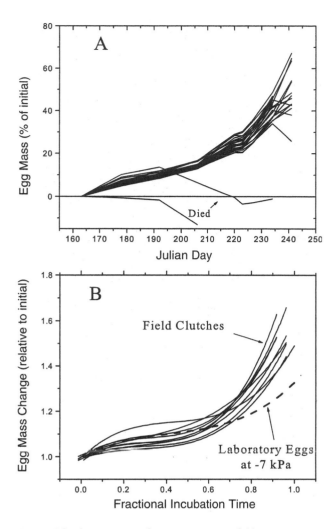

Fig. 13.4. The change in mass of *C. serpentina* eggs in field nests. (A) Eggs in a single nest weighed periodically over the course of incubation. (B) Summary of the polynomial models fitted to the data for each clutch over two seasons. The dashed line is the −7 kPa line from Fig. 13.1.

clutch atmosphere is likely to change as the eggs exchange water during incubation. Eggs are mostly water and therefore are good thermal conductors, and if they swell during incubation (by taking up water), the low thermal conductivity gas phase will shrink over the course of incubation and especially toward the end when egg and clutch heat production will be at its peak. Position of an egg within a clutch could influence egg water exchange but in a multiyear study of water exchange by *C. serpentina* eggs in field nests in Iowa, Lott (1998) found that egg position had no effect on egg water exchange. Hotaling et al. (1985) report small positional effects for *C. serpentina* in New Jersey nests. Ratterman & Ackerman (1989) report no position effects for *Chrysemys* eggs in field nests in Iowa. These observations suggest that, at least with respect to egg water exchange, water vapor is well mixed within the clutch. Nest temperatures inside clutches change diurnally (Packard et al. 1985, 1999) and with depth (Fig. 13.3; Lott 1998); the change in vapor pressure with temperature will induce vapor flow, reducing vapor pressure differences

within the clutch and between the clutch and the surrounding soil. Temperature-induced water vapor flow in soils is well documented (Hillel 1980a; Jury et al. 1991) (Fig. 13.4).

Relatively few observations have been made of the actual patterns of egg water exchange occurring in field nests. Ratterman & Ackerman (1989) provide data for *Chrysemys* eggs in field nests in northwest Iowa, while Lott (1998) and Ackerman & Lott (1999) examined *C. serpentina* egg water exchange in central Iowa. Several other studies have been designed to evaluate the effects of nest water vapor hatchling phenotype (Packard et al. 1993a, 1999) but neither egg water exchange nor soil water potential are typically described. Some egg water exchange data for eggs in field nests in Iowa (Lott 1998) are shown in Fig. 13.4. Field and laboratory (for eggs fully buried in sand [Rimkus et al. 2002]) data are compared in Fig. 13.4B. The patterns are similar but the field eggs appear to take up more water than laboratory eggs at about the same water potential. This may be due to the greater thermal conductivity of a clutch of eggs than of single eggs buried in sand. What is especially interesting here is that the egg water exchange shown in Fig. 13.4 is not strongly correlated with the measured soil water potential (see Table 13.2). Packard et al. (1985, 1993a, 1999) also reported only a weak influence of nest water vapor on hatchling *C. serpentina* phenotype from field nests. These observations may seem surprising at first but water vapor typically is neither measured nor used as a treatment class in the laboratory (Packard et al. 1981, 1980, 1977; Morris et al. 1983; Packard 1999) and thus there is no way of extrapolating to field nests. As we currently understand egg water exchange, variation in egg water exchange can be attributed either to the egg or to the environment. In the case of the egg, the observed variation can be due to changes in either eggshell G_{H2O} or egg vapor pressure. While there is reason to expect that both sources of variation are important, it is difficult to estimate the magnitude of either contribution at this time. Fluctuations in soil water potential are another source of variation. If, as is suggested below, soil water potential varies around some mean value over the period of incubation, then the influence of change in eggshell G_{H2O} and egg vapor pressure may be emphasized and the contribution of soil water potential deemphasized. Another issue, also discussed below, is that measuring nest vapor pressure is technically challenging and may introduce errors of various kinds (Rimkus et al. 2002).

The relationship among the various vapor pressures, conductances, and egg water exchange in natural nests has not been described explicitly. In particular, the relationship of nest water vapor pressure to egg water exchange has not been defined. The relationship can be described in a simple way, by assuming that egg water exchange is described by equation 2 and that the exchanges of water vapor between nest gas and soil and between nest gas and eggs are at steady state. We also assume that the transport of water (liquid plus vapor) in the soil does not limit exchange between the soil and the nest gas (Ackerman 1991). The steady-state exchanges between the egg and nest gas and between nest gas and soil are represented by equations 4 and 5.

$$\dot{M}_{H2O} = k_v \cdot A_v \cdot (P_E - P_N) \tag{4}$$

$$\dot{M}_{H2O} = k_n \cdot A_S \cdot (P_N - P_S) \tag{5}$$

Where $\dot{M}_{H2O}$ is the water exchange of the egg, P_E is the vapor pressure of the egg, P_N is the vapor pressure of the nest gas, and P_S is the soil vapor pressure. A_v and A_S are the areas across which exchange is occurring. k_v and k_n are the permeability of the egg and nest gas, respectively. Note that a permeability for soil is not needed if the soil water transport is not limiting. If the system is at steady state, the left-hand sides of equation 4 are the same and the right-hand sides are equal:

$$k_v \cdot A_v \cdot (P_E - P_N) = k_n \cdot A_S \cdot (P_N - P_S) \tag{6}$$

Solving for P_N:

$$P_N = (k_v \cdot A_v \cdot P_S + k_n \cdot A_S \cdot P_E) / (k_v \cdot A_v + k_n \cdot A_v) \tag{7}$$

and substituting P_N back into equation 4:

$$\dot{M}_{H2O} = (k_v \cdot k_n A_v \cdot A_S / k_v \cdot A_v + k_n \cdot A_S) \cdot (P_E - P_S) \tag{8}$$

If we assume that $A_v = A_S$, then:

$$\dot{M}_{H2O} = (k_v \cdot k_n A_v / k_v + k_n)(P_E - P_S) \tag{9}$$

P_N has disappeared. At steady state, egg water exchange depends only on P_E and P_S as well as a collection of permeability terms and area. Equation 9 can be simplified further by collecting these terms as a conductance $G'_{H2O} = [k_v \cdot k_n A_v / k_v + k_n]$)

$$\dot{M}_{H2O} = G'_{H2O} \cdot (P_E - P_S) \tag{10}$$

This is a familiar form and should make sense. Neither egg nor soil water vapor are influenced by nest water vapor, yet it is the difference between these two variables that determines the direction of water vapor flow and the character of the space between the two that determines the rate of water vapor flow. If an egg at some constant vapor pressure is surrounded by soil at some constant vapor pressure (i.e., both soil and egg have constant but different water potential) and a gas space separates the egg from the soil, then the vapor pressure in the space must be intermediate between the two. Moreover, this space acts as an additional conductance (or resistance) to vapor transport and is in series with the conductance of the eggshell. The only sources and sinks for the water vapor in the space between the egg and the soil are the water in the egg and the water in the soil; that is, egg vapor pressure, P_E, and soil vapor pressure, P_S, determine the nest vapor pressure, P_N, rather than vice versa. P_N is just a point somewhere in the vapor pressure gradient separating the soil from the shell. As such, knowledge of P_N alone yields relatively little information as to either the direction or rate of vapor flow. Among other things, the value of P_N will depend

on where it is measured. If, say, the egg is at -800 kPa and the soil surrounding it is at -30 kPa (see below), the driving force for water vapor exchange is 770 kPa. If an instrument such as a thermocouple psychrometer (Packard et al. 1985, 1992, 1999) is inserted between the egg and the soil, the instrument could read anywhere between about -800 and -30 depending on its position in the space separating the egg and the soil, that is, the space introduced by the psychrometer. One might read -700 kPa and decide that the environment was quite dry or -40 and decide that the environment was quite humid or anywhere in between and be completely misled about the vapor pressure gradient driving water exchange. Any estimate of nest vapor pressure as an isolated measurement is difficult to interpret. There are also technical complications. It is difficult to measure vapor pressure, especially in the field. Thermocouple psychrometry (Rawlins & Campbell 1986), which has been used to estimate P_N (Packard et al. 1985, 1992, 1999), introduces an error when vapor pressure gradients are present; that is, when there is a difference in vapor pressure across the measurement space and which may be the normal state. Because thermocouple psychrometry is an equilibrium technique (Rawlins & Campbell 1986), an error-free value is returned only when $P_E = P_S = P_N$ (i.e., equilibrium). It is hard to know when this occurs in a nest, if indeed it ever occurs. All other conditions (i.e., $P_E \neq P_S$) introduce an error of unknown magnitude. Other sources of error are present as well. The thermocouple psychrometer makes a temperature measurement from which first relative humidity, then ψ are calculated. If there is a temperature difference across or within the psychrometer another error is introduced. For example, if the psychrometer is deployed with an egg (producing heat) on one side and soil then temperature differences will be present. A temperature difference of $0.01\,^{\circ}C$ produces an error of 100 kPa (Rawlins & Campbell 1986), so if an egg is $0.01\,^{\circ}C$ warmer than the soil beneath it (perhaps because of metabolic heating or diurnal temperature fluctuation), it would warm the enclosure of the psychrometer (by conduction) and the device would report a relative humidity that was lower than actual. All of these issues, either separately or together, may contribute to an explanation for the lack of correlation between egg water exchange and nest vapor pressure.

THE CLIMATE FOR WATER EXCHANGE

Clearly, the water exchange of *C. serpentina* eggs ought to be influenced in some way by the state of water in the soil environment surrounding the egg clutch. Numerous studies of *C. serpentina* and other reptile eggs (Packard & Packard 1988a; Packard 1991, 1999) demonstrate that egg water exchange is strongly correlated with proxy measures of environmental hydric state in the laboratory. An implicit assumption in the design of laboratory studies is that water potential is constant over the course of incubation, though the constant value can range from −7 kPa to about −2,000 kPa (e.g., Rimkus et al. 2002). Studies in which eggs are partially buried in vermiculite (Packard 1999) typically use constant water potentials in the range −150 to −900 kPa, though one study (Packard et al. 1980) subjected eggs to a constant −50 kPa environment. The questions of how realistic any of these studies are and to what extent any of them simulate conditions likely to be encountered in the field are not only open, but they are questions that must ultimately be answered. Although there have been a few reports of the soil climate for egg water exchange (Ratterman & Ackerman 1989; Lott 1998; Ackerman & Lott 1999) and some broadly theoretical discussions (Ackerman 1991), the subject has received surprisingly little attention, especially given the extensive interest in reptile egg water. Soil scientists, in general, and soil physicists, in particular, have been deeply interested in soil water (Hillel 1971, 1980a, b; Koorevaar et al. 1983; Campbell 1985; Jury et al. 1991) and provide both insight and the means to generate testable hypotheses that can be of immediate use to researchers interested in the biology of reptilian eggs.

Water in the soil can be assessed by using a field water balance (Jury et al. 1991):

$$P + I - R = ET + D + \Delta W \qquad (11)$$

where the left-hand side is water coming into the soil column (precipitation [P], irrigation water [I], and surface runoff [R]) and the right-hand-side describes water leaving the column (evapotranspiration (ET), drainage (D) downward out of the soil column) as well as water storage (ΔW) in the soil column. The last term, ΔW, represents the water content (θ) of the soil and is the variable that we need. Solving for ΔW and dropping I as irrelevant:

$$\Delta W = -ET - D + P - R. \qquad (12)$$

If ΔW can be estimated or measured in the field at nest depth, then we can estimate both the water potential and the vapor pressure in the soil (Hillel 1971; Koorevaar et al. 1983; Jury et al. 1991; Ackerman 1991) surrounding turtle nests. Fortunately, soil water content and water potential are both (relatively) easily measured in the field (Klute 1986) and can be assessed independently. Jury et al. (1991) point out that, with respect to the soil water content (ΔW): "as long as the input of water from rainfall or irrigation is reasonably regular (i.e. no prolonged drying cycles), the actual soil profile water content in the field will tend to fluctuate about an equilibrium or steady-state value." That is, soil water content (and therefore soil water potential) varies around some constant value when a sufficiently long time base is used. We can extend this hypothesis to the soils around C. serpentina nests and predict that the soil water content and water potential in that soil, when there is regular rain, are relatively constant over the period of incubation. This prediction is eminently testable. The constant water content hypothesis is important because it means that a female turtle, after ovipositing and burying her eggs, has some average expectation that the hydric environment around the eggs will be stable throughout incubation. If the G_{H2O} of the egg can be structured (by selection presumably) to match the expectation of soil water potential then egg water exchange is under selective and therefore evolutionary influence. When there is periodic rainfall, as there is over most of the range of C. serpentina judging from continental rainfall maps, the expected (stable) value is likely to be near the field capacity. Field capacity is the soil water content 2–3 days after a rainfall occurs and is taken to be equivalent to about −33 kPa (Hillel 1980a; Jury et al. 1991) and appears to be relatively independent of soil type. Field capacity is considered to be "quasistable" (Jury et al. 1991) because water loss from the surface (the E in ET of equation 12) and water loss to drainage (D) downward both diminish exponentially after field capacity is reached; that is, within a few days of a rainfall, water loss from the soil column or some section of it will fall to very low values and the remaining water content stabilize. Periodic rainfall at suitable intervals causes the soil water potential to fluctuate at about the field capacity. The average value, taken over 1–3 months (the length of an incubation period for example), will depend on rainfall frequency, but the more frequent the rain, the less the fluctuation at about field capacity. Other factors, such as quantity of rain, soil type, and site drainage and runoff characteristics will also be important. Fig. 13.5 illustrates drainage from soils surrounding a nest, in the absence of evaporation. For soils in which a water table (phreatic water) is present, the gravitational field above the free water inducing flow downward is equivalent to 10 kPa per meter of height above the free water source (the left-hand increasing line). At equilibrium, the gravitational pressure driving drainage is exactly balanced by the matric pressure holding water in the soil capillaries. Thus, if the soil around a nest is 2 meters above a water table, the equilibrium water potential that will be approached as the soil dries by drainage is −20 kPa. Equilibrium is approached (asymptotically) slowly, on the order of weeks to months. Eggs of aquatic turtles, in general, and C. serpentina, in particular, are likely to be in soils in which a water table is present to within a meter or two of the surface. This condition commonly occurs in Iowa and apparently in the Sandhills of Nebraska (Winter 1983, 1986, 1989) and means that soil water potentials in the soil column will tend slowly toward water potentials on the order of −10 to −30 kPa due to drainage alone. Drainage is the dominant process deeper in the soil column, but turtle nests are closer to the surface where evaporation (E) must also be considered. However, evaporation at the surface (Fig. 13.5) is self-limiting because, as the surface layer dries and becomes thicker, the length of the diffusion path for water vapor becomes longer. In essence, the evaporative surface (the drying front in Fig. 13.5) becomes deeper and deeper but it does so at an exponentially decreasing rate such that, after several days (3–10), evaporation has fallen to a very low rate, approaching 0 asymptotically. This depends on such things as surface temperature and soil type. Sand, for example, retains water better than other soils

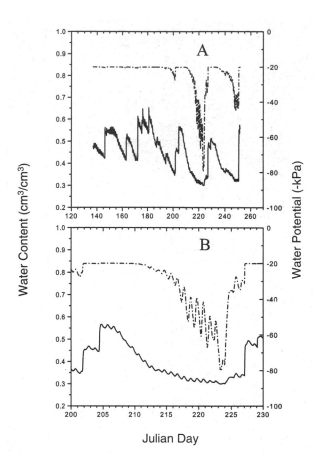

Fig. 13.5. Time-domain reflectometry (TDR) data collected within 1 m of a *C. serpentina* field nest, representing the change in soil water content at nest depth over the course of an incubation period: (A) the full record; (B) the record expanded around a long period without rain. The dashed line represents the calculated soil water potential. The oscillation shown in the TDR data occurs on a daily basis and probably represents the movement of vapor in the soil column due to the daily temperature wave moving through the column (see Fig. 13.2).

(Ackerman 1991) and the drying front would move more slowly downward than in a clay soil, but it appears, based on both theoretical (Hillel 1980; Jury et al. 1991) considerations and field work (Ratterman & Ackerman 1989; Lott 1998), that the drying front would take weeks (without rain) to reach turtle nests at the depth of *C. serpentina* nests. Transpiration is also a potential issue but can be avoided entirely if nests are excavated in open areas away from plant roots. As roots impede the digging of nest cavities, nesting turtles may avoid vegetated areas and thus transpiration may not be an important factor. Packard et al. (1985) report that *C. serpentina* nest in open areas away from vegetation.

Evapotranspiration is not likely to be an issue at all if rain falls at a sufficiently high frequency. Drainage will not approach equilibrium at such rainfall frequencies. Soil water potentials will, as a result, cycle around values between −5 and −60 kPa (i.e., at about field capacity). Table 13.1 summarizes the mean time between measurable rainfalls in central Iowa and several locations in the Nebraska Sandhills. Mean time between rains is on the order of 3–4 days for both

Table 13.1

Summary of rainfall data for central Iowa (Ames) and several locations in the Nebraska Sandhills

Location	Duration of record (yrs)	Mean time* (days)	Longest dry interval (days)
Ames, IA	109	3	33
Valentine, NE[†]	52	4	39
Valentine, NE[†]	52	4	41
Arthur, NE	52	4	52
Crescent Lake, NE	103	4	53

* Mean time is the mean time between measurable rainfall. The data are taken from long-term rainfall records collected by NOAH COOP Weather Stations at the locations and are publicly available.

[†] The two Valentine locations are for an airport and a wildlife refuge, respectively.

Iowa and Nebraska and especially for the Valentine region where some field work on *C. serpentina* nests has been carried out (Packard et al. 1985, 1993a, 1999). The record lengths range from about 50 years to more than 100 years. These rainfall frequencies are sufficiently high that, on average, soil water potential ought to fluctuate at about field capacity. There is simply insufficient time for soils around turtle nests in these regions to dry below field capacity to any significant extent. Table 13.1 also summarizes the longest interval in the record (June through August) without rain; periods of this length are extremely rare in any record but are probably not long enough to lower the soil water potential below −150 to −200 kPa in the absence of significant transpiration. Moreover, this would occur, at worst, over an entire incubation period and the average value over the interval would be higher. The change in soil water content at *C. serpentina* nest depth in central Iowa (Fig. 13.6) during such a long interval (30 days) was measured in the summer of 1998 by using time domain reflectometry (Noborio et al. 1999; Ren et al. 1999). The rate of change could be modeled as a decreasing exponential, consistent with Jury et al. (1991). We estimate that it would take more than 6 weeks without rainfall to reach a

Table 13.2

Summary of soil-water potential at nest depth as measured by tensiometry for *Chrysemys* (Ratterman & Ackerman 1989) and *C. serpentina* (Ackerman & Lott 1999; Lott 1998) over the course of the entire incubation season

Location	Genus	ψ (±sd)	Year	Rainfall* (inches)
Northwest Iowa	*Chrysemys*	−26.7 (22.0)	1985	−1.73
Northwest Iowa	*Chrysemys*	−31.0 (20.0)	1986	+1.42
Central Iowa	*Chelydra*	−29.1 (21.4)	1994	+0.79
Central Iowa	*Chelydra*	−31.8 (26.0)	1995	−1.34

* Rainfall is the departure in inches from the average rainfall for the location.

water potential of as low as −150 to −200 at this site. A duration of this magnitude has not occurred in the past 100 years in central Iowa and it likely to be an infrequent event.

Soil water potentials have been measured around a few turtle nests (Ratterman & Ackerman 1989; Ackerman & Lott 1999; Lott 1998) and are summarized in Table 13.2, together with the departure of the rainfall from average at the location. All the values averaged at about field capacity over the course of incubation, as is expected. These are, in fact, typical values for Iowa. The change in soil water potential near *C. serpentina* nests is shown in Fig. 13.7. The abrupt changes upward are due to rainfall events with slow drying in between. The data for *Chrysemys* are similar (Ratterman & Ackerman 1989). Soils do appear to behave as described by various soil physics texts (Hillel 1980; Jury et al. 1991): The balance of rainfall, surface evaporation, and internal drainage in Iowa is such that soil water content and soil water potential tend to a stable value at about −33 kPa when measured across several months. Moreover, it appears to require what

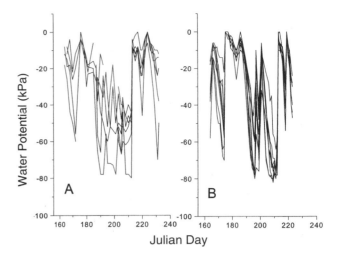

Fig. 13.7. Water potential measured at nest depth measured within 1 m of *C. serpentina* field nests (Lott 1998). The tensiometers were read several times a week and were calibrated to 1 kPa. The abrupt changes are associated with rainfall. Panel A represents six nests in 1994 and panel B represents nine nests in 1995.

seem to be rare conditions to produce soil water potentials lower (drier) than −100 kPa. The expectation that water potentials drier than −150 kPa (i.e., most water potentials used in laboratory studies) occur in the field appears to be low. Packard et al. (1985, 1993a, 1999) report what appear to be dry values for water potential in the Nebraska Sandhills, but these values are for *C. serpentina* nest vapor pressure and not the water potential of the surrounding soil. Interpretation is attendant on all the issues relative to the use of thermocouple psychrometers, which is described above and, in any event, yields little information about the water potential in the sand surrounding the nests or about egg water exchange. In any event, thermocouple psychrometry is not considered sufficiently sensitive to measure soil water potentials around field capacity (Bruce & Luxmoore 1986; Jury et al. 1991) and so is probably not technically suitable for assessing water potentials in the soil surrounding nests. Techniques to measure soil water potential at about field capacity are available (Cassell & Klute 1986), but they have the disadvantage of not being very useful for water potentials drier than about −80 to −100 kPa. In many situations, however, drier values may be very infrequent.

ACKNOWLEDGMENTS

We thank Robert Horton for patiently instructing us in the intricacies of soil physics and for sharing his lab so totally. All misapprehensions, misunderstandings, and oversimplifications remain ours.

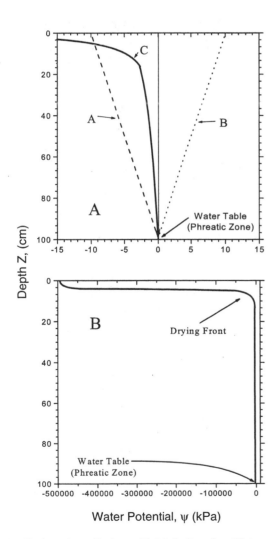

Fig. 13.6. Drainage in a soil column. (A) A is the line of equilibrium potential. B is the line of gravitational potential. C is the drying front imposed on drainage by surface evaporation. (B) The full range of the drying front.

14

Sex Determination in *Chelydra*

FREDRIC J. JANZEN

THE SEX OF AN INDIVIDUAL is a fundamental trait, determining (in animals) whether sperm or eggs are transmitted to form the next generation and thus the pattern of genetic contribution. At the same time, a panoply of behavioral, physiological, and morphological traits intrinsically linked to gonadal sex shape the specific phenotypes of individuals and hence the dynamics of populations. Furthermore, mechanisms of sex determination greatly influence the primary sex ratio and potentially the population sex ratio and effective population size, which are important ecological and evolutionary parameters. Indeed, Fisher (1930) demonstrated that under most circumstances 1:1 primary sex ratios are expected and sex-determining mechanisms that produce such balanced sex ratios should be favored by selection (see also Bull 1983).

For these fitness reasons (and simply based on fair meiotic segregation of chromosomes), one might expect sex determination governed by sex chromosomes to be ubiquitous in dioecious organisms. Remarkably enough however, a wide array of taxa exhibit diverse sex-determining mechanisms, including environmental sex determination wherein the sex of an individual is determined permanently by one or more factors after conception (Bull 1983; Korpelainen 1990). This unusual sex-determining mechanism occurs perhaps most commonly as temperature-dependent sex determination (TSD) in many vertebrates, mainly "reptiles" (Bull 1983; Janzen & Paukstis 1991a). The frequency of TSD raises the question of why so many organisms would leave the important trait of sex determination to the vagaries of the environment.

To resolve this ultimate enigma properly requires clear answers to numerous ancillary questions concerning the basic biology of TSD. Such questions include the following: (1) When and how many times has TSD evolved? (2) When does temperature act developmentally to influence sex? (3) What temperatures produce which sex? (4) How do thermal reaction norms of sex ratio (or responses to them) vary intra- and interspecifically? (5) How does TSD work in nature? (6) What are the underlying physiological and molecular mechanisms? (7) Does temperature influence traits other than gonadal sex in a sex-specific manner? The answers to these questions will also illuminate the conservation implications of TSD under global climate change and human habitat modification, especially given that many species with TSD are already imperiled (Gibbons et al. 2000). Many species have served as subjects for studies of TSD, but common snapping turtles (*Chelydra serpentina*) have played a prominent role in many ways. This chapter will review our current understanding of sex determination in *C. serpentina*, describe key published contributions made to a broader understanding of TSD by research on this species, and suggest promising future directions for investigating TSD using *C. serpentina*.

Temperature-dependent sex determination in vertebrates was first documented clearly in an agamid lizard in a relatively obscure French-language publication more than 35 years ago (Charnier 1966); not long thereafter, this work was extended to testudinid and emydid turtles (Pieau 1971 and numerous pioneering publications since). Common snapping turtles were the fourth vertebrate species for which TSD was described (Yntema 1976). TSD has even been documented in fish (e.g., Conover 1984). Indeed, TSD has now been identified, primarily by laboratory studies, in a diversity of vertebrates including a few fish, some lizards, many turtles, and all tuatara and crocodilians (Bull 1980, 1983; Conover 1984; Ewert & Nelson 1991; Janzen & Paukstis 1991a; Ewert et al. 1994; Lang & Andrews 1994; Viets et al. 1994; Cree et al. 1995). In general, embryonic exposure to specific thermal conditions during a limited period of development (the temperature-sensitive period, TSP) causes male or female differentiation. Few individuals with indeterminate (i.e., incompletely differentiated) or intersex (i.e., a testis and an ovary) gonads are born and there is no evidence that individuals with TSD change sex at a later ontogenetic stage (e.g., Bull 1987). Despite the overall similarity in timing of the TSP, the patterns of TSD vary among vertebrate species. For example, European pond turtles (*Emys orbicularis*) produce female offspring at high incubation temperatures and male offspring at low ones (Pieau 1971; pattern Ia); just the opposite pattern seems to occur in the lizard *Agama agama* (Charnier 1966; pattern Ib). Furthermore, many taxa including *C. serpentina* produce female offspring at high and low incubation temperatures and male offspring at intermediate ones (Yntema 1976; pattern II). Even so, species with the same pattern of TSD may differ dramatically in the pivotal temperature of sex determination (T_{piv} is a temperature at which a 1:1 sex ratio is obtained). To illustrate, the upper T_{piv} of the Amazonian freshwater turtle (*Podocnemis expansa*) in Colombia is 32.6°C (Valenzuela 2001a), whereas the upper T_{piv} of *C. serpentina* in North America is approximately 5°C lower (Ewert et al.

1994). In fact, *C. serpentina* eggs do not even hatch at 32.6°C (Yntema 1978)! The ecology, physiology, molecular biology, and genetics of TSD in vertebrates are more poorly known, so general discussion of these elements will be deferred to the appropriate sections of this chapter.

ORIGIN OF TSD

Given the wide distribution of TSD in vertebrates and the documented great ages of the various groups in which TSD occurs, this sex-determining mechanism is almost certainly very old. But just how old? And did it evolve only once? Since TSD apparently does not leave any trace in the fossil record, the second question is somewhat easier to answer. TSD occurs in at least one species of teleost fish, but is not known in any amphibian, mammal, bird, or snake. This phylogenetic discontinuity almost certainly indicates that TSD has evolved at least twice, once in fish and once in "reptiles." But has TSD evolved more than once in amniotes? This question is highly debatable, but phylogenetic comparative analyses most parsimoniously suggest a single ancestral origin in amniotes and several more recent origins in lizards (Janzen & Krenz 2004). Given the antiquity of the split of "reptile stock" from mammals, the initial origin of TSD in amniotes is ancient (Kumar & Hedges 1998).

All species within the Chelydridae (*Chelydra* and *Macroclemys*) have TSD (Bull 1980; Janzen & Paukstis 1991a). Modern phylogenetic analyses place this family as a close relative to kinosternid turtles (Krenz et al. 2005), with the origin of the clade containing chelydrids being about 85 million years ago (Near et al. 2005). Consequently, TSD has no doubt long accompanied this lineage on its remarkably successful colonization of a diversity of freshwater habitats across a considerable latitudinal gradient from Ecuador to Canada (Ernst et al. 1994). How TSD in snapping turtles has adapted to the concordantly dramatic change in thermal environments across this large geographic range is unknown (but see discussion in "Evidence for TSD" below).

EMBRYOLOGY OF TSD

Perhaps no one has contributed more to our understanding of sex determination in *Chelydra* than Chester Yntema, who discovered TSD in this species more than 30 years ago (Yntema 1976). Indeed, Yntema's pioneering work on the embryology of TSD in *Chelydra* laid an incredibly solid foundation for subsequent research on numerous aspects of TSD in other species as well. Yntema's contributions are particularly notable for identifying the embryonic stages during which gonadal sex is sensitive to temperature (i.e., the TSP) (Yntema 1979) and for clearly illustrating morphological and histological features of the gonads (Yntema 1981). Similar experiments have been conducted to estimate the TSP in other turtles, a lizard, and several crocodilians, but have been far less extensive (reviewed in Janzen & Paukstis

1991a). Even fewer studies have provided detailed information on gross morphology and ultrastructure of gonads and accessory structures. Our understanding of the biological implications of TSD in species without such basic information often rests on the extent to which Yntema's yeomanly embryological work on *Chelydra* can be extrapolated.

Analogous to the famed chicken embryonic series (Hamburger & Hamilton 1951), Yntema developed a valuable embryonic series of stages for *Chelydra* (Yntema 1968). This staging work provided a common language of embryonic development and thereby permitted comparative embryological work at different incubation temperatures. Using this framework, Yntema (1979) showed in a series of exhaustive (and no doubt exhausting!) temperature-shift experiments that the TSP for *Chelydra* occurred approximately between developmental stages 14 and 19 (i.e., from when the forelimb is a simple paddle with a vague digital plate to when digits significantly protrude beyond the edge of the digital plate). Incubation temperatures prior to and subsequent to these developmental stages did not influence sex determination. This TSP corresponds roughly to the second one-fourth of embryonic development (Yntema 1968), which lasts about 1–3 weeks depending on temperature (Yntema 1979), although most turtle researchers have instead referred to this period as the middle one-third (e.g., Janzen & Paukstis 1991a). The TSP was much shorter developmentally (about stages 14–16), but perhaps not temporally, for both sexes when eggs were shifted to an exceptionally cool temperature (20°C) (Yntema 1979). Overall, the developmental stages of the TSP for *Chelydra* are roughly comparable to those of other turtle species examined, although the latter often appear to have a slightly extended TSP to stage 21 or 22 (reviewed in Janzen & Paukstis 1991a).

Equally important has been Yntema's excellent description and plates of gonadal morphology and accessory structures (Yntema 1976) as well as gonadal ultrastructure via histological examination (Yntema 1981). This information has no doubt initially guided numerous researchers studying TSD in amniotes because of its high quality and clarity! Yntema (1976) showed that a neonatal ovary is elongate with follicles present in the cortex, providing a bumpy exterior surface, and is accompanied laterally by a white, threadlike oviduct. In contrast, the neonatal testis is shorter and less elongate, with a smooth surface and no (or only a vestigial) oviduct. Observations on the gonads of unpreserved *Chelydra* neonates indicate that ovaries can be further recognized by their whitish hue, in contrast to the yellowish color and what appear to be transverse capillaries evident in testes (pers. observ.). Yntema's (1981) detailed histological examinations completely confirmed assignment of sex based on assessment of gross gonadal morphology. Testes were covered by a thin squamous epithelial layer and were packed with seminiferous tubules, whereas ovaries were covered by a layer of cuboidal cells and were filled with primary folli-

cles (Yntema 1981). Although ovaries of hatchlings from 20°C and 30°C were similar in length and follicle density, the 20°C ovaries contained less developed germinal epithelium and a higher incidence of epithelial cysts than 30°C ovaries. This observation raises the question of whether individuals of the same sex produced at different incubation temperatures might have different fitnesses in the short term (e.g., Janzen 1995) or in the long term (e.g., Gutzke & Crews 1988).

EVIDENCE FOR TSD
Laboratory Incubation at Controlled Temperatures in Environmental Chambers

The first and foremost evidence for TSD and patterns of TSD derives from laboratory studies in which eggs are incubated at constant temperatures for most or all of embryonic development (Bull 1980; Ewert & Nelson 1991; Janzen & Paukstis 1991a; Ewert et al. 1994; Lang & Andrews 1994; Viets et al. 1994). Such data are particularly abundant for *Chelydra* and originate from several investigators and localities (Table 14.1) (Ewert et al. 2005). Plotting offspring sex ratio (% male) against incubation temperature for the combined data set reveals that *Chelydra* clearly has TSD; some temperatures produce 100% males and others produce 100% females (Fig. 14.1). These results are not caused by embryonic mortality (e.g., see analysis in Paukstis & Janzen 1990). The graphical analysis also reveals that *Chelydra* overall has pattern II TSD, with an average lower T_{piv} at ~21.4°C and an average upper T_{piv} at ~27.8°C. (These values were calculated using inverse prediction after logistic regression analysis of sex ratio data from Table 14.1 for 21 constant incubation temperatures from 20–24.5°C and 54 constant incubation temperatures from 23–31°C, respectively [SAS Institute 2000]). Temperatures below ~21.4°C and above ~27.8°C produce primarily female offspring; males are mostly produced at temperatures between ~21.4 and ~27.8°C.

Figure 14.1 further illustrates a robust fit of the data that might superficially suggest relatively little geographic variation in thermal sensitivity of sex determination in *Chelydra*. Indeed, the upper T_{piv} seems to exhibit some positive latitudinal trend in *Chelydra*, but the variation across ~30 degrees latitude is at most 2–3°C (Ewert et al. 1994). However, a comprehensive study reveals that such seemingly subtle geographic variation in TSD in *Chelydra* is, in fact, both statistically and biologically significant (Ewert et al. 2005). Populations of *Chelydra* arrayed across ~20 degrees latitude exhibit an inverse correlation between lower and upper T_{piv} values that accords with both local climate data and the nesting behavior observed in each population.

This relative invariance of the midrange of the male-producing range of incubation temperatures in *Chelydra* (Ewert et al. 2005) might cast doubt on the traditional utility of T_{piv} as a concept. Perhaps our efforts would, for example,

Table 14.1
Summary of published sex determination data from laboratory constant-temperature incubation studies of *Chelydra*

Males (N)	Females (N)	Intersexes (N)	% Male	Incubation temperature (°C)	Locality	Reference
			84	21.5	MN	Rhen & Lang 1998
32	2	0	94.1	22.5	MN	Ewert & Nelson 1991
134	0	0	100	24	MN	Rhen & Lang 1994
			100	24	MN	Rhen & Lang 1999a
130	1	0	99.2	26.5	MN	Rhen & Lang 1994
8	0	0	100	26.5	MN	Rhen & Lang 1996
			98	26.5	MN	Rhen & Lang 1999a
31	95	0	24.6	29	MN	Rhen & Lang 1994
10	40	0	20.0	29	MN	Rhen & Lang 1994
1	10	0	9.1	29	MN	Rhen & Lang 1996
			36	29	MN	Rhen & Lang 1999a
			24.6	29	MN	Rhen & Lang 1999b
5	22	0	18.5	22.0	no. ON	Brooks et al. 1991
9	0	0	100	25.6	no. ON	Brooks et al. 1991
3	6	0	33.3	28.6	no. ON	Brooks et al. 1991
52	197	0	20.9	20.9	ON	Bobyn & Brooks 1994b
16	85	0	15.8	21.3	ON	Bobyn & Brooks 1994a
51	11	0	82.3	24.8	ON	Bobyn & Brooks 1994a
150	4	0	97.4	25.1	ON	Bobyn & Brooks 1994b
1	42	0	2.3	29.1	ON	Bobyn & Brooks 1994a
0	58	0	0	20	NY/WI	Yntema 1976
0	149	0	0	20	NY/WI	Yntema 1981
19	2	0	90.5	22	NY/WI	Yntema 1976
18	0	0	100	24	NY/WI	Yntema 1976
108	0	0	100	26	NY/WI	Yntema 1976
79	2	0	97.5	26	NY/WI	Yntema 1976
373	3	0	99.2	26	NY/WI	Yntema 1981
17	9	0	63.0	28	NY/WI	Yntema 1976
0	87	0	0	30	NY/WI	Yntema 1976
0	142	0	0	30	NY/WI	Yntema 1981
54	0	0	100	23	NE	Paukstis & Janzen 1990
33	3	0	91.7	25	NE	Gutzke & Bull 1986
30	0	0	100	25	NE	Crews et al. 1989
102	1	0	99.0	26	NE	Packard et al. 1987
36	2	0	94.7	26	NE	Gutzke & Chymiy 1988
50	0	0	100	26	NE	Janzen et al. 1990
1	100	0	1.0	28.5	NE	Packard et al. 1987
0	63	0	0	29	NE	Packard et al. 1984
0	37	0	0	30	NE	Crews et al. 1989
0	33	0	0	31	NE	Gutzke & Bull 1986
0	50	0	0	31	NE	Packard et al. 1987
36	28	0	56.3	21.5	IL	O'Steen 1998
16	4	0	80.0	21.5	IL	O'Steen & Janzen 1999
9	12	0	42.9	21.5	IL	O'Steen & Janzen 1999
154	45	1	77.3	21.8	IL	Janzen et al. 1998
45	0	0	100	24.5	IL	O'Steen 1998
23	0	0	100	24.5	IL	O'Steen & Janzen 1999
36	0	0	100	26	IL	Janzen 1995
50	66	0	43.1	27.5	IL	Janzen 1992
47	47	0	50	27.5	IL	O'Steen 1998
20	24	0	45.5	27.5	IL	O'Steen & Janzen 1999
12	9	0	57.1	27.5	IL	O'Steen & Janzen 1999
91	104	2	46.7	27.6	IL	Janzen et al. 1998
46	178	1	20.7	28	IL	Janzen 1992
37	23	0	61.7	28	IL	Janzen 1995
9	108	0	7.7	28.5	IL	Janzen 1992
0	25	0	0	30	IL	Janzen 1995
0	36	0	0	30.5	IL	O'Steen 1998
0	15	0	0	30.5	IL	O'Steen & Janzen 1999
2	6	0	25.0	21.5	IN	Ewert & Nelson 1991
5	5	0	50.0	22.5	IN	Ewert & Nelson 1991

continued

Table 14.1 continued

Males (N)	Females (N)	Intersexes (N)	% Male	Incubation temperature (°C)	Locality	Reference
7	1	0	87.5	26	DC/VA/MD	Dimond 1983
3	10	0	23.1	28.5	DC/VA/MD	Dimond 1983
0	13	0	0	31	DC/VA/MD	Dimond 1983
0	3	0	0	21.5	TN	Ewert & Nelson 1991
11	2	0	84.6	25	MX	Vogt & Flores-Villela 1992
4	29	0	12.1	28	MX	Vogt & Flores-Villela 1992
0	3	0	0	29	MX	Vogt & Flores-Villela 1992
0	2	0	0	30	MX	Vogt & Flores-Villela 1992

Entries are arranged by population first by descending approximate latitude, then by ascending incubation temperature within each general locality, followed by ascending year of publication within a given incubation temperature.

be more profitably spent on characterizing the shapes of the entire reaction norms of temperature and sex ratio rather than just on analyzing these one or two single T_{piv} values within the reaction norms (sensu Girondot 1999). To illustrate, walking speed (think of this as T_{piv}) is identical between a Congolese and a French population of *Drosophila melanogaster,* but developmental temperature and population exhibit a significant interaction effect (i.e., the shape of the reaction norm for walking speed differs between the two populations; Gibert et al. 2001). Without identifying the interaction, one might logically conclude that walking speed was similar in the two populations. Instead, the interaction reveals that flies from France walked fastest if derived from an 18°C developmental temperature compared to 25 or 29°C, and flies from the Congo walked fastest if derived from a 25°C developmental temperature compared to 18 or 29°C (Gibert et al. 2001). These results suggest local adaptation to different natural thermal environments that would have been obscured by simply focusing on one point (i.e., the mean) of the reaction norm.

In contrast to the plethora of constant-temperature experiments, only two published laboratory studies involving *Chelydra* have even begun to address the effects of controlled diel fluctuations in temperature on sex determination. Fluctuation between 28 and 31°C (12 h at each temperature) produced only females among the offspring sexed from twelve clutches from Illinois (Kolbe & Janzen 2001). The more important contribution in this regard investigated the effects at different exposure times from 22 or 24°C to 30°C in *Chelydra* eggs from New Jersey (Wilhoft et al. 1983). Eggs that spent 4, 8, or 10 h/day at 30°C, and the remainder at 22°C, yielded 100% female offspring; eggs that spent 1, 2, or 3 h/day at 30°C, and the remainder at 24°C, yielded 9.0, 16.6, and 33.3% female, respectively. Thus, only 3–4 h/day incubation at 30°C (with the remainder at a male-producing temperature) tips the balance in favor of female-biased offspring sex ratios. The paucity of such experiments involving *Chelydra* is surprising because major diel fluctuations in temperature characterize the embryonic environment in natural nests (e.g., Packard et al. 1999). As described below, more experiments are critically needed to inform us on the sex-determining relationship between fluctuating temperatures that characterize natural nests and constant temperatures used in most laboratory studies.

Field Incubation at Fluctuating Temperatures in Natural Nests

One of the trickier issues regarding TSD since its discovery concerns its operation under natural conditions. This knowledge is crucial for interpreting some of the most important ecological, evolutionary, and conservation implications of TSD. Does TSD occur in nature or is it a laboratory artifact? How, if at all, do fluctuating temperatures lead to predictable offspring sex ratios? Is pattern II TSD ecologically relevant? Do changing climates alter offspring sex ratios?

The earliest work on TSD in natural nests involved *Emys orbicularis* (e.g., Pieau 1974), yet the first comprehensive research to show clearly that TSD works in nature involved *Chelydra* (Wilhoft et al. 1983; Bull & Vogt 1979; and see be-

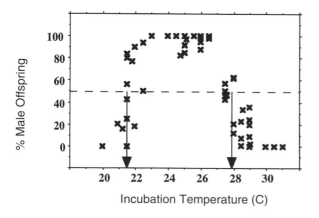

Fig. 14.1. Percent male offspring produced from laboratory constant-temperature incubation studies of *Chelydra*. The graphic of 69 data points (Table 14.1) illustrates that *Chelydra* has pattern II TSD, with a fairly narrow range of lower and upper temperatures tending to produce mixed sex ratios. An even offspring sex ratio (dashed line) is expected at all temperatures in species with genotypic sex determination. Arrows indicate the approximate lower and upper T_{piv}s for *Chelydra*.

low). Since then, a variety of field studies of TSD have been conducted, only a few of which have involved actual experimentation or have measured nest temperatures continuously (also see older literature reviewed in Janzen & Paukstis 1991a). These studies confirm the early field work: TSD occurs in natural nests of many species, as indicated by bimodal distributions of nest sex ratios, and roughly conforms to expected patterns based on laboratory studies.

Despite such results, the arithmetic mean of fluctuating temperatures in a given natural nest is not equivalent in its sex-determining effect to an identical constant incubation temperature used in a laboratory study. The thermal variance clearly matters too. Different statistical models have been successfully developed to equate fluctuating temperature regimes to constant temperature equivalents in the loggerhead sea turtle *Caretta* (Georges et al. 1994) and in *Podocnemis expansa* (Valenzuela 2001a; see also Bull 1985 for map turtles *Graptemys* and Schwarzkopf & Brooks 1985 for painted turtles *Chrysemys*), but these approaches have yet to be applied extensively to *Chelydra*. In fact, their overall utility for understanding the ecology of TSD remains to be shown, but the idea of generating sophisticated statistical/mathematical models holds great promise.

The few published field studies of TSD in *Chelydra* are especially illustrative of the great need for additional work on various issues in this area. Wilhoft et al. (1983) monitored offspring sex ratios and temperatures in eight "reconstructed" *Chelydra* nests in New Jersey. They also recorded offspring sex ratios, but not temperatures, in four natural nests. Sex ratios varied from 100% male to 100% female in the "reconstructed" nests; a similar result was noted for the natural nests. Moreover, in nests with mixed sex ratios, more females were produced in the upper (warmer) levels than in the lower (cooler) levels. These results highlight the potential impact on offspring sex ratios of temperature gradients in the deeper nests of larger species like *Chelydra*.

More recently, Kolbe & Janzen (2002) analyzed data provided in Packard et al. (1999) for 15 *Chelydra* nests in Nebraska for which offspring sex ratios and temperatures (8 weeks at 30-min intervals) at the bottom of nests were recorded. Sex ratios (% males) were negatively correlated with the reported mean temperatures ($r = -0.47$, $P = 0.07$). Similarly, Kolbe & Janzen (2002) detected a negative correlation between mean nest temperatures (middle one-third of development at 72-min intervals) and sex ratio of 10 randomly chosen offspring from each of 14 *Chelydra* nests in Illinois ($r = -0.73$, $P = 0.003$). This finding was mostly confirmed in a smaller-scale study in the same population in a subsequent year (St. Juliana et al. 2004).

These studies illustrate several key points about the ecology of TSD. Note, in particular, that no field study detected convincing evidence of female offspring at lower incubation temperatures, calling into question the ecological relevance of pattern II TSD in *Chelydra* (see Valenzuela [2001a] for a similar conclusion for *P. expansa*). Perhaps *Chelydra* effectively

has pattern Ia TSD under natural conditions (see also Ewert et al. [2005] for a similar conclusion). Second, compared with laboratory temperatures, mean nest temperatures do a poor job of explaining variation in sex ratios ($\leq 50\%$) (see also discussion above). This problem is confirmed by comparing the predicted upper T_{piv} values (± 1 SD) from these field studies (field T_{piv} IL = 25.46 + 0.30°C vs. field T_{piv} NE = 25.40 $\pm$ 0.08°C [excluding the one cold nest because it may, in fact, have involved the lower T_{piv}]) with those calculated for these two populations based on laboratory research (lab T_{piv} IL = 27.52 $\pm$ 0.03°C [14 data points from Janzen 1992, 1995; Janzen et al. 1998; O'Steen 1998; O'Steen & Janzen 1999] vs. lab T_{piv} NE = 27.11 $\pm$ 0.11°C [11 data points from Packard et al. 1984, 1987; Gutzke & Bull 1986; Gutzke & Chymiy 1988; Crews et al. 1989; Janzen et al. 1990; Paukstis & Janzen 1990]), using the maximum likelihood method of Girondot (1999). The field T_{piv} values are sharply lower than the lab T_{piv} values; in fact, incubating eggs in the laboratory at the field T_{piv} values would produce all, or nearly all, male offspring (Fig. 14.1)! This disconnect between mean nest temperatures and constant temperatures clearly indicates the importance of fluctuating temperatures to natural sex determination in species with TSD.

And what of the impact of among-year (or longer-term) climatic variation on cohort sex ratios in amniotes with TSD? Few studies have addressed this question for obvious logistical reasons. The two most extensive investigations have documented clear and strong links between annual variation in local thermal climate and cohort sex ratios in *Caretta* (Mrosovsky & Provancha 1992) and in *Chrysemys* (Janzen 1994a). Furthermore, in both cases, the relationship between climatic temperatures and cohort sex ratios was clearly in accord with the relationship between constant incubation temperatures and offspring sex ratios observed in laboratory studies of these two species. Far less is known in *Chelydra*. Schwarzkopf & Brooks (1985), citing unpublished work, report that "a female bias (sex ratio = 0.31) [was detected] in hatchling snapping turtles from natural nests in Algonquin Park [Ontario, Canada] every year for 3 years." Similarly, Kolbe & Janzen (2002a) noted a heavily female-biased offspring sex ratio (~8.6% male) in 1999 in Illinois, as did St. Juliana et al. (2004) for the same population in 2001 (~22.1% male). These field studies were conducted in warmer-than-average years at the Illinois site (1999 was tied for the third warmest July and 2001 was tied for the eleventh warmest July in Clinton, Iowa, between 1951 and 2004). Much work remains to be done on this important topic, but these studies indicate that among-year variation in climatic conditions may play a substantial role in influencing cohort sex ratios.

MECHANISM OF TSD
Physiology and Endocrinology

One of the hot, and more intriguing, questions involving sex-determining mechanisms concerns the physiological and molecular underpinnings of TSD. This area of research is receiving rapidly expanding attention. Still, we know little about the underlying mechanism(s) of TSD at this point. Much recent research on amniote TSD has focused on molecular bases and physiological pathways (see Lance 1997 and Pieau et al. 1999 for excellent reviews).

Snapping turtles have served as one of the primary models for early research in this area. Injecting or topically treating *Chelydra* eggs near the beginning of the TSP with sex steroids, agonists, antisera, or enzyme inhibitors has revealed important insights into the endocrinological aspects of TSD. One critical result illustrated good concordance between the TSP and the developmental period when differentiating gonads are sensitive to exogenous hormones (Gutzke & Chymiy 1988). This work set the stage for subsequent manipulative endocrinological experiments to ascribe physiological relevance to their results and suggested a causal role for sex steroids in gonadal differentiation. In another experiment, eggs treated with testosterone and then incubated at a female-producing temperature (31°C) yielded no male offspring; interestingly, testosterone-treated eggs incubated at a male-producing temperature (25°C) yielded a preponderance of females (Gutzke & Bull 1986). Furthermore, estradiol completely overrode temperature effects on sex determination at 25°C, leading only to female offspring. These results were largely confirmed in subsequent studies (Crews et al. 1989; Rhen & Lang 1994).

These latter two studies further defined the impact of estrogens on female differentiation in *Chelydra*. Treating eggs at a mostly female-producing temperature (29°C) with an aromatase inhibitor (aromatase converts testosterone to estradiol) had a masculinizing effect (Rhen & Lang 1994). Topical application of estradiol to eggs elevates gonadal aromatase activity at a primarily male-producing temperature (26.5°C), but no information is available to judge this effect at a female-producing temperature (Place et al. 2001). Moreover, Crews et al. (1989) showed that a synthetic estrogen agonist completely overrode temperature effects on sex determination at a male-producing temperature (25°C), concluding that "the classical high affinity, low capacity estrogen receptor" mediates this feminizing effect. Although this information identifies an important, perhaps causal, role for estrogens and aromatase in sex determination of amniotes with TSD, note that administering estradiol to eggs of "reptiles" with genotypic sex determination (GSD) also results in only female offspring (Bull et al. 1988b).

Crews et al. (1989) concluded that androgens are unlikely to be testis-inducing substances because they had negligible effects on sex determination at a female-producing temperature (30°C) (see also Rhen & Lang 1994). These results contrast with an intriguing, albeit weak, positive correlation observed between the concentration of testosterone in yolk at oviposition and the frequency of male offspring produced for 20 clutches of *Chelydra* eggs from Illinois (Janzen et al. 1998). The reason for this difference may reside in the fact that

Janzen et al. (1998) incubated eggs near the upper T_{piv} (27.6°C), which engenders greater sex-determining sensitivity than more extreme temperatures (e.g., Wibbels & Crews [1995] for sliders *Trachemys*). Indeed, lack of expression of gonadal aromatase is insufficient alone for testis development (Place et al. 2001), leaving open the possibility that some androgenic substances may be functionally involved.

As suggested above, recent research has detected considerable concentrations of testosterone and estradiol in egg yolks of turtles at oviposition (Janzen et al. 1998; Elf et al. 2002a) and these substances have been implicated in influencing offspring sex ratio in *Chrysemys* in the laboratory at the T_{piv} (Bowden et al. 2000). For *Chelydra*, yolk testosterone concentrations differed markedly among clutches and less so among the four eggs assayed for each clutch (see fig. 2 in Janzen et al. 1998; Elf et al. 2002a). Yolk estradiol levels for *Chelydra* were low and relatively uniform within and among clutches in the Illinois population (Janzen et al. 1998), but exhibited greater among-clutch variation in a Minnesota population (Elf et al. 2002a). The two turtle species examined by Janzen et al. (1998) with GSD had low, fairly constant levels of testosterone in egg yolks. Although yolk testosterone concentrations for *Chelydra* were positively correlated with the frequency of male offspring near the upper T_{piv} (27.6°C), this relationship did not approach significance in the vicinity of the lower T_{piv} (21.8°C). Moreover, the clutch sex ratio of hatchlings produced from eggs incubated at 28.4°C in this *same* population was unrelated to the concentration of testosterone (or estradiol) measured in the egg yolks (St. Juliana et al. 2004). Thus the biological importance of the positive testosterone result reported by Janzen et al. (1998) is questionable. Similarly, Elf et al. (2002b) noted that estradiol, but not testosterone levels were higher in yolks of embryos developing at warmer incubation temperatures than in yolks of embryos developing at cooler incubation temperatures. In apparent contrast to the lack of relationship between yolk estradiol and clutch sex ratio reported by St. Juliana et al. (2004), this result suggests that yolk estradiol might indeed play an important role in embryonic sexual differentiation in *Chelydra*, as it does in *Chrysemys* (Bowden et al. 2000). Our understanding obviously could benefit greatly from evaluating these endocrinological factors in additional populations, under controlled fluctuating temperatures in the laboratory, and under natural conditions. Future research on clarifying the importance of these maternal physiological effects on sex determination in species with TSD offers great promise.

Overall, these endocrine results have been cleverly co-opted to address other important issues regarding TSD in amniotes (see "Adaptive Significance" below). To be specific, the effects of temperature and sex on potentially fitness-related traits have been investigated by experimentally generating offspring of the opposite sex at otherwise "single sex"–producing temperatures (e.g., Rhen & Lang 1995). Further major advances in clarifying the physiology of TSD are likely to await advances on the molecular front.

Molecular Biology

The molecular basis of GSD in amniotes is increasingly evident (e.g., Ramkissoon & Goodfellow 1996), and our understanding will surely benefit greatly from recent advances in obtaining genomic DNA sequences. In contrast, our knowledge of the molecular basis of TSD currently constitutes a gaping black hole: little has been illuminated and an expanding number of workers are being attracted (inescapably?) to the issue. Indeed, one of the experimental benefits of examining the molecular biology of sex determination by using species with TSD is that individuals with *essentially identical genetic compositions* can be environmentally induced to become male or female. My coverage of this topic will necessarily be brief both because of the vast lacunae in our knowledge base and because of excellent recent reviews (e.g., Pieau et al. 1999; Place & Lance 2004).

Early evidence deriving from mechanistic studies of TSD in "reptiles" has implicated steroid hormones in sexual differentiation of gonads, thus much molecular research has focused on the enzymes responsible for steroidogenesis (reviewed in Lance 1997; Pieau et al. 1999). Of particular interest has been aromatase, which is critical for ovarian differentiation and maintenance. Elevated amounts (but not activity levels) have been detected in gonads of *Emys* embryos from eggs incubated at female-producing temperatures compared with those incubated at male-producing temperatures (Pieau et al. 1999). Accordingly, cloning and expression studies have implicated differential regulation of the aromatase gene in female and male gonads at the level of transcription (Pieau et al. 1999). Of course, other genes are also necessary in such a sexual differentiation cascade. Future research on the molecular genetics of TSD would do well to examine them simultaneously to obtain a complete picture, in particular, if they act upstream of aromatase (e.g., steroidogenic factor 1; Fleming & Crews 2001), and to adopt a rigorous comparative approach in assaying key taxa with and without GSD.

The only published research on the molecular genetic underpinnings of TSD involving *Chelydra* examined tissue of neonates from New Jersey eggs incubated at female-producing (30°C) and male-producing (26°C) temperatures for ZFY- and SRY-like genes (Spotila et al. 1994). The latter genes (but not the former: Bull et al. 1988c; Koopman et al. 1989) have been implicated in the sex-determining pathway for males in all mammals examined (e.g., Sinclair et al. 1990). The turtle ZFY homologue (called Zft) was highly similar to mammal ZFY (167 of 180 amino acids) and even more so to chicken and alligator homologues. Conversely, probes from SRY would not hybridize to turtle DNA under high stringency, suggesting weak similarities. Focusing just on the HMG-box region of SRY revealed multiple unique but related sequences with minimal similarity to mammalian SRY (57–70%). Overall then, an SRY-like gene is unlikely to be involved in TSD.

Despite minimal information at this point, it is difficult to imagine that the entire molecular underpinnings of TSD are fundamentally different from those of GSD, despite their strikingly varied manifestations and biological implications. This hypothesis of a largely conserved molecular genetic pathway underlying both GSD and TSD finds support in the frequency of evolutionary changes between the two categories of sex-determining mechanisms (Janzen & Paukstis 1991a, b) and in the phylogenetic homology of genetic elements in other fundamental traits (reviewed in Gerhart & Kirschner 1997), including in the developmental genetics of testicular differentiation (Smith et al. 1999). In the end, dissecting the molecular bases of TSD will clarify the simplicity or complexity of evolutionary transitions between TSD and GSD and between different types of TSD and could also very well illuminate a general framework for the molecular biology of environmentally regulated systems.

EVOLUTION OF TSD
Quantitative Genetics

Even with knowledge of the physiological and molecular mechanisms underlying TSD, we would still be left with an incomplete understanding of the evolutionary potential of TSD. This is the realm of microevolution. The first quantitative genetic model of TSD (Bulmer & Bull 1982) described its microevolution as being governed by selection on, and heritable variation for, thermal sensitivity of offspring sex determination (e.g., T_{piv}) and maternal choice of thermal qualities of nest sites. Although most empirical effort to date has focused on evaluating the microevolutionary potential of T_{piv} (see below), this model emphasized that nest-site selection might be the more important component (Bulmer & Bull 1982; but see Morjan 2003).

Evaluations of T_{piv} have focused either on (1) geographic variation (interpreted in terms of expected clines in environmental temperatures) or (2) within-population variation (usually interpreted in quantitative genetic terms). In the first case, any geographic variation would embody the collection of evolutionary forces acting on (drift, selection, etc.) or underlying (heritability, genetic covariances, etc.) the pattern (i.e., reaction norm) of TSD. Early studies detected little geographic variation in T_{piv} for *Caretta*, *Chrysemys*, *Trachemys*, and *Graptemys* (e.g., Bull et al. 1982b; Limpus et al. 1985; Mrosovsky 1988) and what little pattern existed tended to contradict perhaps naïve predictions from adaptationist theory. More recently, Ewert et al. (1994) summarized information on geographic variation in T_{piv} in various turtles with TSD. For eight populations of *Chelydra*, the upper T_{piv} increased positively with latitude ($r = +0.69$, $P < 0.05$), contradicting expectations from basic adaptationist theory (see also Ewert et al. 2005). This departure from expectation turns out to be explained well by geographic variation in nesting behavior. The nesting season begins earlier and females choose shadier microenvironments in which to nest in southern populations of *Chelydra* than in northern populations (Ewert et al. 2005). Thus, the thermal environment experience by developing embryos is probably "cooler" in southern populations than in northern ones, consistent with the observation described above of a positive correlation between the upper T_{piv} and latitude. The lower T_{piv} in *Chelydra* exhibits an inverse correlation with latitude ($r = -0.98$, $P < 0.0001$), which reflects a strong inverse correlation with the upper T_{piv} (Ewert et al. 2005), possibly indicating a genetic covariance between the T_{piv} values.

Intrapopulation studies of T_{piv} have detected substantial (among-clutch) variation. As acknowledged in these studies, within-population variation in T_{piv} could be caused by nongenetic maternal effects (e.g., steroid hormone content of egg yolks at oviposition) rather than genetic effects, causing heritability estimates to be at the upper limits. This distinction is important because the evolutionary dynamics of TSD differ dramatically depending on the causal basis of within-population variation in T_{piv} (Morjan 2003).

Bull et al. (1982a) initially described among-clutch variation for *Graptemys* and interpreted it in quantitative genetic terms ($h^2 = 0.82$). Because their laboratory study was conducted at a constant incubation temperature, thereby minimizing environmental variation, they constructed a metric (i.e., the realized heritability) to estimate the quantitative genetic basis of T_{piv} in nature. Temperatures differ substantially among nests; thus, the realized heritability of T_{piv} was considerably smaller ($h^2 = 0.06$) than that of the laboratory T_{piv}, indicating a microevolutionary constraint on T_{piv} under natural conditions.

A similar outcome has been noted for *Chelydra*. Incubating eggs from Illinois at three temperatures near the upper T_{piv}, Janzen (1992) detected substantial among-clutch variation that was interpreted in quantitative genetic terms ($h^2_{27.5} = 0.60$, $h^2_{28.0} = 0.76$, $h^2_{28.5} = 0.34$, $h^2_{combined} = 0.56$); the corresponding realized heritabilities were much lower (0.05, 0.06, 0.03, and 0.05, respectively). Analysis of variance suggested no significant genotype by environment interactions (i.e., clutch sex ratio responses were concordant across temperatures), which was confirmed by high genetic correlations for sex between the temperature treatments ($r^2_{7.5\times28.0} = 0.73$, $r^2_{7.5\times28.5} = 0.52$, $r^2_{8.0\times28.5} = 0.67$). Parallel with the T_{piv} results, this finding suggests that the shape of the TSD pattern in *Chelydra* is also genetically constrained in its microevolutionary potential (see also among-population results described in Ewert et al. 2005).

This conclusion about the microevolutionary potential of TSD in *Chelydra* has been criticized on methodological and sample-size grounds. Incubating eggs from Minnesota at four temperatures near the upper T_{piv}, Rhen & Lang (1998) also detected substantial among-clutch heterogeneity in sex ratio at 27.5, 28.0, and 28.5°C (but not at 29.0°C where nearly all offspring were female). However, using a logistic model of analysis of variance, significant genotype by environment interactions were detected between 27.5 and 28.0°C and between 27.5 and 28.5°C; genetic correlations for sex for these

two comparisons were correspondingly low, suggesting that TSD in *Chelydra* is *not* constrained in its microevolutionary potential. Reanalysis of Janzen's (1992) data using a logistic model of analysis of variance still did not detect a significant genotype by environment interaction ($P = 0.7045$) and using Rhen & Lang's (1998) method for calculating genetic correlations actually strengthens Janzen's original conclusions (cf. $r^2_{7.5\times28.0} = 0.79$, $r^2_{7.5\times28.5} = 0.60$, $r^2_{8.0\times28.5} = 0.83$)! As for sample size, Janzen (1992) used nine fewer clutches (15 vs. 24) than Rhen & Lang (1998), but apparently sexed equivalent or higher numbers of offspring for each clutch and temperature combination. Thus differences between the two studies might be due to different genetic architectures in each population or to relatively small sample sizes (≤ 6/clutch) at 27.5°C for the Minnesota population. Replicate studies in these same or in other populations of *Chelydra* might usefully clarify this important question.

Very little research has focused on nest-site selection in the context of TSD despite its central role in various models concerning the microevolution of TSD (Bulmer & Bull 1982; Roosenburg 1996; Reinhold 1998; Freedberg & Wade 2001). Until recently, evidence for geographic variation in nest-site selection was essentially anecdotal. For *Chelydra*, populations in the southeastern United States regularly nest in the "shade" (Richmond 1945; Ewert et al. 1994) and in southern Mexico often nest "in or beneath vegetation" (Vogt & Flores-Villela 1992), whereas populations in northern Illinois (Kolbe & Janzen 2002), New Jersey (Wilhoft et al. 1983), northern New York (Petokas & Alexander 1980), Quebec (Robinson & Bider 1988), and possibly southern Michigan (Congdon et al. 1987) and northern Minnesota (Rhen & Lang 1995) typically nest in relatively unshaded microenvironments. These natural history observations have been confirmed by a large-scale semiquantitative analysis of vegetation cover around *Chelydra* nests (Ewert et al. 2005): nests in southern populations are highly shaded and those in northern populations are nearly unshaded. More detailed comparative work (e.g., controlled reciprocal transplant experiments) would be helpful to distinguish between local adaptation and phenotypic plasticity as explanations for these patterns, but these studies nonetheless suggest the overriding importance of nest-site selection in affecting the embryonic thermal environment and thus its potential impact on the microevolution of TSD. These conclusions are strengthened by the documented relationships between nest temperatures and nest sex ratios in *Chelydra* (see "Evidence for TSD" above).

But how can females choose the nest thermal environment to possibly manipulate offspring sex ratio when the TSP begins several weeks after oviposition? Indeed, thermal conditions at oviposition can be influenced by numerous transient environmental factors (e.g., cloud cover, proximity to precipitation event, etc.) that may be unlinked to nest temperatures during the TSP. Several studies have noted a qualitative relationship between degree of vegetation cover around nests and nest sex ratio (e.g., Vogt & Bull 1982, 1984) and others have implemented quantitative measures documenting the same result (e.g., Janzen 1994b). Extending this latter approach to *Chelydra*, Kolbe & Janzen (2002) demonstrated that females in an Illinois population overall selected less vegetated sites in which to nest more frequently than expected based on available habitat. Even so, females exhibited substantial heterogeneity in overstory vegetation cover around nests at oviposition, which was strongly negatively correlated with nest temperatures during the middle one-third of incubation ($r = -0.82$, $P = 0.0001$, $n = 16$) and thus with nest sex ratio ($r = +0.98$, $P < 0.0001$, $n = 14$) (see also St. Juliana et al. 2004). Vegetation cover at oviposition can therefore provide a relatively reliable cue to manipulate relative offspring sex ratio.

At the same time, the inheritance of nest-site selection is unclear. Laboratory analyses of geckos with TSD document relatively little among-female variation in nest-site selection in a thermal gradient (Bull et al. 1988a; Bragg et al. 2000). Multiyear field studies of nest-site selection in *Chrysemys* have detected significant individual repeatability for vegetation cover around nests at oviposition (Janzen & Morjan 2001; Valenzuela & Janzen 2001). Repeatability is crucial for a phenotype to be a meaningful target of natural selection and delimits the upper bound of heritability for the trait (see discussion in Janzen & Morjan 2001). Thus, these remarkable field repeatabilities suggest that TSD in *Chrysemys* exhibits microevolutionary potential, providing empirical support for key aspects of models involving nest-site selection. At present, however, we lack any information on the quantitative genetics of nesting biology of *Chelydra*. Creative field experiments, perhaps linking molecular markers and phenotypic variation (e.g., Ritland 1996), on any species with TSD would be extremely valuable for revealing the microevolutionary potential of nest-site selection.

Yet a third possible factor involved in the microevolution of TSD has been almost completely ignored so far. The growing number of studies of a wide variety of taxa has shown that reproductive phenology varies considerably with climatic conditions (e.g., Beebee 1995; Crick & Sparks 1999). Thus timing of the nesting season for amniotes with TSD may also be altered in response to geographic or annual variation in climate (e.g., Congdon et al. 1987; reviewed in Mrosovsky 1994). Indeed, Ewert et al. (2005) find that the nesting season begins and ends earlier in southern populations of *Chelydra* than in northern populations. If this response involves genetic change and not (just) behavioral plasticity, then adaptive evolution of reproductive phenology that potentially maintains offspring sex ratio variation may occur. This reasonable possibility deserves serious quantitative attention and will probably require a consortium of cooperative researchers to evaluate experimentally and over the long term.

Adaptive Significance

Of course, the million dollar question regarding TSD in amniotes is "Why does it exist or persist?" Why indeed do organisms with TSD leave the fundamental trait of sex determination to the stochastic whims of the environment? Despite much attention to this question over the past 30 years, a clear general answer remains elusive (Shine 1999).

The prevailing theoretical framework used to empirically assess the adaptive significance of environmental sex determination (Charnov & Bull 1977) has been successful in non-amniotes (mostly reviewed in Bull 1983; Conover 1984; Korpelainen 1990). The key component of this Charnov-Bull model is the expectation that environmental sex determination is adaptive when some environmental conditions (e.g., cooler incubation temperatures) are better (i.e., increase fitness) for males than for females and vice versa for other environmental conditions (e.g., warmer incubation temperatures) (for a graphical example, see Fig. 3 of Janzen 1995). In other words, the ratio of male to female fitness must vary with temperature in species with TSD. This expectation and theoretical modifications thereof (Roosenburg 1996; Reinhold 1998) have spawned several empirical studies of "reptiles" with TSD seeking thermally dependent traits that have sex-specific benefits (Shine 1999).

Research on *Chelydra* has contributed significantly to our body of knowledge on this topic. Most of this work has centered on the presumed sex-specific fitness advantages of differential posthatching growth rates (mostly summarized in Freedberg et al. 2001). The idea here is that faster growth particularly benefits males in *Chelydra* because they are the larger sex as adults and may defend home ranges (i.e., access to mates?) based on size (e.g., Janzen & O'Steen 1990). Thus, under this sexual size dimorphism hypothesis, males should be produced at incubation temperatures that elicit the fastest posthatching growth rates.

In general, the evidence favors this view. Turtles from Ontario grew fastest over 7 months from 25.6°C incubation as embryos (100% male) compared with turtles from 22.0°C (19% male) and 28.6°C (33% male) (Brooks et al. 1991). A similar result was obtained for four Ontario populations when grown for 11 months posthatching (i.e., 24.8°C turtles [82% male] grew faster than 21.3°C [16% male] and 29.1°C [2% male] turtles) (Bobyn & Brooks 1994a) and for two Ontario populations when grown for 23 months posthatching (i.e., 25.3°C turtles [97% male] grew faster than 21.1°C [21% male] turtles) (Bobyn & Brooks 1994b). Unfortunately, these studies could not separate the potential independent effects of sex and incubation temperature on posthatching growth.

Rhen and Lang adopted a clever method to do just that. By applying estradiol or an aromatase inhibitor to eggs during the TSP, they could generate females at otherwise male-producing temperatures or males at otherwise female-producing temperatures, respectively, that were presumably similar functionally to "normal" turtles of the same sex (Rhen et al. 1996). They found that Minnesota turtles from temperatures that normally produced exclusively males (24 and 26.5°C) grew faster over 6 months posthatching *regardless of sex* than turtles from the otherwise mostly female-producing temperature (29°C) (Rhen & Lang 1995, 1999b). Thus temperature, but not sex per se, influenced posthatching growth rates in accordance with the Charnov–Bull model. Unfortunately, they did not examine the effects of incubation temperature near or below the lower T_{piv}.

On the other hand, O'Steen (1998) incubated Illinois eggs at 21.5°C (56% male), 24.5°C (100% male), 27.5°C (50% male), and 30.5°C (0% male) without the experimental manipulation used by Rhen and Lang. Turtles from 21.5 and 30.5°C were smallest at hatching, but individuals from 21.5 and 24.5°C grew fastest and chose warmer microenvironments over ~10 months posthatching. This result held even after hibernation for 6 months at 7°C! Subsequent work indicated that males generally had higher resting metabolic rates than females, whereas the latter had higher blood levels of thyroxine than the former (O'Steen & Janzen 1999). The contradictory findings of these growth studies, along with comparative analyses, question the broad validity of the sexual size dimorphism hypothesis (Janzen & Paukstis 1991b; but see Ewert et al. 1994). Regardless, further experimental work in *Chelydra* and in other turtles with TSD is needed (Roosenburg & Kelley 1996; Freedberg et al. 2001), in particular, under natural or seminatural conditions (Janzen & Morjan 2002).

The only other experimental study to investigate the adaptive significance of TSD in *Chelydra* adopted a behavioral, performance-based approach recommended for microevolutionary research (sensu Arnold 1983). Eggs from Illinois were incubated at 26°C (100% male), 28°C (62% male), and 30°C (0% male) (Janzen 1995). Hatchlings were weighed, measured, and raced on land and in water in the laboratory to obtain measures of potentially fitness-related traits. Turtles were marked individually, sexed surgically, and then released into a fenced outdoor experimental pond. Survivorship after 6 months was assessed and evaluated in the context of the phenotypic traits measured at hatching. Turtles from the "single sex"–producing temperatures (26 and 30°C) were less likely to be active during performance trials in the laboratory, which translated into *higher* survivorship in the pond. The presumption was that these individuals exhibited a survival advantage over turtles from 28°C because common visual predators at the pond (e.g., bullfrogs) cue on movement of prey. Overall, the results of this study were interpreted as being consistent with the Charnov-Bull model. Although intriguing, this work has yet to be repeated in any other population or species and suffers from having investigated a relatively narrow portion of the thermal range of incubation temperatures.

The search for adaptive explanation for TSD in *Chelydra*

and in other amniotes has been surprisingly frustrating given the success of the Charnov-Bull model for other taxa with environmental sex determination, including a fish with TSD (Conover 1984). Incubation temperature clearly influences neonatal and juvenile phenotypes in *Chelydra* and other species with TSD, including traits that might reasonably be construed to affect fitness (Shine 1999; see above). The critical issue now is not so much more work on the phenotypic effects of incubation temperature, but rather the great need for experimental tests that address possible changes in male-to-female fitness ratios across incubation temperatures (J. Bull, pers. commun.). Until then, we are left with a literature comprising plausible or partial stories consistent with, but not conclusive of, an adaptive explanation for TSD in "reptiles" à la the Charnov-Bull model.

Other models (e.g., Freedberg & Wade 2001) may instead prove to be more fruitful than the Charnov-Bull model in explaining the occurrence of TSD in amniotes. Even so, the fitness benefits of TSD over GSD need only be remarkably minimal in long-lived species to favor the evolution of TSD (Bull & Bulmer 1989), making identification of any adaptive basis for TSD in most amniotes a Sisyphean task. For that matter, TSD originated so long ago in turtles that the conditions for adaptation may no longer obtain, such that TSD may have spread by having been correlated with a different but adaptive trait (Janzen & Paukstis 1991a) or persists simply because it is not maladaptive (Janzen & Paukstis 1988; Girondot & Pieau 1999). Perhaps research on the many related issues raised in this review will shed light on the true explanation for TSD in amniotes.

FUTURE DIRECTIONS

In this review, I have discussed the major issues involving TSD in amniotes, highlighting the many important empirical contributions studies of *Chelydra* have made to a greater understanding of this unusual sex-determining mechanism. Nonetheless, much work remains to adequately evaluate many lingering unresolved questions concerning the basic biology of TSD. *Chelydra* clearly has a strong role to play in successfully tackling these challenges in the future. I see four thrusts of inquiry that are important, in particular, if we hope to develop a complete picture of the biological significance of TSD:

1. Relationships of fluctuating nest temperatures to offspring sex ratios—sophisticated statistical models to successfully predict offspring sex ratios from nest temperature traces. Snapping turtles lay especially deep nests that contain thermal gradients and thus will challenge our ability to construct general statistical models of this sort.

2. Mechanistic underpinnings—detailed studies to clarify the genetic factors and developmental pathways involved in sex determination and sexual differentiation. *Chelydra* can continue to play an important role in such studies because of availability, large clutch

sizes to assess among-clutch variation, and the presence of pattern II TSD.

3. Role(s) of maternal effects—investigations, especially field experiments, of yolk hormones, nesting phenology, and nest-site selection to better elucidate the adaptive significance of TSD. Among-clutch variation in yolk hormones and large population sizes place *Chelydra* in an excellent position to contribute strongly to maternal effects studies involving TSD.

4. Impacts of climate and habitat change—long-term field studies of offspring sex ratios and large-scale quantitative (and perhaps experimental) projects on nesting biology and reaction norms of sex ratio to address increasingly important microevolutionary and conservation issues. Again, large clutch sizes, population sizes, and geographic range position *Chelydra* to serve as an important model for such increasingly crucial research.

ACKNOWLEDGMENTS

My thanks to Gary Packard and Gary Paukstis for getting me started with *Chelydra* research many years ago. Rachel Bowden, Jim Bull, Carrie Morjan, and Tony Steyermark provided constructive comments on the manuscript and Mike Ewert helped with literature pointers and other suggestions. My work on aspects of *Chelydra* sex determination has been supported by National Science Foundation grants DDIG BSR-8914686, DEB-9629529, UMEB IBN-0080194, and LTREB DEB-0089680, and by numerous, generous Turtle Camp and Janzen Lab participants.

15

Physiology and Ecology of Hatchling Snapping Turtles

MICHAEL S. FINKLER
JASON J. KOLBE

OVER THE PAST TWENTY-FIVE YEARS, hatchling snapping turtles have been the subject of many laboratory-based studies, in particular, those examining how maternal effects and variation in environmental conditions can induce variation in hatchling phenotypes and, in turn, how this may affect fitness. In particular, much effort has focused on describing how differences in the environmental conditions in the nest during incubation (e.g., temperature and soil moisture) can induce variation in the sex, size, body composition, and whole-animal physiological performance of the resultant hatchlings (see reviews by Packard & Packard 1988a; Packard 1999; see Ewert, Chapter 10; Janzen, Chapter 14; Ackerman et al., Chapter 13). Although many of these studies have alluded to the ecological significance of phenotypic variation, relatively few have directly measured how variation may influence physiological and behavioral abilities of hatchlings in performing ecologically relevant tasks (e.g., evading predators, acquiring resources, and avoiding or tolerating adverse environmental conditions). Recently, however, a growing number of researchers have begun examining performance and behavior of hatchlings in the field to integrate the findings of laboratory studies with the life history and ecology of these animals. Although the amount of data derived from field-based studies is relatively small, and different studies have often yielded conflicting results, these efforts have been instrumental in shaping our understanding of the interactions between hatchling turtles and their environment.

In this chapter, we provide a descriptive chronological overview of the biology of hatchling snapping turtles from hatching through the first winter of life. We give particular emphasis to environmental influences on three aspects of the early life of hatchling snapping turtles: the process of hatching and nest emergence, the migration of the hatchlings from the nest site to water, and the aquatic biology of hatchlings prior to their first winter. This chapter complements discussions of related topics in the chapters on nesting ecology (see Congdon et al., Chapter 12), thermal biology of eggs and hatchlings (Ewert, Chapter 10; Janzen, Chapter 14; Spotila & Bell, Chapter 7), water relations of eggs (Ackerman et al., Chapter 13), metabolism (Gatten, Chapter 8), growth (Steyermark, Chapter 11), and overwintering biology (Ultsch & Reese, Chapter 9) found in this volume, and integrates these topics into a composite image of the biology of hatchling snapping turtles. We also call attention to aspects of the biology of hatchling snapping turtles that have had minimal or no investigation, yet may ultimately have great importance in our understanding of the life history and ecology of these animals.

HATCHING AND NEST EMERGENCE OF SNAPPING TURTLES

The mechanical processes of pipping, hatching, and nest emergence are the first activities in which the physiological performance of a free-living hatchling can directly influence its survival. In this section, we describe these three processes in snapping turtles based on both laboratory and field observations. We also discuss how maternal and environmental factors can influence hatchling morphology. Finally, where appropriate, we discuss how the phenotypic characteristics of the hatchlings, coupled with existing environmental conditions in the nest, can influence the ability of the hatchling to emerge first from the egg and then from the nest.

Hatching

The incubation period of snapping turtle eggs can be quite variable. Laboratory-based studies have reported incubation durations of anywhere from 54 to 140 days (e.g., Yntema 1968; Mahmoud & Klicka 1971; Ewert 1979; Packard et al. 1980, 1987; Morris et al. 1983; Miller et al. 1987; Janzen 1993a; Finkler 2001; Kolbe & Janzen 2001; Rimkus et al. 2002; Steyermark & Spotila 2001a). Field-based studies, however, have generally reported a somewhat narrower range of incubation periods, lasting from 63 to 119 days in natural nests from northern populations (Obbard & Brooks 1981; Congdon et al. 1987; Robinson 1989; Kolbe & Janzen 2002a). Numerous factors may affect the overall duration of incubation, and subsequently the timing of hatching and nest emergence. Nest temperature clearly influences incubation duration, with increased incubation time resulting from incubation at lower temperatures (see Yntema 1968; Ewert 1985; Packard & Packard 1988; Bobyn & Brooks 1994a, b; Kolbe & Janzen 2002a). In addition, laboratory studies have shown that hydric conditions during incubation influence the duration of incubation, with the incubation period increasing in nests containing wetter substrates (reviewed in Packard & Packard 1988a; Packard 1991, 1999; but see Rimkus et

al. 2002; Ackerman et al., Chapter 13). Other factors influencing incubation duration in snapping turtles include geographic variation (Ewert 1985), clutch identity (Steyermark & Spotila 2001a), egg size (Ewert 1985), and whether pipping (breaking the eggshell and commencement of air breathing) or hatching (emergence from the egg) is used as the end point of incubation (see Gutzke et al. 1984; Miller et al. 1987; Miller 1993).

At present, it is unclear what specific factors are important for inducing the pipping and subsequent hatching of snapping turtles in natural nests. However, a variety of environmental stimuli are able to evoke hatching. In the laboratory, frequent physical disturbance of eggs appears to promote hatching (Finkler, pers. observ.), and it is possible that the movements of clutch mates in the nest stimulate synchronous hatching and subsequent nest emergence in some nests as seen in other turtle taxa (Carr & Hirth 1961; Spencer et al. 2001). Hatchlings from eggs incubated in the laboratory on very dry substrates tend to hatch sooner than do those from wetter substrates (reviewed in Packard 1999), suggesting that osmotic stress to the developing embryo can also induce pipping and hatching. Investigation into the physiological mechanisms that lead to pipping and hatching, such as potential cardiovascular, muscular, and endocrine responses to factors such as physical stimuli or osmotic stress by late-term embryos, is warranted.

Hatchling snapping turtles, like other turtles, possess a keratinous egg caruncle on the tip of the upper tonium that is used to break through the eggshell and shell membranes at hatching (see Yntema 1968, plate 8; Edmund 1969). At pipping, the hatchling pushes forward with the head one or more times, piercing through the shell with the caruncle and extending the head out of the egg. Lateral movements of the forelimbs then expand the aperture, allowing the hatchling room to escape the egg. The total duration of hatching (from pipping to finally escaping from the eggshell) is highly variable, lasting from a matter of minutes to several hours or even days (Finkler & Kolbe, pers. observ.). Although not formally investigated, environmental conditions in the nest may influence the ability of the hatchling to escape from the egg. Hatchlings from eggs incubated under dry conditions in the laboratory may become trapped inside the egg due to drying of the egg contents. The shell membrane may adhere to the integument of the hatchling, and the drying of the shell membrane may make it too tough to penetrate (Finkler, pers. observ.).

At pipping, most chelonians still have appreciable amounts of yolk present in their yolk sac, which is still external to the body (Ewert 1985; Kuchling 1998). Typically, the yolk sac is drawn into the abdominal cavity through the umbilical fontanel in the plastron either during hatching or before emergence from the nest, although some hatchlings may emerge without a fully retracted yolk sac. The duration of this retraction can also be highly variable (from hours to days or even weeks), and is likely related directly to the size of the yolk sac, with larger yolk sacs requiring more time for retraction into the abdomen. Transient skin of the plastron grows over the yolk sac as it is absorbed, eventually enveloping the yolk sac (M. A. Ewert, pers. commun.). This may help to reduce the chance of rupturing the yolk sac during withdrawal into the abdomen or, in the event the yolk sac is not fully withdrawn, during hatching, nest emergence, and overland movement.

The amount of yolk remaining in the yolk sac at hatching can be highly variable and is influenced by many factors. Clutch identity appears to influence yolk sac mass at hatching (Packard et al. 1999; Rhen & Lang 1999b). Yolk sac mass at hatching may also correlate with egg size (Packard et al. 1999; but see Rhen & Lang 1999b). Numerous studies have noted that eggs incubated on vermiculite substrates with low water contents yield hatchlings with larger amounts of yolk remaining than do eggs incubated on wetter substrates (see Morris et al. 1983; Packard et al. 1988; Janzen et al. 1990; Packard 1991, 1999). Packard (1991) suggested that in wet nests, the higher water content of the eggs (due to absorbance of water from the surrounding soil) leads to an increase in overall circulation through the embryonic vasculature and thus increased mobilization of nutrients out of the yolk and delivery to the tissues. The amount of yolk remaining in the yolk sac at hatching also appears to increase with increasing temperature (Packard et al. 1988; Rhen & Lang 1999); thus, alternatively, the lower yolk consumption observed in embryos from eggs incubated in dry nests could be due to the lower thermal conductivity of dry soils and an increase in the temperature of the egg due to metabolic heat production by the embryo (see Ackerman et al., Chapter 13).

Hatchling snapping turtles can vary considerably in overall body size. Congdon et al. (1999) reported a size range of 6–12 g mass and 25–33 mm carapace length for 658 hatchlings from natural nests in a Michigan population, similar to the size ranges for field-incubated hatchlings from Illinois (418 hatchlings, 6–14 g mass and 26–35 mm carapace length; Kolbe unpublished data), New Jersey (55 clutches, 5–10 g mean hatchling mass per clutch; Hotaling et al. 1985), Nebraska (8–9 g mean hatchling mass per clutch; Packard et al. 1999), and Ernst et al.'s (1994) estimates for the species in general (5–9 g mass and 24–31 mm carapace length). Studies using hatchlings from laboratory-reared eggs have likewise yielded a great range of body sizes (e.g., Morris et al. 1983; Wilhoft 1986; Brooks et al. 1991a; Janzen 1993a; Steyermark & Spotila 2001a). Laboratory studies have demonstrated that numerous factors can influence hatchling size. In particular, hatchling mass correlates strongly with egg size (e.g., Ewert 1985; Steyermark & Spotila 2001a). Clutch effects, which may include differences in egg composition and genetic effects (e.g., Rhen & Lang 1999; Packard et al. 1999; Steyermark & Spotila 2001a) can also influence hatchling size. Hatchling size also tends to decrease with increasing temperature (Packard et al. 1988; Brooks et al. 1991a; Steyermark & Spotila 2001a; but see Rhen & Lang 1999), and it tends to increase with in-

creasing soil moisture (reviewed in Packard 1999; but see Rimkus et al. 2002; Ackerman et al., Chapter 13). Finally, hypoxia in the nest during development may reduce the size of hatchling turtles (see Kam 1993 for *Pseudemys nelsoni*).

Nest Emergence

Snapping turtles throughout North America usually emerge from the nest after hatching and migrate to the water before the onset of winter (Costanzo et al. 1995; Sims et al. 2001; see Ultsch & Reese, Chapter 9), although rare incidences of overwintering in the nest have been reported (e.g., Ernst 1966). Congdon et al. (1987) reported that nest emergence in Michigan occurs from late August to early October, with most nest emergence occurring in the month of September. Apparently no data of the time between the hatching of the eggs and the emergence of hatchlings have been published, but in an Illinois population in 1999 the first hatchling from naturally incubated nests emerged 1–8 days after hatching began (Kolbe, unpublished data), comparable to the time of emergence reported for *Malaclemys terrapin* (1–9 days; Burger 1976). Robinson (1989) reported that 96% of hatchlings from 14 nests in a Quebec population emerged in daylight between 7:00 A.M. and 6:00 P.M., and Congdon et al. (1999) found that the majority of nest emergence in a Michigan population occurs during the late morning hours. This is in contrast to observations of nest emergence by sea turtles (Bustard 1967; Mrosovsky 1968; Witherington et al. 1990; Moran et al. 1999) and by *Trionyx muticus* (=*Apalone mutica*) (Anderson 1958) in which hatchlings usually emerge at night, presumably to avoid excessively high temperatures (Moran et al. 1999). Congdon et al. (1999) also noted that nest emergence often coincides with rain. Hatchling snapping turtles thus may tend to emerge and disperse from the nest during periods when temperatures are warm but not excessively high and when the risks of potential dehydration are minimal (see Kolbe & Janzen 2002b). Robinson (1989) observed synchronous emergence of the hatchlings (i.e., all hatchlings emerging within 24 h) in 5 of 14 nests and asynchronous emergence (all hatchlings emerging in >24 h) in the remaining 9 of 14 nests in a Quebec population.

Hatching and nest emergence are particularly interesting mechanical processes as these are the first activities of a free-living neonate whose development has been influenced by various maternal and environmental factors. Thus, they are the first events of the free-living offspring where these factors may influence survival. However, much of our knowledge as to how these factors influence performance during hatching and emergence is largely anecdotal in nature. The physiological and developmental processes that trigger hatching are largely unexplored in turtles. The influence of environmental factors such as temperature and moisture on the duration and success of hatching and nest emergence have not been quantified, nor has the relative importance of synchronous hatching and emergence of clutch mates been addressed. Direct investigation of individual factors and combinations of factors that may affect the ability of the hatchling to hatch and emerge from the nest may yield new insights into the biology of neonate turtles and other precocial oviparous amniotes.

OVERLAND MOVEMENT

The time between emergence from the nest cavity and entrance into the aquatic environment is a crucial stage in the life history of hatchling snapping turtles. Limited resources and harsh environmental conditions may limit the number of hatchlings that successfully complete the journey from nest to water. Numerous factors, both maternal (e.g., direct genetic effects, propagule size, hormone or nutrient provisioning of offspring, and nest-site selection; Bernardo 1996a, b; Mousseau & Fox 1998) and environmental (e.g., predators, temperature, and precipitation) may affect the performance, behavior, and ultimately the survival of migrating hatchlings. Here, we focus on the identified factors that contribute to variation in the ability of hatchling turtles to successfully reach aquatic environments.

Proximity of the Nest Site to Water

Actual distances traveled by hatchling snapping turtles emerging from natural nests are unknown, but the minimum distance a hatchling needs to travel to successfully reach the aquatic habitat can be inferred from nest distances from water. Nest-to-water distances across the species range vary from roughly 1 to 183 m with high variation within populations (reviewed in Iverson et al. 1997). The proportion of nests located close to the water (<20 m) is often relatively high, and decreases with increasing distance from the water (e.g., Congdon et al. 1999). Variation in nest-to-water distance likely results in differential exposure of hatchlings to predators, environmental conditions, and energy requirements for migration. However, evidence that hatchling mortality rates change as nest site distance from the water increases is equivocal (Robinson 1989; Congdon et al. 1987, 1999; Kolbe & Janzen 2002a), and likely depends on environmental characteristics specific to a particular location.

Hatchling Terrestrial Performance and Environmental Tolerance

Because distances from nest site to water and environmental conditions encountered by migrating hatchlings are quite variable both within and among populations, the strength with which environmental factors may exert selective pressure on hatchlings may be highly variable. Under benign environmental conditions and during migrations from nest sites close to water, there may be very little variation in survival of hatchlings based upon phenotype. However, during prolonged overland migration and during instances when hatchlings may be exposed to adverse environmental conditions, the physiological performance and environmental tolerance capacity of hatchlings during the overland migration may affect offspring survivorship by influencing the amount of time spent on land, hydration dynamics, and the ability to avoid predators.

Locomotor performance (speed and endurance) has been the most often studied measure of physiological performance in hatchlings (see Miller et al. 1987; Janzen 1993a; Miller 1993; Finkler 1997, 1999). Although it is unlikely that a hatchling snapping turtle could rely on crawling speed to evade a pursuing terrestrial or avian predator, it has been suggested that hatchlings that have greater physiological capacities (e.g., are faster and/or have greater endurance) could potentially complete the overland migration more quickly, thereby reducing exposure to predators and potentially harsh or variable environmental conditions (but see "Hatchling Behavior during Overland Movement" below).

Laboratory-based studies have demonstrated that many factors, both environmental and those intrinsic to the organism, can influence terrestrial physiological performance. Perhaps the best-studied factor is that of hydric conditions during incubation. Hatchling snapping turtles from eggs incubated on wet vermiculite substrates had forced crawling speeds that were greater than those of hatchlings from eggs incubated on dry vermiculite substrates, both in absolute terms and with body size taken into account (Miller et al. 1987; Finkler 1999). Greater tissue hydration likely accounts for most of the improved performance of wet-incubated hatchlings; Miller et al. (1987) noted that dry-incubated hatchlings improved their performance more than did wet-incubated hatchlings after exposure to water, and Finkler (1999) demonstrated that crawling speed decreases with decreasing relative water content below ~72% of whole body mass. Janzen (1993a) found no difference in locomotor performance between hatchlings from eggs incubated on wet and dry vermiculite substrates, but hatchlings from both treatments had been housed under aquatic conditions prior to testing, which likely alleviated tissue hydration differences between the two treatments. Similarly, physiological endurance may be related to tissue hydration. Miller et al. (1987) found that lactate accumulated at a faster rate with increasing crawling distance in dry-incubated hatchlings than in wet-incubated hatchlings, suggesting that aerobic capacity may be greater in hatchlings from wet nests.

The influence that other environmental and maternal factors may have on the terrestrial locomotor performance of hatchlings has received relatively little attention. Surprisingly, no formal study appears to have been made of temperature effects (either during incubation or posthatching) on terrestrial locomotor performance in hatchling snapping turtles, although in *Apalone mutica* (Janzen 1993b) and *Pelodiscus sinensis* (Du & Ji 2003), warmer incubation temperatures appear to produce hatchlings with faster crawling speeds.

Recently, loss of righting response and righting time (the time necessary for an animal turned on its back to right itself) have also been employed as measures of physiological endurance, in particular, in studies of the tolerance of adverse environmental conditions. Finkler (1999) found that wet-incubated hatchlings tolerate longer periods of desiccation before loss of righting response than dry-incubated hatchlings do, apparently because of their increased body water contents. Both clutch identity and environmental temperature have also been shown to influence righting time (Steyermark & Spotila 2001b).

Both intrinsic and environmental factors may also affect tolerance of adverse environmental conditions during overland movement, in particular, desiccation tolerance. For example, larger hatchlings can survive longer than smaller hatchlings can under desiccating conditions because body water content increases proportionately more with increased body size than the rate of evaporative water loss does (Finkler 1998, 2001). Hydric conditions during incubation may also influence desiccation tolerance, as hatchlings from eggs incubated on wet vermiculite substrates can tolerate greater water loss than can dry-incubated hatchlings (Finkler 1999), because both absolute and relative water contents are greater in these wet-incubated hatchlings. The extent to which dehydration is problematic for hatchlings, however, largely depends on existing environmental conditions during overland movement (Kolbe & Janzen 2002b), and may be tempered by behavioral conservation of water (e.g., reduced activity during unfavorable conditions, microhabitat selection, etc.) (Finkler et al. 2000; Kolbe & Janzen 2002b).

Much of the thermal biology of hatchling snapping turtles during overland movement is unknown. Williamson et al. (1989) found that increased acclimation temperatures elevated the critical thermal maximum (CTM) in 5- to 7-month-old juveniles, but it is unclear whether incubation temperature may similarly influence temperature tolerance in hatchlings. Moreover, the influence of temperature (during incubation and during overland movement) on hatchling terrestrial activity has yet to be explored.

Hatchling Behavior during Overland Movement

Maternal and environmental conditions influencing hatchling snapping turtle performance may also affect hatchling behavior during migration and, in turn, the ability of hatchlings to successfully complete the journey from nest site to water. The time it takes hatchlings to migrate from nest to water (a function of the locomotor activity of the hatchling and the distance the hatchling travels), the direction of dispersal from the nest, and the physical landscape being traversed may influence fitness by causing hatchlings to be more or less susceptible to predation or adverse abiotic conditions as well as by influencing the probability of finding the aquatic habitat.

Snapping turtles are highly aquatic, and as such may benefit from minimizing the amount of time spent in the terrestrial habitat during hatchling overland migration. If time spent on land is indeed risky for hatchling snapping turtles, then orientation mechanisms that increase the chance of finding water likely exist. Such mechanisms could include visual cues, olfactory cues, orientation to magnetic fields, humidity gradients, and/or geotaxis. Noble and Breslau (1938) found little evidence of olfactory orientation, but suggested that humidity gradients, light intensity, and position of the sun might be important cues or orientation. Robinson (1989) found little evidence that magnetic fields, olfactory cues, or humidity gradients influence direction of movement, and suggested that orientation with gravity down a slope is the primary cue directing turtles downhill toward potential water bodies. The influence of visual cues has received relatively little attention in C. serpentina, but Yeomans (1995) found that adult Trachemys scripta out of direct view of water moved directionally toward water bodies under sunny conditions, but randomly under overcast conditions, indicating that visual cues (perhaps polarized light patterns produced in areas with high moisture) are also important in orientation toward water.

Terrain surrounding the nest site also likely influences directional orientation in the hatchlings. Congdon et al. (1999) found the direction of movement was random in two experimental releases where hatchlings emerged from artificial nests on a slope facing the body of water. In a third experimental release, where the topography was uphill in all directions except toward the water, more hatchlings than expected headed downhill toward the marsh. Moreover, Robinson (1989) found a significant negative relationship between the angle of hatchling dispersal from the nest relative to downhill direction and the slope of the nest site; hatchlings from nests on steep slopes moved downhill, whereas those from nests at flat sites move in random directions. The evidence on direction of dispersal from the nest suggests that some hatchling snapping turtles do not orient and disperse toward the water, but the topography surrounding the nest site may influence dispersal direction. Experiments that manipulate physical conditions (i.e., slope, vegetation type, proximity to vegetation, proximity to water, etc.) and environmental conditions (i.e., time of release, temperature, precipitation, etc.) are needed to clarify this component of hatchling snapping turtle behavior.

The distance of nests from water may affect the amount of time spent in the terrestrial habitat, and, in turn, the risks entailed with being on land. However, release-recapture studies of hatchling snapping turtles have not found strong evidence for correlation between linear distance from nest to capture point and the time from release to recapture (Congdon et al. 1999; Kolbe & Janzen 2001). This suggests that measures of locomotor performance under forced conditions (such as those commonly used in laboratory experiments) may not be good predictors of how long it will take a hatchling snapping turtle to travel from nest site to water (see Janzen 1993a). Rather, total distance traveled, duration of activity periods, environmental conditions during activity, performance during voluntary locomotion, and topography of the terrain being traversed need to be taken into account.

Dispersion from the nest is usually measured as the minimum distance a hatchling would have traveled given its recapture location along the drift fence. Congdon et al. (1999) found that hatchling snapping turtles tended to be recaptured symmetrically around the minimum distance from the release point to the fence. Distance from the fence, body size, and time from release to recapture were not related to dispersion distance (Congdon et al. 1999). Kolbe & Janzen (2001) found no effect of body size, clutch, ground vegetation, or aspect on dispersion, but found that hatchlings dispersed farther when released from greater distances and at lower slopes.

Although many laboratory-based performance studies have suggested that timely migration from nest site to water is beneficial to hatchling snapping turtles, field-based studies suggest that periods of reduced activity, which would delay completion of the migration, could also be important for survival. Periods of inactivity or microhabitat selection may reduce predation risk and susceptibility to water loss, or maximize physiological performance (Congdon et al. 1999; Finkler et al. 2000; Kolbe & Janzen 2001, 2002b). Release-recapture studies by Janzen (1993a) and Kolbe & Janzen (2002b) recaptured the majority of hatchlings between 7:00 A.M. and 1:00 P.M., during intermediate temperatures in the diurnal temperature cycle. Moreover, Kolbe & Janzen (2002b) recaptured most of their hatchlings during precipitation events. These findings indicate that hatchlings may limit activity to periods with benign environmental conditions, where thermal and hydric stress is minimized.

There has been relatively little investigation of how topography and microhabitat influence the time needed for a hatchling to complete the nest-to-water migration. However, Kolbe & Janzen (2001) found a strong positive relationship between time from release to recapture and the density of ground vegetation. High vegetation density may enable hatchlings to avoid adverse environmental conditions, but likely presents an obstacle to rapid movement. Other factors, such as slope and substrate consistency, may also influence net overland movement.

Hatchling Survival during Overland Migration

Investigations into variation in hatchling size, morphology, and performance in snapping turtles have often hypothesized how variation in these characters can influence the fitness of the hatchling, in particular, in reference to migration from the nest to water. Although this is a brief stage in the life history of snapping turtles, this period is one where both biotic factors (e.g., predation) and abiotic factors (e.g., excessive temperatures, desiccation) may exert strong selective pressure on the hatchlings. Moreover, because release-recapture studies on land are easier in general to conduct

than are those in water, the period of overland movement is particularly amenable for field testing of hypotheses regarding phenotypic variation on hatchling fitness.

A central theme in many studies of hatchling snapping turtles is the idea that "bigger is better" for hatchling survival during migration from nest to water—that larger hatchlings are better able to tolerate adverse abiotic conditions, evade predators more effectively, and/or complete migration from the nest more quickly than smaller hatchlings can (Janzen 1993a; Congdon et al. 1999). A derivation of this hypothesis, "wetter is better," postulates that the effects on hatchling size, body composition, and physiological performance of increased water availability to the embryo during incubation enhance survival chances during the nest-to-water migration (reviewed in Packard 1999). However, most tests of "bigger is better" or "wetter is better" in snapping turtles are based on laboratory experiments of physiological performance (e.g., Miller et al. 1987; Finkler 1999), which may inadequately reflect the complex array of environmental variables that would ultimately determine survival during the journey from nest site to water.

Three studies that conducted release-recapture experiments in the field (Janzen 1993; Congdon et al. 1999; Kolbe & Janzen 2001) provide some information on the survival of hatchling snapping turtles during the overland migration. All specifically examined "bigger is better"—whether larger hatchlings have better chances of survival—as well as other factors potentially influencing hatchling survival. All three studies had similar recapture rates (59–73%), but their findings are conflicting with regard to selection (based on recapture rates) based on hatchling size. Janzen (1993) detected directional selection favoring larger hatchlings in a single release-recapture experiment. In contrast, Congdon et al. (1999) conducted three separate terrestrial release-recapture experiments that varied in year, terrain, and distance from release site to water. In two of three experiments, there was no evidence of selection based on hatchling size, and in the third, stabilizing selection was detected. Similarly, the release-recapture study conducted by Kolbe & Janzen (2001) found no evidence of survival differences based on the size of the hatchlings, but did find that the probability of survival increased with less ground vegetation and lower slopes at the release site.

An experimental release of hatchlings is a powerful way to ask questions about the early life history of turtles. To evaluate the importance of maternal, environmental, and offspring traits that could affect survival, future experimental releases of hatchling snapping turtles should focus on (1) quantification of the microhabitat through which hatchlings migrate and its effect on offspring survival and behavior, (2) identification of the agent of natural selection and connection with the trait(s) under selection (e.g., Janzen et al. 2000a, b), and (3) identification of the conditions (e.g., predation risk, weather conditions) under which different sources of mortality are likely. These questions may be successfully addressed by comparing multiple populations of snapping turtles with hatchlings that experience different environmental conditions or levels of predation risk. Further, although hatchling body size may be an important phenotype under some conditions, future studies need to move beyond the simple "bigger is better" framework to include other maternal and environmental factors, their interactions, and the conditions in which larger or smaller hatchlings have advantages in performance or survival.

Hatchling snapping turtle performance, behavior, and survival during the overland migration from nest to water are affected by variation in maternal, offspring, and environmental conditions. The complexity of interacting factors is challenging for both interpreting the results of past experiments and designing future experiments. Several aspects of hatchling snapping turtle life history deserve further attention, including the effect of microhabitat conditions, environmental conditions, and population history on survival and behavior of hatchlings. One discrepancy in particular deserves more study: the effect of clutch on survival and behavior in field experiments vs. clutch effects on offspring traits in the laboratory. Clutch effects are often highly significant in the laboratory for offspring traits such as body size, metabolism, survival, thermoregulatory behavior, hormone levels, yolk reserves, and growth (Brooks et al. 1991; Bobyn & Brooks 1994b; Rhen & Lang 1995; O'Steen 1998; O'Steen & Janzen 1999; Packard et al. 1999; Rhen & Lang 1999a, b; Steyermark & Spotila 2000, 2001a, b, c), but experimental releases in the field have rarely found clutch effects on survival and behavior to be important (Congdon et al. 1999; Kolbe & Janzen 2002b). Further investigation of this difference in results between field and laboratory studies may reveal novel interactions among maternal, offspring, and environmental conditions.

AQUATIC PHYSIOLOGY AND ECOLOGY

In contrast to the considerable research conducted on hatchlings during terrestrial dispersion from the nest site, there is a relative paucity of data concerning the physiology, behavior, and ecology of the neonates once they reach the water, in particular, in natural settings (Congdon et al. 1992). As neonatal snapping turtles are small and both morphologically and behaviorally cryptic, they are difficult to study under natural conditions. However, the first year of life is a critical period in the life history of snapping turtles, as survivorship rates are typically lower in this age class vs. subsequent age classes (Brooks et al. 1988; Congdon et al. 1994). Survivorship during this period may depend on the ability of the hatchling to select suitable habitats, avoid predators, and procure enough nutrients to last to the following spring. Herein, we provide an overview of physiological performance, stored energy reserves, feeding behavior, habitat selection, and survival of hatchling snapping turtles in the aquatic environment as they approach the first winter of life.

Aquatic Performance

The physiological performance of hatchling turtles could potentially influence survivorship through the first year of life. Hatchlings that are more robust may be able to procure nutrients more effectively to supplement stored energy derived from the egg, avoid predators more effectively, and tolerate adverse environmental conditions for longer periods. Studies of aquatic physiological performance in hatchling snapping turtles has focused largely on locomotor performance, although tolerance of extreme low temperatures and hypoxic conditions associated with overwintering have also been investigated (see Ultsch & Reese, Chapter 9).

Miller et al. (1987) found that hatchlings from eggs that had been incubated in wet vermiculite substrates had faster burst swimming speeds than did hatchlings from dry-incubated eggs. This difference in swimming speed is apparently sustained at least 50 days posthatching. Finkler (1997, 1998) found that clutch of origin also influenced burst swimming speed. Cooler incubation temperatures also produce hatchlings with greater swimming speeds (Janzen 1995). Swimming speeds increase with time even with only minor growth in animal size (Miller et al. 1987; Miller 1993; Finkler 1998), suggesting that ontogenic changes in performance may occur during the neonatal period.

Locomotor endurance and activity levels have received little attention in hatchling snapping turtles. However, Miller et al. (1987) found that lactate accumulated at a faster rate with swimming in dry-incubated hatchlings than in wet-incubated hatchlings, suggesting aerobic capacity (and perhaps endurance) is greater in wet-incubated hatchlings.

Stored Energy Reserves

Wilhoft (1986) estimated that approximately 49% of the lipids and 47% of the protein initially present in snapping turtle eggs remain stored in the bodies of the hatchlings at hatching. These nutrients, derived primarily from the yolk of the egg, are available to the hatchling in the form of residual yolk (i.e., yolk still present in the yolk sac at the time of hatching) as well as somatically stored metabolic substrates such as triglycerides in fat bodies and glycogen in the liver. With opportunities to feed prior to winter likely limited in northern populations of snapping turtles, hatchlings may rely heavily on the remainder of nutrients originally provisioned in the egg for sustaining them throughout the first autumn and winter of life.

As previously noted (see "Hatching" above), the amount of yolk remaining in the yolk sac at the time of hatching can be highly variable and is influenced by both maternal investment in the eggs and by environmental conditions during embryonic development. Because the amount of yolk remaining in the yolk sac at hatching can vary, several studies have discussed the importance of variation in this nutrient source on the energetics of hatchling turtles during the first autumn and winter of life. In aquatic species with delayed emergence from the nest (e.g., *Chrysemys picta, Trachemys scripta*), yolk may support long-term energetic needs of the hatchling and may support soft-tissue growth (Congdon & Gibbons 1990; Tucker et al. 1998; but see Packard & Packard 1986). These species tend to have proportionally higher lipid contents in their eggs (Congdon & Gibbons 1985) and also tend to have larger amounts of yolk in their yolk sacs at hatching than do species that emerge from the nest soon after hatching. Thus, factors such as nest environment that can induce variation in the amount of yolk remaining at the end of incubation may have long-term influences on the energetics of the hatchlings. Snapping turtle hatchlings, however, have proportionately less residual yolk at the time of hatching than do species with delayed emergence, and that yolk is almost completely consumed within the first three weeks posthatching (Wilhoft 1986; Finkler & Kressley 2002). The decrease in yolk sac content is not accompanied by increases in somatic stores of triglyceride or glycogen over the same period (Finkler & Kressley 2002), suggesting that in snapping turtles most of the lipids absorbed from the yolk are used for maintenance metabolism in the immediate neonatal period rather than stored. However, having increased residual yolk could reduce the amount of somatically stored energy substrates that need to be mobilized for maintenance metabolism during the immediate posthatching period. Subsequently, variation in yolk sac content could influence the amount of stored fat and glycogen present in the hatchling at the onset of winter and the degree to which tissues atrophy during winter (Mahmoud & Klicka 1971; Costanzo et al. 2000; Finkler et al. 2002).

Rhen and Lang (1999b) found that incubation temperature influences hatchling energy reserves, with hatchlings from eggs incubated at cool temperatures having less residual yolk, less fat body mass, and a greater loss in whole body mass, than did hatchlings from eggs incubated at higher temperatures. This is presumably due to higher resting metabolic rates at a given ambient temperature in hatchlings from eggs incubated at lower temperatures (O'Steen & Janzen 1999). As hatchling metabolic rate also increases with increasing ambient temperature (O'Steen & Janzen 1999), high temperatures may also induce a faster rate of stored energy depletion in hatchlings (see Mahmoud & Klicka 1971).

Feeding Behavior

In the laboratory, hatchlings will begin feeding within two to four weeks of hatching (McKnight & Gutzke 1993; Miller 1993; Bobyn & Brooks 1994b; Steyermark & Spotila 2001c). However, feeding activity is strongly influenced by temperature, and apparently little feeding occurs at temperatures below 15–18°C (Obbard 1983; Williamson et al. 1989; see Mahmoud & AlKindi, Chapter 6). Thus, regional climate may have an appreciable influence on the duration of feeding before hatchlings overwinter, in that hatchlings at higher latitudes may only have a short period to feed before overwintering (perhaps a matter of a few weeks, if at all), but hatchlings at lower latitudes may have ample opportunity to feed (several weeks) prior to overwintering.

Hatchlings in the laboratory will take live food, carrion, and commercially processed feed (see Burghardt & Hess 1966; Miller 1993; Finkler 1997; Steyermark & Spotila 2000). Movements from live animal prey seem to stimulate feeding: three-week-old hatchlings held at 25°C and fed crickets as their first meal ($n = 24$) made their first strikes at the prey in a median time of 18 minutes whereas those fed commercial turtle feed ($n = 24$) made their first strikes at the pellets in a median time of 170 minutes (Finkler, unpublished data). Burghart and Hess (1966) and Burghart (1967) observed that neonates appear to develop preferences for food (meat or earthworm pieces) based on first feedings, but it is not clear how long such preference is sustained, or the ultimate impact on the nutrition and energetic budgets of the neonates.

The behavior and mechanics of feeding differ between hatchlings and adults. Whereas adults are typically ambush predators, hatchlings will actively pursue prey. Older juveniles and adults strike with high speed to grasp prey in their jaws (see Lauder & Prendergast 1992), whereas the attack of a hatchling is much slower, and usually the neck is somewhat extended and the mouth open before the hatchling actually strikes at the prey. Once the prey is grasped, the hatchling will use its forelimbs to assist in tearing its food into smaller pieces.

A single study on aggressive interactions among juvenile snapping turtles competing for a stationary food source (Froese & Burghardt 1974) suggests the development of a social hierarchy early in the life of the individuals that correlates with body size. Animals with high ranks in the social hierarchy also grew at a faster rate than did lower-ranking individuals. Thus, social interactions among hatchlings may also influence their future fitness.

Habitat Selection

Juvenile snapping turtles (including hatchlings) select shallower aquatic habitats than do older animals (Hammer 1971; Congdon et al. 1992). Several possible reasons for this preference have been proposed. Selection of shallow areas may enable juveniles to escape predators found in deeper water, such as predatory fish and older turtles (Congdon et al. 1992). Moreover, shallow areas may have a higher density of vegetation cover in which hatchlings can evade predators such as wading birds and mammals (Congdon et al. 1992; Froese 1978). Hatchlings in shallow areas may also have greater opportunity to forage, as presumably the same characteristics that attract small turtles to shallow areas may also attract smaller organisms of other species, including those that might serve as prey, to those same areas (Congdon et al. 1992; Froese 1978). Furthermore, as shallow water tends to warm more quickly than does deeper water, hatchlings may be able to remain more active, forage more effectively, and grow more quickly in these areas with higher temperatures (Congdon et al. 1992).

The influence of most habitat characteristics on habitat selection has yet to be investigated, but several studies have examined temperature selection in hatchling and juvenile snapping turtles. Juvenile snapping turtles preferentially select warm temperatures (mean, 28–30°C) within aquatic thermal gradients (Williamson et al. 1989; Bury et al. 2000) regardless of acclimation temperature (Williamson et al. 1989). Hatchlings from eggs incubated at high temperatures select areas with cooler water temperatures, whereas hatchlings from eggs incubated at lower temperatures appear to select warmer areas (O'Steen 1998; Rhen & Lang 1999a). Moreover, Rhen and Lang (1999a) found that incubation temperature, time of day, and ingestion of food interact to influence temperature selection. Since metabolism is strongly influenced by ambient temperature, the influence of incubation temperature on behavioral thermoregulation by the hatchlings may have lasting effects on hatchling growth and development (O'Steen 1998; Rhen & Lang 1999a).

The propensity for juvenile turtles to select shallow aquatic habitats, however, may also present some risks. Snapping turtles in general select shallow water for overwintering (see Ultsch & Reese, Chapter 9). In small, still bodies of water, this may make the hatchlings more susceptible to freezing, anoxia, and/or drying of the overwintering location. Ultsch & Reese (Chapter 9) noted that young juveniles have lower anoxia tolerance than do adults. Thus, overwintering may be an important source of mortality for juvenile turtles, ultimately affecting demographics of snapping turtle populations (Ultsch & Reese, Chapter 9).

Survivorship through the First Year

Few data for survivorship of hatchling snapping turtles through the first winter of life are available, but those that are suggest that survivorship from nest emergence to the end of the first year is low compared with other age classes (Brooks et al. 1988; Congdon et al. 1994). In a life table constructed for females of an Ontario population, Brooks et al. (1988) estimated annual survival probability for snapping turtles of all juvenile age classes (hatchling to 17 years) to be 75.4%. A recalculation of these values by Galbraith et al. (1997) adjusted juvenile survival probability to be 82.8%. These values were based on observations of adult female survivorship (estimated from recapture rates) and nest success, and did not directly measure juvenile recapture rates or distinguish among juveniles of different age classes. Thus, the values presented in this life table may overestimate survivorship of younger juveniles and underestimate survivorship of nearly mature juveniles. Congdon et al. (1994) estimated a survivorship rate of 9% for Michigan hatchlings through the first year, but believed this to be an underestimate of actual survivorship, as hatchlings were small enough to escape the traps used to recapture animals. For life table

analyses, Congdon et al. (1994) adjusted the probability of survivorship from hatching to one year of age to 47%. However, in a later study of the same population, Congdon et al. (1999) found survivorship from hatching through at least one year of age to be, again, approximately 9%. Further direct measurement of hatchling survivorship is clearly needed to ascertain whether survival probabilities estimated in life table analyses are accurate. Moreover, similar, long-term investigations into the life history and demography of other snapping turtle populations are needed to evaluate potential geographical variation in juvenile survivorship in this wide-ranging species.

No study has been conducted to examine the factors causing mortality in hatchling snapping turtles after entering the water. However, potential major contributors to mortality include predation, adverse environmental conditions (e.g., freezing or hypoxia during overwintering), and malnutrition due to a lack of environmentally derived resources. Investigation into the relative influence of these factors on survivorship during different intervals in the first year of life is warranted.

Two studies have examined the influence that attributes of individual hatchlings have on their survivorship during the first year of life. In an experimental pond, Janzen (1995) observed that survivorship through the first year was higher for hatchlings from eggs incubated at temperatures that generated all males or all females than for hatchlings from eggs incubated at an intermediate temperature that produced mixed sexes; he ascribed this to a tendency for intermediate-temperature hatchlings to run away from nearby predators as opposed to remaining stationary. In a long-term study, Congdon et al. (1999) tested the influence of hatchling size on survivorship in a natural population in Michigan but found no evidence that hatchling size influenced survival of the hatchlings to one year of age.

The time spent in the aquatic environment is by far the longest of the three periods in the early life history of hatchling snapping turtles that we have addressed in this chapter, and it is likely that events during this period have more influence on survivorship of the hatchlings than do events occurring during hatching or overland migration. However, our understanding of how various characteristics of the hatchling and factors present in the environment influence survival is arguably less for this period than for the other two. The long duration of this period, the cryptic nature of the hatchlings, and the complexity of the habitat make exploration of the aquatic biology of hatchling turtles particularly challenging. However, further exploration of this life stage is sorely needed to ascertain whether phenotypic variation in the hatchlings does indeed lead to variation in fitness.

SUMMARY AND FUTURE DIRECTIONS

The hatchling stage is one of the most intensively studied stages in the life history of *Chelydra serpentina*. Indeed, much of what we know about this species and the physiology and ecology of turtles in general has been gleaned from research on hatchling snapping turtles. However, as we have frequently noted, many aspects of the biology of hatchlings are still largely unexplored. As we close this chapter, we wish to highlight three areas of study that merit particular attention in future research.

1. *Hatchling performance in complex environments.* Most studies that have investigated factors that influence hatchling performance and survival have been laboratory-based experiments focusing on one or a few potential variables. The few studies that have examined performance in the field have in most cases also focused on one or a few potential factors that may influence that performance. Rarely is an extensive array of potentially important maternal and environmental factors included in analyses of performance under natural conditions, nor are potential interactions among these factors regularly addressed. As our knowledge of the effects environmental and maternal factors independently exert on performance increases, research should begin focusing more on potential interactions among these various factors and how the synthesis of complex arrays of factors ultimately influence performance and survival.

2. *Aquatic ecology and survivorship of hatchling turtles.* The importance of phenotypic variation in hatchling snapping turtles on fitness, to which so many researchers have alluded, must ultimately be assessed in the context of long-term survivorship, as snapping turtles, like other turtles, have delayed sexual maturation.

However, direct measurements of survival rates of hatchlings beyond the immediate posthatching period are largely unavailable. Moreover, the relative importance of various factors that may influence survival once the hatchling enters the aquatic environment is largely unexplored. Such information is critical to evaluating the importance of phenotypic variation in the early life history of this organism.

3. *Interpopulational and geographic variation in hatchling physiology and ecology.* Chelydra serpentina is a wide-ranging species (see Ernst, Chapter 1, and Moll & Iverson, Chapter 17), inhabiting many different types of aquatic environments in diverse climates. However, variation in hatchling phenotypes among populations and differences in the ecological importance of phenotypic variation among populations have rarely been addressed, despite the potential for selective pressures exerted on the hatchlings to differ markedly among populations. Such comparisons among populations may help to resolve issues of the importance of phenotypic variation on fitness, and provide insight into the adaptive capabilities of this wide-ranging turtle species.

ACKNOWLEDGMENTS

We thank M. Ewert, K. Miller, and an anonymous reviewer for their constructive comments on previous drafts of this manuscript. We also thank A. Steyermark for his vision of assembling this important volume, and for inviting us to be part of it.

16

Population Biology and Population Genetics

DAVID A. GALBRAITH

A PRODUCTIVE DEFINITION OF POPULATION biology might be that it is the study of structure, or nonrandom distributions, in space and time of interacting groups of individuals of the same species. In other words, population biology consists of observation and hypothesis tests regarding phenomena above the level of individual but below the level of species, and it includes the examination of varying scales ranging from demes and breeding assemblages to patterns detectable in comparisons among populations separated widely by geography, habitat, or climate. These phenomena include spatial and temporal ecology, the results of interactions of groups of individuals, and the distribution of genetic diversity, among other things.

This rather dry definition belies the intellectual and practical challenges of studying the population biology of a long-lived, cryptic vertebrate like *Chelydra serpentina*. And it does not impart a sense of the importance of developing and testing quantitative hypotheses about patterns of survivorship, reproduction, and genetic diversity within and among populations for our understanding of evolutionary ecology and for our ability to come to terms with the effects of the activities of our own species on the survival of others.

Few species present as many contrasts and opportunities in this regard as *C. serpentina* does. One of the most widespread of nonmarine reptiles, snapping turtles have a phenomenal geographic distribution extending over 50 degrees of latitude in the Americas, being found in Ecuador, Colombia, Costa Rica, Honduras, Mexico, 41 of the 48 continental United States, and at least six Canadian provinces (Ernst & Barbour 1972), a span of more than 6,000 km. This taxon is also widely recognized by the general public, at least in temperate climates, and much of its range coincides with the world's largest population of herpetologists (Brooks 1992).

For all of this range, familiarity, and opportunity, rigorous understanding of the population biology of *C. serpentina* across its range remains elusive. Even in well-studied habitats only a fraction of individuals are encountered for more than a few moments per year. Despite dozens of single-capture or mark-recapture studies of various aspects of the biology of *C. serpentina* since the middle of the twentieth century, to date only two populations have been studied in sufficient detail, number, and duration to result in the publication of life tables. These two long-term studies, at the University of Michigan's E. S. George Reserve in Livingston County, Michigan (ESGR, 42°28' N, 84°00' W Congdon et al. 1987, corrected) and at the Algonquin Provincial Park Wildlife Research Station in the District of Nipissing, Ontario (WRS, 45°35' N, 78°30' W, Galbraith 1986), are separated by less than 600 km in total, some 350 km north-south and 450 km east-west. Both populations are within the northernmost portion of this species' geographic range and both are of the northernmost species, *C. serpentina serpentina*.

In this chapter we review the population biology of snapping turtles with a focus on what is and is not known about both the demography and genetic diversity of those populations. First, we consider temporal structure of populations as we look at studies of the demographics of snapping turtle populations and the geographic variation in the components of the demographic environment. Second, we examine what is known about the genetic diversity and structure of snapping turtle populations and present eight predictions regarding that diversity relative to other aquatic turtles. Finally, we outline the use of snapping turtles as model organisms with which to test hypotheses about population biology and suggest new avenues for research on this fascinating species.

Several topics that are within the sphere of population biology are not considered in this chapter. Readers are referred to other chapters in this volume for reviews of environmental sex determination (Janzen, Chapter 14), energetics (Spotila & Bell, Chapter 7, and Gatten, Chapter 8), overwintering biology (Ultsch & Reese, Chapter 9), growth (Steyermark, Chapter 11), activities of hatchlings (Finkler & Kolbe, Chapter 15), and movements and activity (Congdon et al., Chapter 12).

WHAT IS A POPULATION OF SNAPPING TURTLES?

Although populations are considered to be a basic unit of biological integration, there is no single definition of a population that covers all purposes and situations. Elements of several definitions are important in understanding snapping turtle populations. Krebs' 1972 definition of a population as a group of individuals of the same species occupying a particular space at a particular time is often the most reliable and practical. Cole's 1957 definition that a population is a biological unit wherein the demographic parameters of birth rate, death rate, sex ratios, age structure, and so forth, can be described as applying to the unit is also important, and shapes the assumptions underlying most field studies.

Testing the assumption that an assemblage of individuals within a defined area at a defined time (population *sensu* Krebs 1972) actually consists of a single biological unit (population *sensu* Cole 1957) is highly problematic. It is now possible to dissect membership of genetically distinct but spatially overlapping populations through the use of assignment tests, but this process has not been applied to turtle populations.

Gibbons (1990) presented a coherent definition of a population of aquatic turtles, concluding that a population is bound by the geographic constraints of living within a body of water (and adjacent habitat components like nesting areas and overland migration/dispersal routes) such that proximity of individuals at the time of mating permits regular gene flow. Most researchers studying aquatic turtles define the population under study solely on the basis of geography, usually as one or more bodies of water in which turtles of interest reside most of the year and additional terrestrial space used by individuals as deduced from studies of movements. A further means of defining population membership may be to consider that group of female turtles making use of a particular location for nesting.

In his consideration of the problems of defining a population, Gibbons (1990) also noted that the demographic and other characteristics of actual populations are highly dependent on the contingent history of the populations under study. Determination of the characteristics of a life table for one population of one species, for example, should not be interpreted as providing a "typical" life table for that species, at least not until comparable observations have been made of other populations and used to test for significant differences.

The geographic bounds of snapping turtle populations have not yet been well defined, but approaches in other turtle populations are suggestive. To define the characteristics of a metapopulation of slider turtles (*Trachemys scripta*), Burke et al. (1995) analyzed mark-recapture records at eight aquatic habitats at the Savannah River Site in South Carolina. A total of 26 years of mark-recapture study was necessary to detect movements of marked individuals in all eight areas studied, representing a total terrestrial area defined by a convex polygon enclosing all eight water bodies within 841 hectares of terrestrial habitat. A landscape ecology approach was taken by Joyal et al. (2001) to document habitat use in spotted turtles (*Clemmys guttata*) and Blanding's turtles (*Emydoidea blandingii*) in Maine, by tracking individual movements among wetland and upland habitat components. Approaching turtle movements in real landscapes that are patchy in time and space emphasized that the turtle populations could not be localized to single patches of wetland. Furthermore, drying of wetland patches may take place frequently enough that "extinction" and "recolonization" are not primarily due to death of individuals and repopulation through arrival of propagules and subsequent reproduction. Instead, in long-lived, mobile species, individuals may experience several such cycles during their lives by serially moving among relatively ephemeral bodies of water as they dry and flood (Burke et al. 1995).

Operationally, populations of snapping turtles have usually been defined by researchers as those individuals occupying one or more bodies of water at a particular time or times, or those making use of one or more nesting sites, or a combination of the two. These two components of capture location correspond to different habitat uses and they may be separated by considerable distances. For example, making use of radiotelemetry, Obbard & Brooks (1980) found that adult female *C. serpentina* at the WRS undergo an annual nesting migration averaging 10.6 km, with some females regularly making overland movements of 500 m during the migration. It is thought that the extent of the migratory distance in this migration is a result of the scarcity of suitable nesting sites in the watershed.

DEMOGRAPHIC STRUCTURE
Age Estimation

Few characteristics of individuals are as important to our understanding of the dynamics of populations as is individual age. It is through estimates of individual age (combined with observations of fecundity) that growth rates of populations can be calculated through the use of life tables and other methods (see below) and that physical growth rates may be reconstructed. Testing hypotheses concerning life-history evolution and reproductive investment (see Congdon et al., Chapter 12; Congdon et al. 2001) usually requires individual age as an important parameter.

As with every other datum that may be generated in field studies, estimates of individual age are subject to error. In long-lived species, the interpretation of subsequent analytical results rests on the assumptions of the effects of measurement error on the observed variable and the assumption that errors are truly random. Consideration of the detection and quantification of measurement error in age estimates is therefore vital in population studies of snapping turtles.

In an ideal population study of long-lived organisms like snapping turtles, individual age would be determined, not estimated. If all hatchlings could be permanently marked at hatching, at least to year-of-hatching or cohort identity, then age at subsequent recapture could be determined with reliability. Very few population studies have attempted to mark entire cohorts of hatchling turtles over several years, but it has been done.

For example, despite a long-term field study of snapping

turtle hatchling cohorts marked by toe clips, few hatchlings or juveniles have ever been recovered in the Algonquin Park population (Cunnington & Brooks 1996). Note that such marking techniques themselves must be considered as an anthropogenic interference in the cohort marked.

Only a few studies have been able to follow cohorts of free-living aquatic turtles of any species from hatchling through to the onset of reproduction by known individual. In the long-term study of Blanding's turtles (*Emydoidea blandingi*) at the E. S. George Reserve, the first capture of a truly known-age first nesting female was recorded after more than two decades of study (Congdon et al. 1993).

In most population studies, individual age must be estimated because we do not have access to true age. Techniques for estimating individual age may be grouped into two categories: estimates that count discrete natural markers of temporal events (quantum aging), and estimates calibrated against a continuous biological, chemical, or physical process such as the accumulation of biochemical substances. In studies of mammals, age has been estimated by calibration of the rate of racemization of aspartic acid in insoluble proteins in the lens of the eye, assayed after death (e.g., Bada et al. 1980; George et al. 1999). It does not appear that this method has been applied to turtles.

In many organisms, growing seasons, dormancy periods, or other discrete seasonal or life-history events leave phenotypic marks in some hard part of the anatomy. Most age estimation techniques that have been developed for turtles rely on such natural temporal markers in bone or keratin. In turtles, natural markers of temporal events are deposited on two tissues: as growth rings, annuli, or lines visible ex-

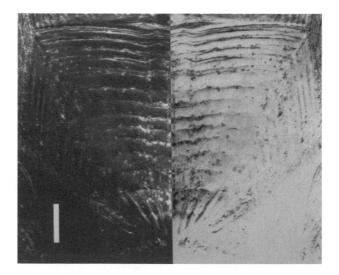

Fig. 16.1. Photograph of the left side of the fourth vertebral scute of a female snapping turtle captured in the Welland River, Ontario, Canada (mass, 6.6 kg; carapace length, 31 cm), an example of an adult snapping turtle displaying "clear" annuli or growth rings. Cranial is toward the top of the page. The white scale bar indicates 1 cm along the cranial-caudal axis. The left side of the figure is the original photograph of the left side of the scute. The right side of the photograph is also the left side of the scute, reversed left-for-right and printed as a negative to highlight annuli.

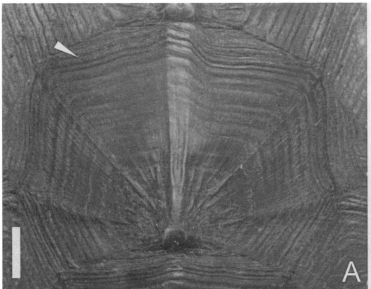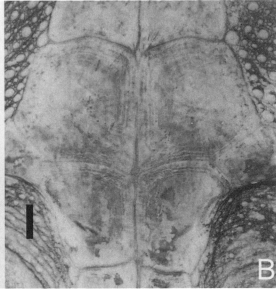

Fig. 16.2. Photographs of portions of the carapace and plastron of a juvenile snapping turtle captured in Algonquin Provincial Park, Ontario, Canada, illustrating the difference in the retention of growth-related markings on the two surfaces. On both panels, cranial is toward the top of the page. (A) The fourth vertebral scute of this juvenile has retained at least 11 clear growth lines as the scute has enlarged. The margin between the third and fourth vertebral scutes has been marked by the right-most point of the white arrowhead. The white scale bar indicates 1 cm along the cranial-caudal axis. (B) The central portion of the plastron of the same juvenile snapping turtle, showing very poor retention of growth-related markings on the softer underside. The black scale bar indicates 1 cm along the cranial-caudal axis.

ternally on the keratinacious scutes of carapace or plastron, and as growth lines in compact bone that can only be revealed postmortem through histological preparations.

Annuli or growth rings on the carapace are attractive age markers in snapping turtles because the large size of the carapace and the relatively rectangular shape of the vertebral scutes on this species present very regular-appearing rings on many individuals. An annulus is so called because of its ringlike appearance, not because it is an annual structure. Because of the presence of growth-related annuli on the carapacial scutes (Fig. 16.1) age estimation is a commonly attempted procedure in field studies of snapping turtles. Although some annuli may be formed on the plastron of younger individuals, the thicker, harder keratin of the carapace provides a better record of annual variation in growth (Fig. 16.2).

In a recent critical review of the use of carapacial annuli to estimate individual age in turtles, Wilson et al. (2003) found that of 150 case studies drawn from the literature, less than one third of studies provided any test at all of whether the technique was reliable. Of these, in only a fraction of studies did the authors believe that the method was reliable. Furthermore, of 22 studies that actually tested for an annual deposition of annuli, less than one quarter found that these features were added consistently on an annual basis. Citing widespread difficulties ranging from unjustified assumptions of the frequency of deposition of annuli to problems in interpretation, Wilson et al. (2003) concluded that this method should not be used without case-specific justification. Researchers must accept that validation is necessary on

a population by population basis, including documenting the frequency with which these features are formed and remain observable.

Three sources of measurement error are readily apparent in age estimates based on counts of annuli on carapacial scutes. First, any count of rings on a scute depends on the observer being able to accurately count the number of visible rings. Observer error can be further resolved into error among observers and error among counts by the same observer. Second, if the annuli are to be interpreted as truly representing individual age, it is assumed that the individual in question is still growing throughout life and that deposition of the rings has been annual without interruption or changes in depositional frequency. Major changes in growth rate at various times in life can result in slowing or stopping apparent ring deposition, leading to a systematic error underestimating true age in many individuals. Third, older annuli (those deposited when the individual was young) may become obscured or be completely lost with age or damage to the tissues involved. In extreme cases of large adults, and especially in large males encountered in central Ontario, Canada, most or all of the area of the carapacial scutes can appear completely smooth.

The magnitude of these errors was first determined quantitatively by Galbraith & Brooks (1989), who found that age estimates of larger adult snapping turtles at the WRS population may have uncertainties exceeding 100% of the estimated ages (see Fig. 16.3). In addition, between-observer biases or errors were a significant source of error in age estimates. Similar cautionary results have been obtained by

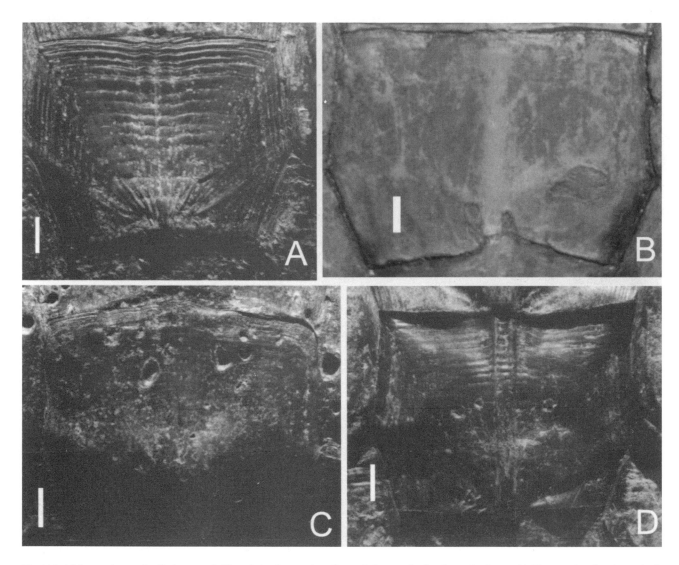

Fig. 16.3. Adult snapping turtles display remarkable variation in retention of growth lines on the fourth vertebral scute. (A) The complete fourth vertebral scute of the turtle shown in Fig. 16.1. (B) The fourth vertebral scute of a large adult male snapping turtle from Algonquin Provincial Park, Ontario, Canada, in 1983 (carapace length [CL] = 36 cm). (C) Worn and damaged fourth vertebral scute of an adult snapping turtle (location of capture and sex unknown; CL = 27 cm). In addition to the typical appearance of an older turtle that has lost many older markings, this individual displayed numerous lesions, perforating the keratin and invading the bony layer of the carapace, that may have been caused by a bacterial infection. (D) Fourth vertebral scute of a snapping turtle that displays both substantial areas of wear and also retention of some growth-related surface features (location of capture and sex unknown; CL = 27 cm). On all panels, cranial is toward the top of the page and the white scale bar indicates 1 cm along the cranial-caudal axis.

Brooks et al. (1997) and Wilson et al. (2003). Given these warnings the reliability of any conclusion based on adult age estimates must be considered low, even with some form of calibration.

Estimating age through counts of carapacial annuli of juvenile snapping turtles still appears justifiable, at least in strongly seasonal climates, but some form of calibration of the marks being counted is needed. Field researchers should take care to clarify just what it is that each observer is counting and to pay attention to the topography of the scutes, especially in the areas represented by the first two years of life.

A third type of chronological marker in age-related studies is the use of mark-recapture over a known span of time to gauge changes in individuals related to that known dura-

tion. This involves the application of an artificial marker that is recognizable at some future capture date, and testing for phenotypic changes (typically in body size or reproductive performance) taking place subsequent to the first marker. Such chronological markers may be individual-specific (tags, carapace notches) or they may be temporal in the case of the injection of tetracycline or some other biochemical substance that will leave a permanent mark in bone, visible postmortem.

Survival and Mortality

Few studies have directly quantified sources of adult mortality in *C. serpentina*. As recovery of individuals postmortem is difficult or impossible without a previously attached radio transmitter, samples of individuals that have died of known causes are rare. In the absence of observed incidents of natural mortality, adult survivorship in all studies to date has been estimated by mark-recapture.

In the Algonquin Park long-term study, two causes of natural adult mortality were documented. Some adults appear to die of bacterial infections during or after winter seasons. The rate at which this cause of death occurs is not known, nor is there any information to suggest that some individuals are more susceptible to this form of mortality than others.

The other known cause of mortality in adult snapping turtles is predation by mammals. A catastrophic increase in adult mortality was observed in the Algonquin Park population between 1987 and 1989 (Brooks et al. 1991b). During this relatively short interval, approximately 50% of the known adult snapping turtles were killed outright or severely wounded some time during aquatic overwintering torpor. Although no direct observations of the mammal species responsible were made, circumstantial evidence indicated that the predators were river otter (*Lontra canadensis*). This observation of a significant increase or pulse in adult mortality by nonprimate mammalian predation has not been repeated by other workers. It is therefore impossible to judge anything about the frequency with which this particular type of catastrophe might affect snapping turtle populations.

Another indication of the low predation risks faced by adult snapping turtles under most conditions is the limited degree to which adults that display limb amputations have been encountered, injuries presumably suffered during predation attempts by mammals. Saumure (2001) documented one limb amputation among 229 snapping turtles captured in Quebec and cited other reports of limb loss in Ontario as less than 1% of the adult population. Reported frequencies of healed injuries must be interpreted with caution, as more severe injuries do happen and may be fatal (Brooks et al. 1991b).

Patterns in Survivorship

Note that humans have been a factor in the demographic environment of snapping turtles for thousands of years, but the intensity and form of the interactions that affect present *C. serpentina* populations across their North American range are likely increasing. These interactions, including habitat alteration and fragmentation, increases in adult mortality from trapping and road kill, and the effects of pollutants (organochlorides and hormone mimics among others) are not expected to be neutral with respect to the persistence of turtle populations.

At the same time as human activities and elements of the built environment have had negative effects on turtle populations, some human activities have been of local benefit to snapping turtle populations. For example, in at least some areas human disturbance of the terrestrial substrate, from the preparation of gardens and fields for planting to the construction of gravel-shouldered roads and railway embankments, have increased nesting opportunities for snapping turtles. Indeed, the very existence of the WRS population and the initial phases of the long-term population study of snapping turtles there are linked to females nesting on a gravel and sand artificial dam. Given that this northern population is in fact situated on the Canadian shield, little natural nesting substrate is available in this area. Most known nests throughout the duration of this study were constructed on the dam, along roadsides, or along an abandoned gravel railway embankment.

It is also possible that interactions with researchers change the demographic environment of turtle populations, although the effects of such activities have not been quantified by comparative study. Field work by researchers should not be assumed to be benign and should always be subject to appropriate controls. For example, after examining the data from previously published mark-recapture studies of toads and frogs, Parris & McCarthy (2001) claim to have detected a significant decline in the probability that an animal would be recaptured after being subjected to a toe clip for marking or tissue-sampling purposes.

Fecundity

Because detailed reviews of the reproductive biology and life-history evolution in snapping turtles are presented elsewhere in this volume (e.g., Congdon et al., Chapter 12), we will refer to variation in clutch size and other reproductive variables only to draw attention to the major trends that appear present and that would be of importance at the level of demography.

Iverson et al. (1997) conducted a detailed review of the characteristics of female snapping turtles and reproduction as they vary geographically. In addition to contributing observations on such parameters as clutch size, egg dimensions, and relative clutch mass from a study site in Nebraska, Iverson et al. (1997) assembled literature reports of female reproductive characteristics from 48 other locations ranging from 50° N in Manitoba to 8° N in Colombia (Table 16.1; Iverson et al. 1997). Unfortunately many of the literature accounts of reproduction in this species are statistically inadequate. Of the 48 records cited by Iverson et al. (1997), only eight are of observations on at least 20 individuals. Even the field results of Iverson et al. (1997) are based on a sample of 22 clutches.

Clutch size varies enormously in *C. serpentina,* with the smallest clutches of less than 20 eggs and the largest on record as being more than 100. This variation is present among females within populations and among populations (Iverson et al. 1997).

Table 16.1

Summaries of values used in construction of life tables for *C. serpentina* at the E. S. George Reserve (ESGR), southeastern Michigan, and the Algonquin Park Wildlife Research Station (WRS), central Ontario

Population	ESGR	WRS
Reproduction		
Mean clutch size	28.0	34.0
Clutch frequency	0.85	0.72
Primary sex ratio	0.5	0.66
Annual fecundity	11.9 eggs	16.16 eggs
Emigration rate	0.005	unknown
Age at first report	11	17
Survivorships (l_x)		
Nest (age 0)	0.23	0.0635
Mean juvenile (ages 1–12)	0.745	0.766
Adult females	0.93	0.966
Age at $R = 1$	/6	infinite
Source	Congdon et al. 1994	Brooks et al. 1988

The number of eggs produced by a female snapping turtle is strongly correlated with body size. Importantly, Iverson et al. (1997) detected significant correlations between body size and latitude, longitude, and elevation among 20 temperate populations in North America. In a multivariate analysis of data from 19 studies, Iverson et al. (1997) found that 81% of the variation in clutch size among populations is explained by a multivariate model including carapace length and latitude of the study site. In this analysis clutch size was positively correlated with both carapace length and latitude.

The relationship between clutch size and body size appears to be different in tropical populations. Female snapping turtles in tropical locations are much larger than females in temperate climates but they also appear to produce smaller clutches than their temperate counterparts (Iverson et al. 1997). It is possible that fewer eggs per clutch in the tropics is offset by production of more clutches per year, but this remains to be demonstrated in this species.

Ewert (2000) demonstrated that Florida snapping turtles, *C. serpentina osceola*, nest as early as February each year and that they produce more than one clutch per season, possibly as many as three or four. The number of eggs produced per season per female was estimated by Ewert (2000) to be between 19 and 36, with a mean of 27.6, indicating that multiple clutches are not a means of producing many more eggs per season in this species. Although individual clutches as large as 52 eggs have been reported in *C. serpentina serpentina* in Florida, clutches no larger than 31 eggs have been reported in *C. serpentina osceola* (Jackson & Ewert 1997). The few reports available suggest that the clutches of the Florida subspecies appear to average less than 20 eggs.

Bobyn & Brooks (1994) demonstrated that hatchling snapping turtles from the WRS population and from another

further south in Ontario, located at Cootes Paradise, Hamilton, displayed genetic differences in growth rates, with hatchlings from the more northerly population growing more slowly than those from the south when exposed to the same environment.

Contributions of Males to Reproduction

The contribution of males to reproduction is not often factored into demographic analysis. Males simply "don't appear" on life tables as they are often constructed, because their participation is not a major factor in determining the net reproductive rate of turtle populations. This is not to suggest that males could not be represented in demographic studies of turtles. For the most part, however, they have not been.

The fact that the females of many turtle species are able to store viable sperm for several seasons is one factor that limits the need to include information about males in life-table analysis. In general, it is assumed that male gametes are not a limiting factor to females, such that either mating before a particular nesting season, or stored sperm from previous seasons, will be available to fertilize every clutch females are able to produce. These assumptions have not been tested in the field.

The presence of environmental sex determination and nonequal sex ratios in *C. serpentina* also affects the ability to include information about males in life tables in a meaningful fashion. As unequal primary sex ratios have been demonstrated in this species, by definition it's not possible to make the usual assumption that half of all offspring at birth are females and half are males for the purposes of estimating m_x.

Life Tables and Population Growth

A primary objective of demography is to estimate the intrinsic rate of increase of the population under study. This rate, r, is usually defined as the change in population size, N_i, over time interval t, in the familiar equation for exponential growth,

$$N_t = N_0 e^{rt}$$

Thus, if $r = 0$, then $N_t = N_0$; if $r < 0$, then $N_t < N_0$ and the population is declining. Similarly if $r > 0$, then $N_t > N_0$ and the population is growing in size.

For long-lived, sometimes cryptic organisms capable of extended dispersal and migratory movements like *C. serpentina*, directly measuring population size at any given time is almost impossible, unless a whole population can be defined as inhabiting a single small area and then all individuals can be captured and counted. Even more challenging for such taxa is seeking to understand the dynamics of a population by directly deriving rates of change in population size, as this would require repeated determination of population size without the observer interfering in the very dynamic of interest.

Conventionally, two types of life tables have been constructed. A life table consisting of a schedule of survival and mortality rates for all ages of a single cohort is termed a vertical life table (Deevey 1947). Vertical life tables express the various risks and reproductive probabilities as experienced by an individual organism through its life. In contrast, life tables can also be constructed that express the demographic environment of all age classes of a population at any given time; this is termed a horizontal life table (Deevey 1947).

A review of the development of the use of life tables in turtle populations is provided by Frazer et al. (1990), along with a presentation of a complete life table of the well-studied *Pseudemys scripta* population in Ellenton Bay, South Carolina. In brief, a life table is a "digestible summary of vast amounts of information on survivorship, fecundity and age at maturity for a particular population" (Frazer et al. 1990).

The conventional formulation of a life table relates age-specific survivorship, l_{x2}, defined as the probability of survival of an individual from age 0 to age *x*, and age-specific fecundity, m_{x2}, defined as the average number of female-producing eggs laid per individual female at age *x*. If l_x and m_x can be determined over all ages then the net reproductive rate, R_0, of the population can be calculated from

$$R_0 = \sum l_x m_x$$

Thus, if $R_0 = 1$, then $N_t = N_0$; if $R_0 < 1$, then $N_t < N_0$ and the population is declining. Similarly if $R_0 > 1$, then $N_t > N_0$ and the population is growing in size. The same information can be used to obtain the intrinsic rate of increase of the population, *r*, through an analytical solution to

$$1 = \sum l_x m_x e^{-rx}$$

(Ricklefs 1979). Although simple in concept, deriving l_x and m_x for real populations is often difficult. In theory the simplest way to determine l_x is to mark all individuals of a particular cohort at natality and then census the population each year from $x = 0$ to $x = \infty$, or at least until *x* is equal to the oldest known age class for the population. In real populations there is always risk of failure to recapture all individuals of a particular cohort at any particular year.

In long-lived organisms, where median life expectancy might be equal to or greater than the length of an average academic career, construction of a true vertical life table is practically impossible. Similarly, deriving a true horizontal life table, with the demographic characteristics of all age classes in a population at a single time, would require far more data than has ever been recovered from a single population. As noted above, a key concern in population studies of long-lived species is deriving estimates of individual age. Unless age can be determined reliably through some means, such as recovery of an individual marked as a hatchling, then the age of each individual can only be estimated.

The pragmatic approach that has been taken to date in preparing life tables of turtle populations is to construct horizontal life tables through estimated rates of mortality and reproduction, either for discrete individual ages or, more commonly, by using some version of life stages defined operationally depending on the researcher's ability to classify individuals by age or stage. Stage-specific rates have usually been estimated from data collected for other purposes throughout several years of study. While allowing a tractable life table to be constructed, this technique depends on the crucial assumption that the demographic environment is stable enough that annual or longer changes in that environment can be safely discounted.

Life tables have been published for two *C. serpentina* populations. Brooks et al. (1988) and Galbraith et al. (1997) published similar horizontal life tables for female snapping turtles at the WRS. Brooks et al. (1988) developed a life table as a tool to inform management decisions and indicate that protection of adult females is critical to population survival. Galbraith et al. (1997) revised the earlier life table of Brooks et al. (1988) and used it as the basis for estimating the costs of four different management scenarios involving manipulation of various life stages.

The long-term study of snapping turtles at the University of Michigan's E. S. George Reserve, in southeastern Michigan, was used as the basis for a life table constructed by Congdon et al. (1994). Unlike the Algonquin Park population, where few juveniles have ever been captured, the critical annual rate of survivorship for juveniles at the E. S. George Reserve was estimated from the capture of more than 1,000 hatchlings and juveniles between 1975 and 1992. The demographic parameters of these two populations are summarized in Table 16.1.

Although widely used in demography, the life-table model of population growth cannot be used in all situations. As noted above, in real field populations of turtles the age of any individual can only be estimated, not determined, at this time. Those age estimates may include significant sources of error due to both the biological basis of the technique and the repeatability of human observation (Galbraith & Brooks 1987, Brooks et al. 1997, Wilson et al. 2003), and thus may have dubious value as the source of demographic data.

Two alternative approaches to studying population growth make use of either size classes or life-history stages as the temporal variable, rather than year class as in the life table or the Leslie matrix. Developed as a matrix, stage-based models also permit that the intrinsic rate of increase of the population can be estimated.

Cunnington and Brooks (1996) used a stage-based model to test for sensitivity of population replacement to changes in survival of various life stages in *C. serpentina*. They noted that the key differences between the ESGR and WRS studies were the higher annual survival of adult females in the northern WRS population (0.96) relative to the ESGR population (0.95), and the lower probability of hatching and emergence at WRS (0.065) relative to ESGR (0.50). If the goal is to find management recommendations to increase the probability of population survival, then all options can ultimately be reduced to changes in survival of one or more life stages. The high adult female survivorship documented at WRS means that there is very little "room" in which to increase survivorship further. Therefore protecting females is not an option that will increase population survival. Instead, low survival from eggs through to emergence of hatchlings presents a much greater opportunity to change population growth at WRS. Although Brooks et al. (1988) and Galbraith et al. (1997) had previously predicted that "head starting" programs for *C. serpentina* might be ineffective or too expensive as conservation options for northern populations, the findings of Cunnington and Brooks (1996) suggest otherwise. The situation appears different at the ESGR population. The lower female survival and higher survival in the first year suggest that increasing survivorship of adult females may be more effective under conditions like those experienced by the ESGR population than those at WRS. Using a matrix-based approach similar to the demography of a population of diamondback terrapins (*Malaclemys terrapin*) in Rhode Island, Mitro (2003) found that protecting nests from predation would increase population growth rate, but also that populations that were experiencing high mortality at other critical life stages would not benefit greatly from nest protection.

Some population biologists have developed demographic models on the basis of size classes rather than age. In part, these models were developed to reflect the biological reality that reproduction in many species is more a function of size than of age. These models also permit some examination of demography in field populations when age determination is not possible. To date a size-based model approach has not been developed for snapping turtles, but the stage-based approach of Cunnington and Brooks (1996) suggests that the approach is possible.

POPULATION GENETIC STRUCTURE IN *C. SERPENTINA*

Like the demographic study of populations, much of the effort to understand genetic diversity within populations rests in uncovering structure in the distribution of elements of that diversity. Consideration of population genetics together with a more general review of population biology is not to suggest that population genetics is often considered as an adjunct of the broad topics encountered in population biology. However, many of the basic themes encountered at the population level when considering whole organisms (especially testing for structure, association, and the aggregate effects of individual behavior) also apply to the distribution of genetic diversity below the species level.

The overall life-history patterns of snapping turtles allow for a basic set of predictions about relative levels of genetic diversity and of structure in that diversity. Many factors can affect genetic diversity at the population level. Feedback between population genetic diversity and demographic characteristics should be considered when studying either. Population-level phenomena such as rates of multiple insemination and environmentally mediated dispersal in turtles affect genetic diversity and structure sufficiently that it is possible to detect them using genetic markers (Scribner & Chesser 2001).

The expected variance in reproductive success among both females and males suggests that effective population size in most populations of snapping turtles might be relatively low, promoting genetic drift and random fixation of alleles. At the same time, the relatively high mobility and potential for dispersal in this species suggest that gene flow may also be relatively high. Thus, we predict that snapping turtles may display relatively low genetic diversity within populations because of the effects of stochastic variation in reproductive success. Four predictions can thus be made about the direction of effects of demographic factors on genetic diversity within snapping turtle populations, and four more can be made about such effects on among-population diversity.

Prediction 1. High Variance in Reproductive Success among Females Predicts Relatively Low Levels of Genetic Diversity

Great variation in clutch size occurs among female snapping turtles (Iverson et al. 1997), as noted above. Environmental stochasticity affects incubation conditions and rates of nest predation. This suggests a high variance in a female's reproductive success in any given year in a species that lays dozens of eggs in its annual clutch. In accordance with the bet-hedging model of reproductive effort we should expect the year-to-year variance in clutch success to be averaged over the reproductive life of the individual, somewhat decreasing the net variance in numbers of surviving hatchlings. These sources of variance in reproductive success among females suggest that effective population size is relatively small and genetic drift should tend to reduce allelic diversity within snapping turtle populations.

Prediction 2. High Variance in Reproductive Success among Males Predicts Relatively Low Levels of Genetic Diversity

Among males it has been suggested that only the largest or most aggressive of individuals are able to secure home ranges in the most reproductively attractive locations along the annual nesting migration routes of females (Galbraith 1991). This in turn suggests a high variance in the number of males achieving reproduction each year. This variance might be expected to be offset somewhat by the effects of multiple paternity, as a male holding the most adventitious mating location cannot exclude other males in other locations from achieving at least some inseminations. As above, this predicted variance in male reproductive success would tend to reduce effective population size and promote genetic drift, reducing allelic diversity.

Prediction 3. Variance in Reproductive Success between Males and Females Predicts Differences in Nuclear and Mitochondrial DNA Diversity

It is possible that male snapping turtles experience a higher variance in lifetime reproductive success than females do. The strong sexual size dimorphism in this species, the apparent system of forcible insemination and other factors noted under Prediction 2 suggest that variance in reproductive success among males may be very high. If male snapping turtles experience higher variance in reproductive success than females do, this is yet another factor predicting a reduction in effective population size relative to a population without such a difference. However, this difference is expected to have different effects on nuclear and mitochondrial genes. It is predicted that nuclear genomes should experience higher genetic drift over time than mitochondrial genomes in this species.

Prediction 4. Environmental Sex Determination Predicts Relatively Low Levels of Genetic Diversity

If sex ratios regularly depart from 1:1 in snapping turtle populations, we expect that such a skew will decrease effective population size relative to a population with an unbiased sex ratio. It is possible, although admittedly speculative, that skewed sex ratios resulting from environmental sex determination (ESD) might therefore reduce effective population size, increasing genetic drift and thus decreasing expected genetic diversity within populations.

Prediction 5. Geographic Variation in the Demography of Snapping Turtles Predicts Geographic Variation in Population Genetic Diversity

Clutch size is known to vary with latitude among temperate populations of *C. serpentina* (Iverson et al. 1997), as do productive attributes of females such as age at first reproduction. Each of these suggests that variance in reproductive success may vary geographically, with females at the northerly portions of the distribution experiencing the highest such variance. If so, then we predict that genetic drift should be relatively higher in northern populations, decreasing genetic diversity relative to populations to the south.

Prediction 6. Postglacial Recolonization Predicts a Founder Effect among Northerly Populations

Because the habitats of northern populations of snapping turtles in North America were recolonized from the south at the end of the Wisconsinan ice age, we predict that genetic diversity within northern populations should be influenced by the alleles brought into those populations by their founders as populations expanded northward. This founder effect is predicted to have two effects. First, it should mean that different alleles may be found among northern populations (typical of founder effects), and second, that northern and southern populations may display differences in mean heterozygosity.

Prediction 7. Long-Distance Dispersal Capability Predicts Low Levels of Genetic Structure among Populations

The capacity of adult female snapping turtles to make extended annual nesting migrations, including long overland movements, suggests that gene flow among populations may be higher in snapping turtles relative to more sedentary taxa. High gene flow will reduce population genetic structure, and thus we predict that, relative to turtles that do not make extended movements or that may be restricted by terrestrial or other geographic barriers, genetic diversity among snapping turtle populations may be relatively low.

Prediction 8. Longer Mean Generation Times of Some Populations of Snapping Turtles Predict Slowing in Rate of Molecular Evolution

In a review of phylogenetic patterns across several reptilian taxa indicated by the rate of molecular evolution in three coding genes (cytochrome *b* and $NADH_4$ in the mitochondrial gemone and c-*mos* in the nuclear genome), Bromham (2002) found a significant negative relationship between body size and rate of molecular evolution. This relationship was interpreted as being the result of an increase in generation time among larger organisms relative to smaller species. If the same effect holds true at the scale of comparisons among populations of the same species (by no means assured), then snapping turtle populations that experience significantly longer generation times than others (as indicated in more northerly populations) may also experience slower rates of molecular evolution.

Studies to Date

Although there are still some data with which to test these predictions, available studies seem to be in accordance with the expectation of relatively little structure and low diversity in snapping turtles compared with other turtle species, at scales ranging from within and among local populations to comparisons on a regional or phylogeographic level.

In one of the few published studies of genetic variation within and among populations of snapping turtles, Phillips et al. (1996) examined mitochondrial DNA variability and allozymes in a total of nineteen specimens of *Chelydra serpentina* drawn from all four recognized subspecies. The objective of Phillips et al. (1996) was to test for the degree of genetic divergence among the four recognized subspecies. On the basis of mitochondrial DNA haplotypes, they suggested that the two Central American subspecies (*C. s. rossignonii* and *C. s. acutirostris*) be considered as species distinct from the two North American subspecies (*C. s. ser-*

pentina and *C. s. osceola*), but this interpretation was challenged by Sites and Crandall (1997).

At the level of geographic differentiation in genetic markers within species (phylogeography), recent studies have included snapping turtles among the turtle assemblages in the southeastern United States assayed for diversity. Of ten turtles surveyed by Walker et al. (1998) from Louisiana and Arkansas east to Virginia, the Carolinas, Georgia, and Florida, nine show some structuring into major phylogeographic units (Avise 2000). Only one species, *C. serpentina,* displayed no significant substructure across the range studied (Walker et al. 1998, Avise 2000). Note that the range of snapping turtle populations sampled by Walker et al. (1998) is only a small fraction of the range of this species, comprising specimens from just 10 of the 41 states within the U.S. range, and none outside of the United States. Nevertheless this sample should have been sufficient to detect such structure if it were present, given that the geographic region sampled includes Pleistocene refugia.

The lack of a phylogeographic structure observed in *C. serpentina* across the southeastern United States contrasts sharply with findings in other turtle species, including closely related alligator snapping turtles (*Macroclemys temminckii*). Analysis of variation in the control region of mitochondrial DNA from 158 specimens of *M. temminckii* from 12 drainages from across the range of this species revealed significant population structure among river systems (Roman et al. 1999).

At the end of the Wisconsinan ice age, snapping turtles and painted turtles (*Chrysemys picta*) were among the first species to recolonize formerly glaciated areas (Holman & Andrews 1994). Snapping turtles were present in 13 of 19 known Pleistocene glacial age turtle faunas across North America, interpreted by Holman & Andrews (1994) as an indication of their adaptation to cold conditions relative to other turtles. The apparent cold tolerance of snapping turtles, combined with their likely capacity for relatively high dispersal relative to other turtles, led Walker & Avise (1998) to suggest that snapping turtles may not have been confined to warmer refugia during the late Tertiary and Quarternary. Gene flow among populations spread over this wide range would have minimized subsequent founder effects as northerly habitats were recolonized.

Bobyn & Brooks (1994b) concluded that the duration of the incubation season is one of the most important factors governing the ability of snapping turtles to exist in colder climates. The ability of female snapping turtles to survive and reproduce at the northerly WRS site does not appear to be affected by average ambient temperatures. Successful incubation and emergence of hatchlings instead appears to be limited by the length of the incubation season and incubation temperatures (Bobyn & Brooks 1994b).

Phillips et al. (1996) also surveyed their nineteen specimens of snapping turtles for variation at 27 enzyme loci. As with some other turtle populations, little diversity was de-

tected among the proteins assayed. Of the 27 enzyme loci assayed, only 8 were polymorphic among the specimens sampled (average polymorphism = 0.296; calculated from Table 4, Phillips et al. 1996). An average of 1.333 alleles per locus were detected across all allozymes sampled (calculated from Table 4, Phillips et al. 1996). Heterozygosity averaged 0.0196 across all populations and loci in this study (calculated from Table 4, Phillips et al. 1996). Most of the allozyme diversity detected in this study was found among common snapping turtles, but as this study was based on only nine snapping turtles collected from known locations and one collected from the pet trade, very little can be inferred about the distribution of variation among sites. Of the ten heterozygous loci detected among all specimens, seven were detected within common snapping turtles and only one of these individuals had more than a single heterozygous locus. The reported mean heterozygosity of allozyme loci in snapping turtles was just one seventh that reported in sliders, *P. scripta,* at the Savannah River Site in South Carolina ($H =$ 0.148, Scribner et al. 1984). Scribner et al. (1986) found that populations separated by terrestrial barriers had lower heterozygosity and greater interpopulation divergence than those separated by water.

An unpublished study of genetic diversity among snapping turtles from four lakes in Nova Scotia using random amplified polymorphic DNA (RAPD) markers found no significant structure among sites, interpreted as either due to high gene flow or insufficient time since postglacial colonization to allow for differentiation in these neutral markers (Marlene Snyder, pers. commun.).

C. SERPENTINA AS A MODEL ORGANISM FOR POPULATION BIOLOGY STUDIES

Although snapping turtles have been frequently used as model organisms in the study of physiology, relatively few areas of population biology have used snapping turtles as models to characterize general trends or to test hypotheses that may be applicable to other organisms. However, they have been used to illustrate life history and reproductive trade-offs in long-lived organisms, and as models of the behavior of turtle populations (and of that of other long-lived species) in conservation and management scenarios. Both Congdon et al. (1994) and Brooks et al. (1988) made use of data from long-term ecological studies of *C. serpentina* to predict that snapping turtle populations would be vulnerable to removal of adult females.

Still relatively unexploited is the potential for snapping turtles to serve as a model for testing the limits of local adaptation in demographic characteristics across a wide range of habitats and climates in a reptile. This species is a generalist with a broad range. Some effort has gone into discussions as to factors that limit the species' distribution at its northern boundary (usually considered to be the bounds of the thermal regime necessary for successful egg incubation and hatchling emergence). The degree to which snapping turtles could be considered as a single wide-ranging species or should be divided into more than one species is, of course, of paramount importance in considering the utility of this taxon as a single unit for comparative purposes.

CONCLUSIONS

Snapping turtles are a familiar and widespread species of aquatic turtles and yet only a handful of field studies have attempted to follow the basic population processes of this species in sufficient detail to resolve survivorship. Nowhere has any field study of snapping turtles extended long enough to surpass the theoretical time needed to accommodate two generations. As a result, and because of the difficulties in documenting survivorship, in particular, among younger life stages, changes in the demographic environments of single populations over time have not yet been documented sufficiently to produce, for example, comparative longitudinal life tables by which mortality and fecundity of different cohorts could be examined.

Long-term studies of turtle populations are particularly valuable for both our academic understanding of these species and also for our ability to manage such populations. Such studies also present significant challenges to the management of the research programs themselves and also to careers of academic researchers in an impatient era. As the planning of most research projects depends on individual university faculty, researchers who are engaged in long-term field studies may have to consider 20 to 30 years of active research to encompass even a single generation of a species like *C. serpentina* within their careers. Relatively few researchers have the patience and persistence to justify this kind of work to funding agencies. Projects that have the scope to extend beyond the career horizon of a single researcher are even rarer and are often based at a field research station where continuity of field methods and sites, marking schemes and records can be established.

Snapping turtles present several advantages and challenges as the subjects of population study. Of particular and recurring interest is the use of carapacial annuli as a means for estimating individual age. Although this method may be of use in estimating the ages of juveniles the great sources of error that are inherent in the process of "aging" this species makes conclusions based on population samples of "aged" adults of dubious value at best. The mobility of adults (especially females) and the physical advantages of the large carapace as a platform for equipment attachments make radiotelemetry of *C. serpentina* a popular option for a variety of purposes.

The great north-south geographic range of *C. serpentina* also presents unexplored opportunities to examine demographic variation among populations of a single reptile species inhabiting greatly divergent habitats and climate zones. Although several studies have reported elements of fecundity data across a portion of the range, the only published studies to date that have generated sufficient survivorship data to attempt to formulate a life table have both been within the northernmost 10% of the range. In reviewing geographic variation in reproductive characteristics of snapping turtles, Iverson et al. (1997) concluded that "we still do not have sufficient data to permit a complete analysis of geographic variation in its life-history traits and a full investigation of its causes." Even for reproductive variables such as clutch size and female body size, published observations are largely lacking for the southern 60% of this species' range.

The substantial east-west range of this species in North America also presents considerable diversity among habitats and climates that could be investigated further. For example, both water content and dry mass of snapping turtle eggs have been shown to be correlated with longitude among four studies at approximately 42° latitude across the northern United States, which may be related to differences in climate in the east-west direction across the continent (Finkler et al. 2004).

In many respects the population biology of *C. serpentina* is unique among aquatic chelonians, and stands in strong distinction to the closely related *M. temininikii*. Although all aquatic turtles are capable of overland movement to some degree, terrestrial barriers that have allowed for phylogeographic differentiation among other taxa have not proven effective as barriers to *C. serpentina*. It is possible that anatomical adaptations of this species—reduced carapace and pastron and limbs capable of effective terrestrial locomotion and clearance of obstacles—are part of a suite of adaptations that facilitate dispersal and migration, whether along aquatic routes or over land.

This capability for movement between watersheds has implications for both the demography and the genetics of snapping turtles. If snapping turtles are more able to move between bodies of water than are painted turtles, sliders, or kinosternids, for example, then short-term changes to single bodies of water, such as occasional drying, should pose less of a threat to the long-term persistence of snapping turtle populations than to others. However, if snapping turtles are more regularly exposed to terrestrial environments than are other turtles, the risks presented by such exposure would be expected to influence survivorship of the affected life stages.

At the level of population genetics, the few studies of *C. serpentina* to date indicate little structure among or within populations, interpretable as relatively high levels of gene flow. At phylogeographic levels of resolution, where studies have demonstrated genetic differences among populations of several aquatic turtle species in North America at the level of major watersheds, no substructure among *C. serpentina* populations has been found.

ACKNOWLEDGMENTS

I extend my appreciation to R. J. Brooks, P. T. Boag, B. N. White, M. E. Obbard, E. G. Nancekivell, J. D. Congdon, and the many colleagues with whom I have had the privilege of collaborating on studies of *C. serpentina*. I thank C. A. Bishop for the carapace specimens used in some of the illustrations in this chapter, and G. P. Brown, M. W. Bruford, and J. D. Litzgus for helpful suggestions and encouragement. This chapter is Royal Botanical Gardens contribution no. 105.

17

Geographic Variation in Life-History Traits

DON MOLL

JOHN B. IVERSON

T HE SNAPPING TURTLE (*Chelydra serpentina*) has the greatest distribution of any North American freshwater turtle, ranging from Florida to Texas to Saskatchewan to Nova Scotia (Iverson 1992a). Snapping turtles are also the second largest North American freshwater turtle (Conant & Collins 1991). However, despite its range and body size, and the fact that more than 30 natural history studies of snapping turtles have been published (Tables 17.1 and 17.2), a relatively complete suite of life-history data is available for only three northern populations, one in Ontario (Loncke & Obbard 1977; Obbard 1983; Brooks et al. 1986; Galbraith 1986; Galbraith et al. 1989; Brown et al. 1994a, b), one in Pennsylvania (Ernst, in Iverson et al. 1997), and one in Nebraska (Iverson et al. 1997). Other comprehensive studies either lacked data on egg size (Hammer [1969] in South Dakota; White & Murphy [1973] in Tennessee; Kiviat [1980] in New York; Congdon et al. [1987, 1994] in Michigan) or age and/or size at maturity (Petokas & Alexander [1980] in New York; Congdon & Gibbons [1985] in North Carolina; Steyermark & Spotila [2001a] in Pennsylvania) or lumped data from distant populations (Yntema [1970] for New York and Wisconsin). However, sufficient data on life-history traits of snapping turtles are available from across its range to permit the preliminary study of geographic variation in the life-history traits of *Chelydra serpentina* (Iverson et al. 1997). This chapter reviews those data (including material published subsequent to Iverson et al. 1997) and discusses patterns in their variation across latitude, longitude, and elevation. When relevant we have also made comparisons of snapping turtle reproductive characteristics with patterns characteristic of other turtle lineages, and with general vertebrate and turtle life-history theory.

Table 17.1

Published and unpublished life-history data for *Chelydra serpentina*

Location	Lat	Long	Elev	n	CL	PL	GBM	CS	EL	EW	EM	CM	CS/GBM ×100	EM/GBM ×100	CM/GBM ×100	CM/GBM ×100	Source
Manitoba	50	96–99	~300	9	–	–	–	65.4 (49–80)	–	26	est. 10.1	–	–	–	–	–	Norris-Elye 1949; Berezanski 1999, pers. commun.
Ontario	45.5	78.5	400	63	284	est 222	5,170	34.0 (18–66)	est 27.7	est 26.9	11.6	394.4	0.658	0.224	7.63	8.26	Obbard 1983; Galbraith 1986; Brooks et al. 1986; Galbraith et al. 1989; Loncke and Obbard 1977
Quebec	45	72	200	1	362	est 283	est 10,965	83	–	–	–	–	0.760	–	–	–	Bleakney 1957
Quebec	45	73.5	50	28	239	est 187	est 3,073	–	–	–	–	–	–	–	–	–	Mosimann and Bider 1960
Minnesota	45	93.5	300	1	254	est 199	est 3,703	24	–	–	–	–	0.648	–	–	–	Breckenridge 1944
New York	44.5	75	500	28	259	196	4,370	30.9 (16–59)	28.4	est 27.5	11.1	308	0.707	0.254	7.05	7.58	Petokas and Alexander 1980
New York	44	74.5	500	1	285	est 223	est 5,269	36	–	–	–	–	0.683	–	–	–	Weber 1928
Ontario	43.5	80	300	24	284	est 222	5,840	48.1	est 26.6	est 25.9	10.4	500.2	0.824	0.179	8.57	9.37	Brooks et al. 1986
Wisconsin	43.5	88.5	300	70	–	–	–	44 (?–85)	–	–	–	–	–	–	–	–	Mathiak 1966 in Vogt 1981
Wisconsin	43.5	89.5	300	4	296	est 231	est 5,918	35.5 (26–40)	–	–	–	–	0.600	–	–	–	E. Moll, pers. commun.
New York & Wisconsin	43–45	–	–	255	263	est 206	4,400	36.7 (11–73)	26.8	est 26.2	10.6	410	0.834	0.241	9.20	10.13	Yntema 1970
New York	43	75.5	200	53	–	–	–	43.7	–	–	–	–	0.665	–	–	–	Yntema 1979
South Dakota	43	101.5	950	291	319	247	7,364	49 (31–87)	–	–	–	–	0.365	–	–	–	Hammer 1969
New York	42.5	76.5	125	2	279	est 218	est 4,937	18	–	–	–	–	–	–	–	–	Hamilton 1940
New York	42.5	76.5	300	9	237	est 185	est 2,995	20.8 (12–30)	28.5	–	est 12.54	est 260.8	0.694	0.419	8.71	9.90	Pell 1941
New York	42	74	<10	27	262	est 205	est 4,072	29.6 (16–54)	–	–	–	–	0.727	–	–	–	Kiviat 1980
Michigan	42.5	84	300	78	258	201	est 3,884	27.9 (13–49)	–	31.0 (x-ray)	–	–	0.718	–	–	–	Congdon et al. 1987
Wisconsin	42.5	88	300	1	–	–	–	–	25.9	est 25.4	est 9.62	–	–	–	–	–	Edgren 1949
Nebraska	42.5	100.5	900	20	–	–	–	55.6 (39–80)	est 28.5	est 27.7	12.54	697.2	–	–	–	–	Packard and Packard, pers. commun.
Nebraska	42.5	100.5	900	1	–	–	–	104	–	–	13.89	1444	–	–	–	–	Miller et al 1989
Nebraska	42.5	100.5	900	1	–	–	–	109	–	–	12.42	1354	–	–	–	–	Packard et al. 1990
Nebraska	42	102.5	1165	77	325	256	7,878	46.8 (20–73)	27.6	26.8	11.44	526.9	0.594	0.145	7.00	7.10	Iverson et al. 1997
Illinois	42.5	38	300	3	238	est 186	est 3,033	23 (16–37)	–	–	–	–	0.758	–	–	–	E. Moll, pers. commun.
Illinois	42	90	300	17	–	–	–	48.5 (25–83)	–	12.3	–	–	–	–	–	–	Janzen 1993, 1992, pers. commun.
Illinois	42	90	300	3	est 281	220	est 5,046	40 (34–51)	est 28.3	est 27.4	12.27	481.3	0.793	0.244	9.54	10.78	Janzen, pers. commun.
Iowa	41–43	94	300	9	255	194	est 3,748	36.4 (21–52)	–	–	–	–	0.971	–	–	–	Christiansen and Burken 1979; Christiansen, pers. commun.
Connecticut	41.5	72.5	100	1	est 389	est 304	13,636	52	24	–	–	–	0.381	–	–	–	Finneran 1947
Colorado	41	~103	~1200	1	260	est 203	est 3,977	61	–	–	–	–	1.534	–	–	–	Hammerson 1999
New Jersey	40.5	74.5	50	117+	261	202	4,220	32.3 (15–52)	25.7	est 25.4	9.69	309.8	0.765	0.230	7.34	8.11	Hotaling 1990
Pennsylvania	40–42	75–80	–	9	–	–	–	–(28–60)	–	–	–	–	–	–	–	–	Hulse et al. 2001
Pennsylvania	40.5	76.5	150	44	295	208	est 5,857	30 (15–50)	26.2	26.0	10.3	310	0.512	0.176	5.29	5.69	Ernst, pers. commun.
Illinois	40.5	90	150	4	–	–	–	32 (25–46)	–	–	–	–	–	–	–	–	D. Moll, pers. commun.

Location	Lat	Long	Elev	N	CL	PL	GBM	CS	EL	EW	EM	CM					Reference
Indiana	40	85	300	2	224	158	2,387	25 (23–27)	26.6	26.0	9.75	242.4	1.066	0.400	10.27	12.04	Iverson, unpublished
Illinois	39–40	88.5	300	4	246	est 192	est 3,336	35.0 (18–48)	–	–	–	–	1.049	–	–	–	E. Moll, pers. commun.
Missouri	39	94	300	1	298	est 233	est 6,041	33	–	–	–	–	0.621	–	–	–	Smith et al. 1983
Virginia	39	77	~10	21	–	–	–	23.8 (12–44)	–	–	–	–	–	–	–	–	Ernst et al. 1997
Kansas	38.5	95	300	1	–	–	–	–	28.4	28.1	est 12.84	–	–	–	–	–	Gloyd 1928
Virginia	38.5	77	30	22	260	est 203	3,963	ca. 30 (10–47)	est 27.8	est 27.0	11.71	est 304	0.656	0.295	7.68	8.55	Wilgenbusch, pers. commun.
Missouri	37	93	400	6	268	204	4,441	31.8 (29–37)	27.5	26.8	11.63	385.6	0.716	0.262	8.68	9.47	Thomas and D. Moll, pers. commun.
Virginia	37	77	50	85	–	–	–	28.6 (13–48)	–	–	–	–	–	–	–	–	Ash 1951
Virginia	36.5	76	50	1	281	211	5,300	55	27.7	–	12.6	693	1.038	0.238	13.08	15.38	Mitchell and Pague 1991; Mitchell, pers. commun.
Virginia	36.5	76	50	13	–	–	–	27.0 (7–55)	–	26.3	10.9	est 294.3	–	–	–	–	Mitchell 1994
Tennessee	36.5	89.5	75	2	–	–	–	28.5 (25–32)	27.95	–	est 11.92	est 339.9	0.693	–	–	–	E. O. Moll, pers. commun.
Tennessee	36	86.5	200	25	est 224	175	2,870	19.9 (12–42)	–	–	–	–	–	–	–	–	White and Murphy 1973
North Carolina	36	80	300	4	est 221	173	2,856	23.6	25.8	25.8	9.63	225.7	0.826	0.337	8.30	8.58	Congdon and Gibbons 1985
North Carolina	36	81	400	5	261	est 204	4,295	30.4 (10–43)	–	–	–	–	0.708	–	–	–	Brown 1992
North Carolina	–	–	–	4	251	est 196	est 3,570	27.5 (N = 2) (25–33)	28.2	–	12.16	est 282.5 (N = 2)	0.637	0.343	7.83	8.77	Palmer and Braswell 1994
Arkansas	35	92	65	2	240	189	3,163	29 (25–33)	28.1	27.3	12.29	349.6	0.956	0.389	11.29	13.27	Iverson, unpublished
Louisiana	31.5	91.5	50	6	232	est 181	est 2,798	20.5	27.5	27.5	est 11.42	est 234.1 (16–31)	0.733	0.408	8.37	9.50	E. Moll, pers. commun.
Florida	30.5	84	17	1	–	–	–	54	–	–	–	–	–	–	–	–	Jackson and Ewert 1997
Florida	30	85	<10	46	–	–	–	33.2 (17–52)	–	–	–	–	–	–	–	–	Ewert and Jackson 1994
Florida	29.5	82.5	50	8	–	–	–	16.6 (14–20)	est 27.1	est 27.1	est 11.87	est 197	–	–	–	–	Iverson 1977
Florida	29.5	82.5	50	1	243	168	est 3,233	15	27.7	27.7	13.17	197.6	0.464	0.407	6.13	6.73	D. R. Jackson, pers. commun.
Florida	29.5	83	–	2	–	–	–	30.5 (30–31)	–	–	–	–	–	–	–	–	Jackson and Ewert 1997
Florida	29	81	2	1	–	–	–	23	–	–	–	–	–	–	–	–	Jackson and Ewert 1997
Florida	27	82.5	50	25	216	est 169	est 2,254	15.3 (6–21)	est 26.2	est 25.6	9.9	151.8	0.678	0.441	6.74	7.58	Punzo 1975, pers. commun.
Honduras	16	87.5	200	1	370	270	est 11,724	34–35	34–35	est 32–33	est 19.3	–	–	0.165	–	–	Medem 1977
Costa Rica	10	84	758	3	–	–	–	27.7 (20–38)	–	–	–	–	–	–	–	–	Flausin et al. 1997
Costa Rica	10.5	83.5	10	6	350	265	10,752	35.2	35.4	34.3	20.35	716.3	0.327	0.189	6.66	7.14	D. Moll, pers. commun.
Colombia	8	77	50	1	312	245	est 6,953	27	36.3	33.0	est 20.42	est 551.3	0.388	0.294	7.93	8.61	Medem 1962, 1977

Missing data were estimated when possible (indicated with "est"). Carapace length (CL, in millimeters) or plastron length (PL, in millimeters) were estimated from the mean of the PL/CL ratio (0.782) from turtles from Michigan (Congdon et al. 1987), South Dakota (Hammer 1969), and this study. Body mass (in grams) was estimated from carapace length (in millimeters) based on the data reported in Lagler and Applegate (1943) as presented in Iverson (1984; grand body mass [GBM] = $0.00015853CL^{3.064}$). Egg length (EL) and/or egg width (EW) were estimated from egg mass (EM); and egg mass was estimated from mass (EM); based on the regressions relating those variables for eggs from Nebraska (Iverson et al. 1997; EM = $0.510EL + 0.722EW - 11.936$). Other abbreviations are latitude (Lat, in ° N), longitude (Long, in ° W), elevation (Elev, in meters), clutch size (CS), clutch mass (CM, in grams), and spent body mass (SBM = GBM − CM − 100 g).

Table 17.2.

Variation in age and size at maturity in *Chelydra serpentina*

Location	Age	CL	PL	BM	Source
Ontario	18*	249	est 195	est 3,484	Obbard 1983; Galbraith 1986; Galbraith et al. 1989
Quebec	–	200	est 156	est 1,780	Mosimann & Bider 1960
New York	–	220	167	2,950	Petokas & Alexander 1980
New York and Wisconsin	–	200	est 156	1,900	Yntema 1970
New York	–	216	est 169	est 2,254	Kiviat 1980
South Dakota	–	254	197	est 3,703	Hammer 1969
New York	–	200	est 156	est 1,900	Pell 1941
Michigan	12*	206	159	est 1,949	Congdon et al. 1987
Nebraska	11	285	225	6,000	Iverson et al. 1997
Iowa	9–12	229	162	est 2,696	Christiansen & Burken 1979; Christiansen, pers. commun.
New Jersey	–	200	143	1,700	Hotaling 1990
Pennsylvania	–	208	est 163	2,008	Hulse et al. 2001
Pennsylvania	10–12*	200	160	est 1,780	Ernst, pers. commun.
Virginia	–	est 198	155	est 1,732	Mitchell 1994
Tennessee	–	185	145	est 1,402	White and Murphy 1973
North Carolina	–	187	est 146	1,470	Brown 1992

Populations in order of decreasing latitude (see Table 17.1). Carapace length (CL) or plastron length (PL) in millimeters, body mass (BM) in grams, and age in winters (except asterisk indicates age "in years"). For samples with a range for age, the mean of the maximum and minimum values was used for interpopulation comparisons. Missing data estimated (est) as in Table 17.1.

METHODS

Reproductive data were compiled from published literature and from unpublished data supplied by colleagues (Table 17.1). Abbreviations used include carapace length (CL), maximum plastron length (PL), gravid body mass (GBM), spent body mass (SBM), egg mass (EM), and clutch size (CS). All statistical analyses done across populations include only those samples representing at least three females. Relationships between variables were evaluated using least-squares linear regression. All means are followed by ±1 SD.

FEMALE SIZE AND AGE AT MATURITY

The largest snapping turtles are found at high latitudes, longitudes, and elevations. For example, female snapping turtles in the Sandhills of Nebraska and South Dakota are the largest on average (mean, 322 mm CL) of all populations in the United States and Canada (mean for 26 samples, 262 mm). Average body size is positively correlated with latitude (for CL, PL, and GBM, $P = 0.007$, 0.007, and 0.017, respectively, for 24 populations) and elevation ($P = 0.001$, 0.0005, and 0.0003, respectively), but not longitude ($P = 0.069$, 0.045, and 0.059, respectively). More than 53% of the variation in mean carapace length (CL) across populations can be explained by the multivariate model: $CL = 2.47Lat + 0.054Elev + 147.07$; $r = 0.73$; $P = 0.0004$; $n = 24$.

Females in populations with large average body sizes also mature at larger sizes. For 14 populations, average female body size is positively correlated with estimated (usually minimum) size at maturity (e.g., for CL, $r = 0.79$; $P = 0.0008$). Body size at maturity was also correlated with latitude (e.g., for CL, $r = 0.52$; $P = 0.047$; $n = 15$), longitude ($r = 0.69$; $P = 0.007$; $n = 15$), and elevation (e.g., for CL, $r = 0.81$, $P = 0.0004$; $n = 14$).

Age at maturity (Table 17.2) apparently varies from about 9–12 years in Iowa (Christiansen, pers. commun.; Christiansen & Burken 1979) to as high as 17–19 years in Ontario (Galbraith et al. 1989), and is significantly correlated with latitude ($r = 0.93$; $P = 0.024$), but not with longitude or elevation ($P > 0.40$) or mean female body size ($P > 0.84$ for CL, PL, and GBM). However, data on age at maturity are available for only five populations ranging across only about five degrees of latitude.

Of the five populations for which age and size at maturity data are available (Table 17.2), it is not surprising that the population in Ontario (at the highest latitude and with the shortest growing season) requires the longest time to reach maturity (Galbraith et al. 1989). However, it was not expected that snapping turtles in Nebraska and Iowa would have the fastest rates of growth to maturity (Table 17.2). For example, although snapping turtles in Nebraska and Michigan lie at nearly the same latitude and mature at nearly the same age (11 vs. 12 winters), Nebraska turtles reach maturity when the carapace is 38% longer and the body is over 200% heavier than in Michigan (Table 17.2). Given the much shorter growing season in Nebraska (approximately 145 frost-free days) compared with Michigan (172 frost-free days), it seems likely that genetic differences in juvenile growth rate must exist between at least those populations.

It is also surprising that Ontario female snapping turtles delay maturity for 18 years rather than mature much earlier, at sizes comparable to those in populations from the midwestern and eastern United States (see Table 17.2). Although common garden experiments are needed to test for genetic differences among snapping turtle populations (e.g., see Niewiarowski & Roosenburg 1993), the preliminary data suggest (see also Galbraith et al. 1989; Brown et al. 1994a) that selection at high latitudes, longitudes, and elevations favors increased growth rates, and hence increased adult body size, with a concomitant allometric increase in clutch size (Iverson et al. 1993—review the advantages of this strategy) rather than relative increases in egg size (see below). Perhaps relative increases in clutch size without increases in body size are not possible in snapping turtles (e.g., if egg size is already at or near the minimum size for viability).

As pointed out by Charnov (1990), Charnov et al. (1993), and Congdon et al. (2003) a strong positive correlation exists between delayed sexual maturity and longevity. The increased likelihood of death before maturity is attained may be compensated for by increased fecundity and/or by increased probability of offspring survivorship through enhanced offspring quality or size in such species that do not practice parental care. Snapping turtle life-history strategy seems to mainly favor the former solution for coping with this phenomenon, but as discussed below (e.g., see the summer length hypothesis) advantages may be accrued from maintaining smaller egg size at some or all latitudes beyond the general clutch size–egg size trade-off explanation (see below) as well (i.e., there may be advantages in maintaining smaller egg sizes and possibly smaller hatchlings as well as increasing clutch sizes rather than in the need to simply direct energy toward the clutch).

Iverson et al. (1993) reviewed the possible advantages of increasing clutch size and/or egg size with increasing latitude for vertebrates in general. These include (1) the long-day hypothesis, (2) the spring productivity hypothesis, (3) the juvenile competition hypothesis, (4) the size-selective predation hypothesis, (5) the nest loss hypothesis, (6) the climatic uncertainty or seasonal hypothesis, and (7) the summer-length hypothesis. With the exception of the long-day hypothesis (which involves maternal care of the young—a non-snapper trait), one or a combination of these hypotheses may explain selective forces that influence latitudinally (and possibly longitudinally and elevationally) related snapping turtle reproductive patterns (also see Iverson et al. [1993] for more detailed analyses, pros and cons, original references, and suggested approaches for testing these hypotheses).

The spring productivity hypothesis argues that, due to the great pulse of productivity during the spring months at temperate latitudes, reproduction may not be as resource limited as it may be in the tropics. As a result, temperate species may be able to convert more energy into more young during the spring, resulting, in snapping turtles (and in turtles in general), in greater clutch sizes.

The juvenile competition hypothesis is based on the concept that in temperate regions the burst in spring productivity follows an extended harsh-winter period in which populations of potential competitors are reduced. Under such circumstances the most effective strategy for a species would be to produce many small young that exploit these resource-rich, low-competition environments and grow very quickly to larger sizes.

As part of the size-selective predator hypothesis, diversity theory predicts that temperate zone predators will in general be less diverse and less specialized than in the tropics. If relatively few size-selective predators exist at high latitudes, and predation is no higher on smaller juveniles than on those somewhat larger, there may be some advantage to be gained at higher latitudes in producing more and relatively smaller eggs that have nearly the same level of fitness as would fewer, larger eggs. Iverson & Smith (1993) suggested that size-selective predators feeding only on very small turtles is probably a very rare occurrence (at least in the temperate zone), and that an egg or hatchling even twice as large or larger would probably remain equally vulnerable to the normal array of predators. Although many studies of hatchling turtles suggest that larger hatchlings may enjoy greater survivorship (e.g., Miller et al. 1987; Janzen 1993a; Janzen et al. 2000a; Tucker, 2000) perhaps because of greater running and/or swimming abilities, other studies by Congdon et al. (1999) and Filoramo & Janzen (2002) found no such advantages associated with larger size. Filoramo & Janzen (2002) thought that size advantages related to greater mobility and increased protection from bird predation were probably negated in drier conditions when red-eared slider hatchlings buried themselves and moved erratically toward water on their initial movement from the nest. Congdon et al. (1999) conducted both short- and long-term field experiments with hatchling snapping turtles at Michigan's E. S. George Reserve and found no evidence that greater size conferred any survivorship advantages there at all under any environmental circumstances in which the tests were conducted. As extra size may be of little advantage to young snapping turtle survivorship, and survival rate may often be very low (e.g., release of hatchling and one-year-old juvenile snapping turtles in Algonquin Park, Ontario, indicates a survival rate of <1% during the first 5 years after release [Galbraith et al. 1989; Cunnington & Brooks 1996; R. J. Brooks, unpublished]), then increased clutch sizes would seem to be a key to snapping turtle survival (and increasingly so at higher latitudes).

The nest loss hypothesis, slanted here toward turtle reproduction patterns, predicts that selection favors the production of smaller clutches of eggs in the tropics (the type II reproductive pattern—Moll 1979, and below). If deaths in nests of high-latitude overwintering species that produce fewer clutches (only one in high-latitude snapping turtles) occur at greater frequency due to exposure to autumn and winter weather, as seems likely, then it may be advantageous to produce more offspring to compensate for losses due to low nesting frequency and higher nest mortality. Some evidence from Algonquin Park, Ontario, shows that this may be the norm in high-latitude snapping turtle populations. Of 257 clutches of snapper eggs studied here in a four-year experiment, 215 did not emerge before fall, and only one of these successfully overwintered (Obbard & Brooks 1981b; Bobyn & Brooks 1994b). In the more than 20 years this population has been studied no "bumper crops" of juveniles have been observed entering the breeding population (Cunnington & Brooks 1996; Brooks, unpublished data).

The climatic uncertainty or seasonality hypothesis argues that high seasonality of resources and/or climatic uncertainty characteristic of high latitudes may cause reductions in population size and consequently reduce competition, predation, and other density-dependent factors during parts of the year. This would result in the population's access to temporarily resource-rich environments during these periods. Under such circumstances selection for large clutch sizes and presumably smaller eggs would be favored to maximize the use of the resources that become available during these "windows of opportunity."

The summer-length hypothesis reasons that because summers are shorter at high latitudes, and offspring may need to develop quickly to a critical size before the onset of cold weather, egg size may be under intense latitudinal selection. For some vertebrates (e.g., frogs) larger eggs may be favored, but available evidence for turtles suggests that smaller eggs with their shorter incubation times may be more advantageous at higher latitudes because of the shorter incubation season and cooler temperatures characteristic of such regions.

NESTING SEASON

The timing of nesting within snapping turtle populations is inversely correlated with spring temperatures (Congdon et al. 1987; Obbard & Brooks 1987; Hotaling 1990). In addition, the average commencement of nesting in snapping turtle populations is correlated with latitude across populations (Table 17.3, Fig. 17.1 [upper]; e.g., for earliest nest date, $r = 0.83$, $P = 0.0002$, $n = 15$; for latest nest date, $r = 0.76$, $P = 0.0025$; $n = 13$). The latest reliable reports of nesting are in early July (Table 17.3), and the reports of eggs retained until July 22 (Gloyd 1928; captive animal), August 4 (Galbraith et al. 1988; dead on road), and September 1 (Weber 1928; captive) must be considered anomalous. The average length of the nesting season is similar at several localities at high latitude: South Dakota (13.5 days; calculated from Hammer 1969), Michigan (18 days; Congdon et al. 1987), New Jersey (15 days; Hotaling 1990), and Nebraska (16 days; Iverson et al. 1997). However, the season may be longer at lower latitudes (Table 17.3). For example, in northern Florida, gravid females have been encountered as early as March 16 (Jackson & Ewert 1997), and fresh nests found as late as June 9

Table 17.3

Geographic variation in the timing of nesting season in snapping turtles

Location	Earliest nest	Latest nest	Average first nest	Source
Manitoba	June 15	June 29	–	Berezanski 1999, pers. commun.
Ontario	May 29	–	June 10	Obbard & Brooks 1987
Quebec	–	July 5	–	Bleakney 1957
Quebec	June 16	July 9	–	Robinson & Bider 1988
Maine		July 1	mid-June	Coulter 1958
South Dakota	June 5	June 30	June 9	Hammer 1969
Michigan	May 22	June 28	June 1	Congdon et al. 1987
Nebraska	June 1	June 28	June 8	Iverson et al. 1997
New York	May 28	June 21	–	Petokas & Alexander 1980
New York	–	–	June 10–20	Hamilton 1940
New York	June 4	June 30	–	Kiviat 1980
New Jersey	May 17	June 21	May 24	Hotaling 1990
Indiana	May 30	June 10	–	Minton 2001; Iverson, unpublished
Virginia	May 17	June 29	–	Gotte 1988; Mitchell 1994; Ernst et al. 1997
Tennessee	mid-May	–	–	White & Murphy 1973
North Carolina	May 21	–	–	Brown 1992 ($n = 1$)
North Carolina	May 30	July 1	–	Palmer & Braswell 1994
Texas	late May	–	–	Vermersch 1992
Northwest Florida	April 28	May 9		Ewert & Jackson 1994
North Florida	May 18	June 9		Iverson 1977
North Florida	–	–	March	Jackson & Ewert 1997
South Florida	–	–	mid-June	Punzo 1975
South Florida	–	–	late Feb–March	Ewert 1976
South Florida	–	–	early Feb–April ?	Ewert 2000

Populations are arranged in order of decreasing latitude.

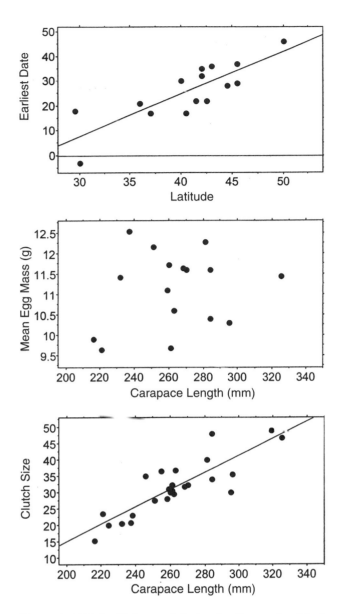

Fig. 17.1. (*top*) Relationship between latitude and earliest nest date (May 1 is day 1) for 15 populations of snapping turtles (with $n \geq 3$) across the species' range; regression equation is Date = 1.72Latitude–43.99 ($r = 0.83$; $P = 0.0002$). (*middle*) Relationship between mean carapace length (in millimeters) and mean egg mass (in rams); regression is not significant ($r = 0.26$; $P = 0.46$; $n = 16$). (*bottom*) Relationship between mean carapace length (in millimeters) and mean clutch size; regression equation is CS = 0.26CL–37.69 ($r = 0.85$; $P \leq 0.0001$; $n = 25$).

(Iverson et al. 1997). Furthermore, in March in south Florida, Ewert (2000) found gravid females with one to three sets of corpora lutea and up to two sets of enlarged preovulatory follicles, suggesting that nesting might extend from early February through at least April. For comparison, the Central American snapping turtle (*Chelydra rossignoni*) apparently nests from April to June (Alvarez del Toro 1960). *Chelydra acutirostris* nests at least from late February through April in Costa Rica (Moll 1997, and unpublished).

Snapping turtles typically nest in the early morning or early evening (Hammer 1969; Petokas & Alexander 1980; Congdon et al. 1987; Iverson et al. 1997); however, the fre-

quency of morning nesting seems to be related to average morning temperatures (i.e., inversely related to latitude and longitude). Nesting frequency in the morning for a population in Florida (27° N) was 100% (Punzo 1975), whereas that in New Jersey (40.5° N) was 63% (Hotaling 1990), that in Michigan (42.5° N) was 63% (Congdon et al. 1987), that in New York (44° N) was 44% (Petokas & Alexander 1980), and that in western Nebraska (42° N, but 102.5° W) was 33% (for Lat, $r = 0.89$, $P = 0.0454$, $n = 5$).

In Nebraska, there was a tendency for large females to nest in the morning (Iverson et al. 1997); however, Petokas & Alexander (1980) found no significant difference in body size between females nesting in the morning versus the evening. Larger female snapping turtles apparently nest earlier in the season than smaller females in South Dakota (Hammer 1969) and New York (Petokas & Alexander 1980), but no such pattern exists in Ontario (Obbard 1983), Michigan (Congdon et al. 1987), or Nebraska (Iverson et al. 1997). The definitive relationship between temperature and daily nesting patterns must await further study.

Snapping turtles typically produce a single clutch per year, although preliminary data from two long-term studies (Obbard 1983; Congdon et al. 1987, 1994) suggest that some females may not nest every year. There is some evidence that other freshwater turtles, particularly larger species, may not nest every year either. Based on ovarian analyses, Kuchling (1998) thought that Madagascan big-headed turtles (*Erymnochelys madagascariensis*) might be biennial breeders, and Junk & da Silva (1997) and Dobie (1971) suggested that giant South American river turtles (*Podocnemis expansa*) and alligator snapping turtles (*Macrochelys temminckii*) might not nest annually. Furthermore, the production of as many as four clutches per year in snapping turtles from extreme southern Florida has been verified (Ewert 2000). Clutch frequency data from other southern populations are sorely needed. Multiple clutches per year may also be produced by the Central American snapping turtle, which nests from April to June (Alvarez del Toro 1960). Painted turtles (*Chrysemys picta*) and sliders (*Trachemys scripta*) are examples of other wide-ranging turtles that are known or suspected to produce more clutches per year during the extended nesting periods possible in the more southern portions of their ranges also. Reliable data on clutch frequency are extremely difficult to obtain for this (or any) turtle species and estimates of this life-history trait may be the most significant gap in our knowledge of the life history of this species (Gibbons 1982).

The characteristics of snapping turtle nesting sites have been reviewed by Ernst et al. (1994). Snapping turtles usually select open areas as nesting sites when such locations are accessible to them. Substrates consisting of loose sand, loam, vegetable debris, sawdust piles, muskrat and beaver lodges, forest clearings, and even soil accumulated in the cracks of granitic rocks on shorelines may be utilized. Human-altered areas such as gravel and dirt roads, dams, and railway embankments may also be used when suitable natural sites are lacking. Moll (1997) found *Chelydra acutirostris* at Tortuguero, Costa Rica, to nest in abandoned agricultural fields near shorelines in the earliest stages of tropical rain forest succession near shorelines.

EGG SIZE

Mean egg mass (EM) in snapping turtles in the United States and Canada ranges from 9.63 to 12.54 g (Fig. 17.1 [middle] mean of 24 populations from Table 17.1 = 11.26 g), but averages about half that for Central American (19.3–20.4 g; Medem 1977; Moll, in Iverson et al. 1997) and South American snapping turtles (20.4; Medem 1962, 1977; Iverson et al. 1997). Individual egg size within a population of *Chelydra serpentina* can vary greatly (Table 17.4), possibly due in part to the large range in adult body size in this species, along with the usual pattern of a correlation between egg size and female body size within a population (e.g., Obbard 1983, in Ontario; Congdon et al. 1987, in Michigan; Hotaling 1990, in New Jersey; and Iverson et al. 1997, in Nebraska; but see Steyermark & Spotila 2001a, in Pennsylvania). However, no such correlation exists across temperate populations (Table 17.1) despite great variation in average female body size (e.g., for GBM and EM, $r = 0.10$, $P = 0.72$, $n = 16$). Mean egg size across populations is not significantly correlated with latitude ($P = 0.24$–0.56 for EL, EW, and EM), longitude ($P = 0.07$–0.39), or elevation ($P = 0.06$–0.41).

Mean egg size in Nebraska declined through the nesting season, despite the fact that body size of nesting females did not (Iverson et al. 1997). No other study of snapping turtles has examined this relationship, but perhaps females carrying large eggs, regardless of their body size, are stimulated by their more distended oviducts to deposit their eggs earlier. It is also possible that egg size and clutch size variation throughout the season (see also the discussion of clutch size below) may be in part determined by the amount of energy accrued at a given moment and the amount of time remaining in the nesting season. One would expect that there would be individual variation in the amount of yolk within follicles at emergence and variation among individuals in the amount of time it takes to complete the production of a clutch. An individual that emerges from overwintering with large ovaries and/or accrues energy rapidly early in the season would be able to produce relatively large (perhaps maximal) eggs and clutches early in the year. At the other extreme, another individual might have small ovaries/follicles upon emergence and/or accrue energy slowly and therefore shell up smaller clutches of small eggs late in the year. A continuum between the extremes would probably exist within the population.

Relative egg mass (REM = mean egg mass/mean gravid female body mass × 100) in snapping turtles averages 0.28 (range, 0.145–0.441; $n = 16$). REM in *Chelydra rossignoni* (0.189; Moll, in Iverson et al. 1997) is not significantly different ($t = 1.04$, $P = 0.16$) from that in *C. serpentina*, perhaps because much of the variation in REM is related to body size across populations. REM decreases with female size across populations (for CL, $r = -0.90$; $P < 0.0001$; for GBM, $r = -0.92$, $P < 0.0001$; $n = 16$), as well as within at least two of them for which the data are available (Obbard 1983; Iverson et al. 1997). This pattern of a relative decrease in egg size with body size seems to be typical of freshwater turtles (Rowe 1994).

Average REM in snapping turtles (0.28) is well below the average values for turtles in general (1.6 ± 1.5 for 169 populations in Table A1 in Iverson et al. 1993), but similar to that in other large turtles with large clutches of small, round eggs (e.g., 0.25 for the alligator snapping turtle, *Macrochelys*; 0.04–0.10 for sea turtles; 0.13–0.24 for larger South American sidenecks, *Podocnemis*; and as low as 0.02 for some trionychid softshell turtles). Among other things, this indicates that snapping turtles probably experience no pelvic canal constraints on egg size like some turtles do with smaller clutches of relatively big eggs (Congdon & Gibbons 1987).

REM in populations of snapping turtles does not vary with longitude or elevation ($P = 0.74$ and 0.11, respectively), although it decreases with latitude (for 14 samples, $r = 0.69$, $P = 0.007$). On the contrary, partial correlation analysis of these factors with egg size across populations (with the removal of GBM effects) suggests that there are no latitudinal ($r = 0.34$, $P = 0.24$, $n = 14$), longitudinal ($r = 0.23$, $P = 0.44$), or elevational effects ($r = 0.15$, $P = 0.61$) on relative egg size.

Table 17.4

Variation in egg size within populations of snapping turtles with large samples

Location	n	Range	Source
Wisconsin/New York	3,976	7.0–14.7	Yntema 1970
Nebraska	1,112	8.65–16.95	Packard & Packard, pers. commun.
Nebraska	2,165	6.42–19.07	Iverson et al. 1977
New Jersey	3,779	6.20–13.76	Hotaling 1990
Virginia	553	6.06–22.76	Wilgenbusch, pers. commun.
Missouri	191	9.5–13.9	Thomas, pers. commun.
Florida	~100	5–13	Punzo 1975

Range is given in grams.

The absence of a latitudinal decrease is contrary to the general pattern of a decrease in size-adjusted egg size with latitude across turtle species (Iverson et al. 1993). This may be because egg size is already near the lowest level necessary for hatchling viability.

Only two studies (Obbard 1983; Iverson et al. 1997) have examined variation in egg size between years in a snapping turtle population. Neither could identify differences among years.

CLUTCH SIZE

Average clutch size (CS) for snapping turtles is 32.9 eggs (Table 17.1; range of means, 10.7–65.4). The largest clutches are produced by those from Manitoba (65) or the Nebraska Sandhills (means, 47–55; Iverson et al. 1997; Finkler 1998; Packard & Packard, pers. commun.; Table 17.1). The two largest recorded clutch sizes for the species (104 and 109 eggs) were from the latter region (Miller et al. 1989; Packard et al. 1990). Average clutch sizes for *Chelydra rossignoni* (27.7 and 35.2; Flausin et al. 1997; Moll, in Iverson et al. 1997) and *Chelydra acutirostris* (27 and 35.2; Medem 1962, 1977; Moll 1997) are similar to those for *Chelydra serpentina*.

Clutch size also varies extensively both within and across populations (Table 17.1; e.g., 20–73 in a single population in Nebraska). This pattern is primarily a result of the highly significant correlation between body size and clutch size observed in all populations in which the relationship has been examined (Fig. 17.1 [bottom] and Pell 1941; Ash 1951; Yntema 1970; White & Murphy 1973; Obbard 1983; Wilhoft 1986; Congdon et al. 1987; Hotaling 1990; Iverson et al. 1997; Steyermark & Spotila 2001a), as well as among populations. Body size explains 23–77% of the variation in CS within seven populations with data (Pell [1941], using CL; White & Murphy [1973], using PL; Obbard [1983], using CL; Congdon et al. [1987], using PL; Hotaling [1990], using PL; Iverson et al. [1997], using CL, PL, and GBM; Steyermark & Spotila [2001a], using BM) and 73% (for CL; 74% for GBM) of that variation across 25 populations (Table 17.1 and Fig. 17.1 [bottom]; $r = 0.85$, $P < 0.0001$).

Because body size in snapping turtles is correlated with latitude and elevation (see above), CS is also positively correlated with those factors (for latitude, $r = 0.68$, $P < 0.0001$, $n = 38$; for longitude, $r = 0.55$, $P = 0.0004$, $n = 38$; for elevation, $r = 0.58$, $P < 0.0001$, $n = 38$). Across populations, 75% of the variation in clutch size is explained by the multivariate model (CS = 0.24CL + 0.29Lat - 42.37; $r = 0.87$; $P < 0.0001$; $n = 23$).

Clutch size declined through the nesting season in the population in western Nebraska (Iverson et al. 1997), despite the fact that body size did not. A possible explanation for this phenomenon was discussed above in relation to seasonal changes in egg size. Although this pattern has not been reported previously, it is expected to be the case in South Dakota (Hammer 1969) and New York (Petokas & Alexander 1980), where nesting female body size was reported to decrease during the season, and clutch size is assumed to vary with body size. A decrease in clutch sizes laid later in a nesting season has also been observed in *Chrysemys picta* from the Nebraska sandhills, a population known to produce multiple clutches (Iverson & Smith 1993).

Relative clutch size (RCS = mean CS / mean gravid female body mass × 100) in *C. serpentina* averaged 0.73 (range, 0.51–1.05; $n = 25$), much higher than values for *C. rossignoni* in Central America (0.33; Moll, in Iverson et al. 1997) or *C. acutirostris* in northern South America (0.39; Medem 1962, 1977; estimated by Iverson et al. 1997). This pattern may be related to variation in body size across populations, since RCS tends to decrease slightly with female size across populations of *C. serpentina* (for CL, $r = 0.37$, $P = 0.068$; for GBM, $r = 0.38$, $P = 0.06$, $n = 25$). Relative clutch size in snapping turtles (0.73) is near the average for turtles in general (0.74 ± 0.66 for 166 populations in Table A1 in Iverson et al. 1993).

Relative clutch size in snapping turtle populations does not vary with latitude, longitude, or elevation ($r = 0.01–0.13$, $P = 0.57–0.95$, $n = 23$), and partial correlation analysis of each of those three independent variables versus clutch size across populations (with the removal of GBM effects) also shows no significant relationships ($r = 0.10–0.31$, $P = 0.15–0.65$, $n = 23$). This is contrary to the general pattern of an increase in size-adjusted clutch size with latitude across turtle species (Iverson et al. 1993).

Congdon et al. (1987) reported that clutch size varied among years in Michigan even when female body size did not. However, neither Obbard (1983) nor Iverson et al. (1997) could identify such a pattern in Ontario or Nebraska, respectively. Brown et al. (1994a) determined that environmental factors such as levels of productivity and/or ambient temperatures may influence growth rates and reproductive output (i.e., clutch size) in Ontario snapping turtles, however.

EGG SIZE VERSUS CLUTCH SIZE

There was no indication of a negative correlation between unstandardized egg size and clutch size in western Nebraska (Iverson et al. 1997), Ontario (Obbard 1983), or Pennsylvania ($r = 0.15$, $P = 0.56$; calculated from fig. 1 in Steyermark & Spotila 2001a), whereas Congdon et al. (1987) found a positive correlation between egg width measured from x-rays and unstandardized clutch size in Michigan. Across populations, mean egg size (as EM) per clutch is not correlated with mean clutch size ($r = -0.04$, $P = 0.86$, $n = 24$). However, this relationship may be complicated by the fact that both egg size and clutch size are correlated with body size within populations (see above). When the effect of body size (GBM, CL, or PL) was removed by partial correlation analyses, there was still no evidence of a negative correlation between egg and clutch size across the populations (e.g., for GBM, $r = -0.01$, $P = 0.97$, $n = 16$) or within the Nebraska (Iverson et al. 1997) or Pennsylvania populations (Steyermark & Spotila 2001a). This is contrary to the interspecific pattern of a negative correlation for turtles in general as demonstrated by Moll (1979) and Iverson et al. (1993), as well as the intraspecific data available for painted turtles (Iverson & Smith 1993; Rowe 1994), razorback musk turtles (Iverson 2002b), red-eared sliders (Tucker et al. 1998), rough-footed mud turtles (Iverson et al. 1991), Sonoran mud turtles (van Loben Sels et al. 1997), and yellow mud turtles (Iverson et al. 1991), but not wood turtles (Brooks et al. 1992) or giant river turtles (Valenzuela 2001).

It may not always necessarily be important to recognize correlations between egg size and number. In optimal propagule size theory, an "offspring size and number trade-off" refers to a trade-off in the number of offspring that survive as a consequence of variable offspring size and number relationships. Because a finite quantity of energy is assumed, a negative correlation between offspring number and size would be expected. This assumes, however, that there is flexibility in both offspring size and number. If one variable, say egg size, is less variable than the other (clutch size), which varies largely as a consequence of resources (as is true in *Chelydra*), then a negative correlation might not be detectable. Therefore, as clutch size varies, egg size doesn't vary very much, which obscures the detection of a negative correlation. A demonstrated lack of a negative correlation between egg and clutch size therefore does not necessarily reflect a lack of a trade-off between egg size, number, and survival.

CLUTCH MASS

The heaviest clutches of *Chelydra serpentina* are produced by turtles from Manitoba and the Nebraska Sandhills (mean CM, 503–697 g; Table 17.1), and the maximum known clutch mass for the species (1,444 g) was also recorded from the Sandhills (Miller et al. 1989). However, mean clutch mass in *C. rossignoni* in Costa Rica is greater (716 g; Moll, in Iverson et al. 1997). CM in populations of *C. serpentina* other than in Nebraska and Manitoba ranged from 152 to 500 g (mean for 19 samples, 318 g; Table 17.1), significantly lighter than Nebraska clutches ($t = 4.46$, $P < 0.0001$) as well as those for *C. rossignoni* ($t = 4.25$, $P = 0.0002$). However, as for egg and clutch size, these patterns of variation in CM are primarily the result of the highly significant correlation between body size and CM in all four populations in which the relationship has been examined (Obbard 1983; Hotaling 1990; Iverson et al. 1997; Steyermark & Spotila 2001a), in combination with significant geographic variation in body size.

Within populations, body size typically explains more than 50% of the variation in CM (Obbard 1983; Hotaling 1990; Iverson et al. 1997; Steyermark & Spotila 2001a), and across populations it explains more than 74% of clutch mass variation (for CL, $r = 0.86$, $P < 0.0001$, $n = 16$; for GBM, $r =$

0.86, $P > 0.0001$; $n = 16$). Thus, because body size in populations of snapping turtles is correlated with latitude and elevation (see above), unstandardized clutch mass was also correlated with these factors (for latitude, $r = 0.62$, $P = 0.004$; for longitude, $r = 0.61$, $P = 0.004$; for elevation, $r = 0.73$, $P = 0.0002$; $n = 20$). Across populations, more than 83% of the variation in CM is explained by the multivariate model (CM $= 15.94$Lat $+ 8.66$Long $- 1000.22$; $r = 0.86$; $P < 0.0001$; $n = 20$).

Although clutch mass is the product of clutch size and egg mass, across populations it is correlated with the former ($r = 0.97$, $P < 0.0001$, $n = 22$), but not with the latter ($r = 0.24$, $P = 0.29$, $n = 22$). In addition, partial correlation analysis to remove the effects of GBM revealed that adjusted clutch size was still correlated with clutch mass ($r = 0.50$, $P = 0.048$, $n = 16$), but that egg mass was not ($r = 0.18$, $P = 0.51$, $n = 16$). Increases in clutch mass are thus due primarily to increases in clutch size and not egg size. This would be expected from optimal life-history theory due to the intense selection on egg size and less on clutch size (Gibbons et al. 1982; Elgar & Heaphy 1989; Iverson & Smith 1993).

Clutch mass in Nebraska declined through the nesting season, even though body size of nesting females did not (Iverson et al. 1997). This pattern has not been reported in other populations. However, Hammer (1969, in South Dakota) and Petokas & Alexander (1980, in New York) reported that body size decreased during the nesting season, and since clutch mass was presumably correlated positively with body size, clutch mass may have declined through the season at those sites.

Relative clutch mass (RCM = mean CM / mean GBM × 100) averaged 7.84 (range, 5.3–9.5; Table 17.1) across 16 populations of *C. serpentina*, very similar to values for *C. rossignoni* (6.7; Moll in Iverson et al. 1997) or *C. acutirostris* (7.9; estimated from Medem 1962, 1977). In addition, RCM in *C. serpentina* did not vary with body size in Nebraska (Iverson et al. 1997), New Jersey (Hotaling 1990), Ontario (Obbard 1983), or New York/Wisconsin (Yntema 1970). Furthermore, RCM does not vary with female size across populations (for CL, PL, or GBM, $r = 0.04–0.23$, $P = 0.40–0.88$, $n = 16$). This allocation of a constant proportion of body mass to reproductive output suggests that total output may be under primarily genetic control, and that selection may be intense on the maintenance of that proportion. Relative clutch mass in snapping turtles (7.8) is typical of other turtles studied (7.9 ± 3.7 for 166 populations in Table A1 in Iverson et al. 1993).

RCM does not vary with latitude, longitude, or elevation ($r = 0.01–0.26$, $P = 0.38–0.96$, $n = 14$), and partial correlation analysis of each of those three independent variables versus clutch mass (with the removal of GBM effects) also demonstrates no significant relationships (e.g., for GBM, $r = 0.19–0.51$, $P = 0.19–0.51$, $n = 14$). This is consistent with Iverson et al. (1993), who found no significant correlation between latitude and RCM across turtle species. Moll & Moll

(1990) thought that RCM values might be more useful indicators of adaptations of species to localized environmental conditions than as responses to larger regional trends.

No evidence of annual variation in RCM was found in Ontario (Obbard 1983) or Nebraska (Iverson et al. 1997). In addition, Brown et al. (1994b) could detect no effect of leech infestation on RCM in snapping turtles in Canada. Although these preliminary data suggest that proximate levels of resource availability may have very little effect on reproductive output in snapping turtles, differences in size-adjusted clutch size and clutch mass between two Ontario populations may indicate otherwise (Brown et al. 1994a).

GENERAL PATTERNS

Geographic variation in female size across the range of *Chelydra serpentina* is significant, and mean female size is correlated with latitude, longitude, and elevation, with the largest females found in western Nebraska and South Dakota. Variation in growth rate to maturity among populations suggests that at northern, western, and high-elevation locations a greater selective advantage accrues from maturing at a larger body size than from maturing at an earlier age, perhaps because of the strong body size–clutch size correlation.

Tropical snapping turtles (*C. rossignoni* and *C. acutirostris*) differ from those in the United States and Canada in their production of larger eggs, relatively smaller clutch sizes, and perhaps more clutches per year. However, they do not differ in raw clutch size, relative egg size, or relative total clutch mass. Perhaps these differences represent the earliest stages of adaptation toward the type II reproductive pattern characteristic of long-term tropical residents such as the semiaquatic and terrestrial batagurids (Moll 1979). These turtles produce very large eggs and hatchlings, very small egg clutches, and lay their eggs asynchronously or throughout extended nesting seasons during the year. Possibly these tropical snappers are responding to a different array of specialized size-specific hatchling and egg predators than those that plague them in the temperate zone. If so, they are showing more evolutionary initiative than the other relatively recent tropical colonists, the sliders. These generalists have shown little inclination to vary (other than displaying extended nesting seasons) from their ancestral temperate (type I) reproductive pattern in which multiple large clutches of relatively small eggs are produced annually (Moll & Legler 1971; Moll & Moll 1990).

The reproductive strategy in *Chelydra serpentina* seems to be very conservative. Most of the geographic variation in their life-history traits appears to reflect variation in female body size. For example, despite significant geographic variation in body size (with larger turtles at high latitude, longitude, and elevation), neither mean egg size, average relative egg size, nor relative clutch size or relative clutch mass varies with latitude, longitude, or elevation. However, within populations, egg size and clutch size (and therefore clutch mass) increase with body size, although clutch size increases much more rapidly and significantly with body size than does egg size. As an example, in Nebraska a 10% increase in female CL above the population mean produces (on average) a 12.4% larger clutch and only an 8.0% heavier egg. In addition, the largest mean egg mass among samples with $n > 3$ is only 54% larger than the smallest mean mass, whereas the largest mean clutch size is 259% larger than the smallest. This suggests that, despite the supposed advantages of larger body size to hatchling turtles (Swingland & Coe 1979; Janzen 1993; Haskell et al. 1996; Janzen et al. 2000a; Tucker 2000; but see Congdon et al. 1999 and Filoramo & Janzen 2002), selection may be operating on snapping turtles to maintain a minimum viable egg size (i.e., 9–13 g) and to maximize growth and/or clutch size, as discussed above.

Snapping turtle reproductive patterns seem in general to conform to the predictions of bet-hedging theory, especially at higher latitudes (Cunnington & Brooks 1996). According to the theory, long adult lives and relatively low annual reproductive output may minimize the effect of normally low annual juvenile survivorship on an individual's reproductive success over its life. Continued reproduction over the long term by adults will eventually result in enough juveniles surviving to adulthood (perhaps in the occasional "bumper crop" year when normal sources of juvenile mortality may be reduced for some reason) to maintain or even increase the population. Cunnington & Brooks (1996) supplied evidence that bumper years probably don't occur very often, however, and population models they generated suggest the great importance of adult survivorship in maintaining healthy populations.

The advantages of large body size to painted turtles at higher latitudes have been reviewed previously (Iverson & Smith 1993). Of those supposed advantages, three may apply to snapping turtles. First, larger turtles are slower to lose heat than smaller turtles. Second, a larger body can store more catabolic or anabolic products necessary for survival through a longer winter. Finally, larger turtles have more absolute reproductive output (in particular, clutch size in snapping turtles) than smaller turtles with the possible advantages discussed above. Determining the relative importance of these advantages as selective forces for the evolution of large body size in northern, western, and high-elevation snapping turtle populations presents a real challenge for future work.

In conclusion, despite the number of studies undertaken with snapping turtles, sufficient data are still not available to permit a complete analysis of geographic variation in its life-history traits and a full investigation of its causes. Indeed, no life-history data are available from the southwestern part of the species' range nor for most of the Gulf Coastal region. In addition, many studies have reported only selected reproductive parameters when others were clearly available (e.g., egg size and not clutch size or body size; see Table 17.1).

Future work should strive to report sets of life-history traits that are as complete as possible, and in particular, to quantify the most elusive of those traits, clutch frequency. It should also focus on sorting out the relative influence of genetic versus local environmental effects on observed variation in life-history traits.

ACKNOWLEDGMENTS

Support for this work was provided by Earlham College and the Joseph Moore Museum of Natural History. J. L. Christiansen, C. Ernst, D. R. Jackson, F. J. Janzen, J. C. Mitchell, E. O. Moll, G. C. and M. J. Packard, F. Punzo, R. B. Thomas, and J. C. Wilgenbusch very generously shared unpublished data as noted in the tables. J. K. Tucker and two anonymous reviewers provided valuable comments on early drafts.

REFERENCES

Ackerman RA. 1977. The respiratory gas exchange of sea turtle nests (*Chelonia, Caretta*). Respir. Physiol. 31:19–38.

———. 1980. Physiological and ecological aspects of gas exchange by sea turtle eggs. Am. Zool. 20:575–583.

———. 1981a. Oxygen consumption by sea turtle (*Chelonia, Caretta*) eggs during incubation. Physiol. Zool. 54:316–324.

———. 1981b. Growth and gas exchange of embryonic sea turtles (*Chelonia, Caretta*). Copeia 1981:757–765.

———. 1991. Physical factors affecting the water exchange of buried reptile eggs. In Egg Incubation: Its Effects on Embryonic Development in Birds and Reptiles, ed. DC Deeming and MWJ Ferguson. Cambridge University Press, Cambridge, U.K., 193–211.

———. 1994. Temperature, time, and reptile egg water exchange. Isr. J. Zool. 40:293–306.

Ackerman RA, R Dmi'el, and A Ar. 1985a. Energy and water vapor exchange by parchment-shelled reptile eggs. Physiol. Zool. 58:129–137.

Ackerman RA and D Lott. 1999. Water exchange by snapping turtle (*Chelydra serpentina*) eggs in field nests. Am. Zool. 39:551.

Ackerman RA, R Seagrave, R Dmi'el, and A Ar. 1985b. Water and heat exchange between parchment-shelled reptile eggs and their surroundings. Copeia 1985:703–711.

Adams LA and HT Martin. 1931. An addition to the urodele fauna of Kansas from the Lower Pliocene. Univ. Kans. Sci. Bull. 19:289–297.

Agassiz L. 1857. Contribution to the Natural History of the United States of America, Vol. 2. Little, Brown, Boston, 541–640.

Albrecht PW. 1967. The cranial arteries and cranial arterial foramina of the turtle genera *Chrysemys, Sternotherus,* and *Trionyx:* A comparative study with analysis of possible evolutionary implications. Tulane Stud. Zool. 14(3):81–99.

———. 1976. The cranial arteries of turtles and their evolutionary significance. J. Morphol. 149:159–182.

AlKindi AYA, IY Mahmoud, and F Al-Siyabi. 2001. Circulating catecholamines in the loggerhead sea turtles, *Caretta caretta* during nesting season. In Proceedings: Perspective in Comparative Endocrinology: Unity and Diversity, ed. HJTh Goos, RK Rastogi, H Vaudry, and R Pierantoni. Monduzzi Editore S.p.A.-Medimond Inc., 325–330.

Al-Nakshabandi G and H Kohnke. 1965. Thermal conductivity and diffusivity of soils as related to moisture tension and other physical properties. Agric. Meteorol. 2:271–279.

Altland PD. 1951. Observations on the structure of the reproductive organs of the box turtle. J. Morphol. 89:599–621.

Alvarez del Toro M. 1960. Los reptiles de Chiapas. Tuxtla Gutierrez, Chiapas, Mexico.

Anderson JF and GR Ultsch. 1987. Respiratory gas concentrations in the microhabitats of some Florida arthropods. Comp. Biochem. Physiol. A 88:585–588.

Anderson PK. 1958. The photic responses and water approach behavior of hatchling turtles. Copeia 1958:211–215.

Andrews R. 1982. Patterns of growth in reptiles. In Biology of the Reptilia, ed. C Gans and FH Pough, Vol. 13 Physiology D. Academic Press, London, 273–320.

Ar A, C Paganelli, R Reeves, D Greene, and H Rahn. 1974. The avian egg: water vapor conductance, shell thickness and functional pore area. Condor 76:153–158.

Aresco MJ. 2005. Mitigation measures to reduce highway mortality of turtles and other herpetofauna at a north Florida lake. J. Wildl. Manage. 69:549–560

Arnold SJ. 1983. Morphology, performance and fitness. Am. Zool. 23:347–361.

Arslan M, P Zaidi, J Lobo, AA Zaidi, and MH Quzi. 1978. Steroid levels in preovulatory and gravid lizards *(Uromastix harkwicki)*. Gen. Comp. Endocrinol. 34:300–303.

Ash RP. 1951. A preliminary report on the size, egg number, incubation period, and hatching in the common snapping turtle, *Chelydra serpentina*. Virginia Acad. Sci. 2:312.

Ashley LM. 1955. Laboratory anatomy of the turtle. Wm. C. Brown, Dubuque, IA.

Attaway MB, GC Packard, and MJ Packard. 1998. Hatchling painted turtles (*Chrysemys picta*) survive only brief freezing of their bodily fluids. Comp. Biochem. Physiol. A 120:405–408.

Auffenberg WA. 1957. The status of the turtle *Macroclemys floridana* Hay. Herpetologica 13:123–126.

———. 1974. Checklist of fossil land tortoises (Testudines). Bull. Fla. State Mus., Biol. Sci. 18:121–251.

Auth DL. 1975. Behavioral ecology of basking in the yellow-bellied turtle, *Chrysemys scripta scripta* (Shoepff). Bull. Fla. State Mus. Biol. Sci. 20:1–46.

Avery HW. 1988. Roles of diet protein and temperature in the nutritional energetics of juvenile slider turtles, *Trachemys scripta*. Masters thesis, State University College at Buffalo, Buffalo, NY, 64 pp.

Avery HW, JR Spotila, JD Congdon, RU Fischer, EA Standora, and SB Avery. 1993. Roles of diet protein and temperature in the growth and nutritional energetics of juvenile slider turtles, *Trachemys scripta*. Physiol. Zool. 66:902–925

Avise, JC. 2000. Phylogeography: the History and Formation of Species. Harvard University Press, Cambridge, MA.

Babcock HL. 1932. The American snapping turtles of the genus *Chelydra* in the collection of the Museum of Comparative Zoology, Cambridge, Massachusetts, U.S.A. Proc. Zool. Soc. Lond. 44:873–874.

Bada JL, SE Brown, and PM Masters. 1980. Age determination of marine mammals based on aspartic acid racemization in the teeth and lens nucleus. Report Int. Whaling Comm. Special Issues 3:113–118.

Bagatto BP and RP Henry 1999. Aerial and aquatic respiration in the snapping turtle, *Chelydra serpentina*. J. Herpetol. 33:490–492.

Baird II. 1960. A survey of the periotic labyrinth in some representative recent reptiles. Kans. Univ. Sci. Bull. 41:891–981.

———. 1970. The anatomy of the reptilian ear. In Biology of the Reptilia, ed. C Gans and TS Parsons, Vol. 2. Academic Press, New York, 193–275.

Baker PJ, JP Costanzo, JB Iverson, and RE Lee. 2003. Adaptations to terrestrial overwintering of hatchling northern map turtles, *Graptemys geographica*. J. Comp. Physiol. B 173:643–651.

Bakken GS. 1976. A heat transfer analysis of animals: unifying concepts and the application of metabolism chamber data to field ecology. J. Theor. Biol. 60:337–384.

Bakken GS and DM Gates. 1975. Heat-transfer analysis of animals: some implications for field ecology, physiology and evolution. In Perspectives in Biophysical Ecology, ed. DM Gates and RB Schmerl. Springer-Verlag, New York, NY, 255–290.

Baldwin FM. 1925. The relation of body to environmental temperatures in turtles, *Chrysemys marginata belli* (Gray) and *Chelydra serpentina* (Linn.). Biol. Bull. 48:432–445.

———. 1926a. Some observations concerning metabolic rate in turtles, *Chrysemys marginata belli* and *Chelydra serpentina* (Linn.). Am. J. Physiol. 76:196.

———. 1926b. Notes on oxygen consumption in turtles, *Chrysemys marginata belli* and *Chelydra serpentina* Linn. Proc. Iowa Acad. Sci. 33:315–323.

Bara G and WA Anderson. 1973. Fine structural localization of 3β-hydroxysteroid dehydrogenase in rat corpus luteum. Histochem. J. 5:437–449.

Bartholomew GA 1982. Energy metabolism. In Animal Physiology: Principles and Adaptations, ed. MS Gordon, GA Bartholomew, AD Grinnell, CB Jorgensen, and FN White. Macmillan, New York, 46–93.

Baumgardt LL. 1997. Temperature-dependent sex determination: factors affecting sex allocation in a northern population of snapping turtles. Unpublished thesis, University of North Dakota, Grand Forks, ND.

Beaupre SJ and D Duvall. 1998. Variation in oxygen consumption of the western diamondback rattlesnake (*Crotalus atrox*): implications for sexual size dimorphism. J. Comp. Physiol. B 168:497–506.

Beebee TJC. 1995. Amphibian breeding and climate. Nature 374:219–220.

Beggs K, J Young, A Georges, and P West. 2000. Ageing the eggs and embryos of the pig-nosed turtle, *Carettochelys insculpta* (Chelonia: Cartettochelydidae), from northern Australia. Can. J. Zool. 78:373–392.

Bell T. 1832. Zoological observations on a new species of *Chelydra* from Oehningen. Trans. Geol. Soc. Lond. Ser. 2 4:379–381.

Bellairs Ad'A. 1949. The anterior brain-case and interorbital septum of Sauropsida, with a consideration of the origin of snakes. Zool. J. Linn. Soc. 41:482–512.

Bellairs Ad'A and AM Kamal. 1981. The chondrocranium and

the development of the skull in recent reptiles. In Biology of the Reptilia, ed. C Gans, Vol. 11. Academic Press, London, 1–264.

Bennett AF. 1982. The energetics of reptilian activity. In Biology of the Reptilia, ed. C Gans and FH Pough, Vol. 13. Physiology D. Academic Press, London, 155–199.

Bennett AF and WR Dawson. 1976. Metabolism. In Biology of the Reptilia, ed. C Gans and WR Dawso, Vol. 5. Physiology A. Academic Press, London, 127–223.

Bentley CC and JL Knight. 1998. Turtles (Reptilia: Testudines) of the Ardis local fauna late Pleistocene (Rancholabrean) of South Carolina. Brimleyana 25:3–33.

Bentley PJ and K Schmidt-Nielsen. 1970. Comparison of water exchange in two aquatic turtles, *Trionyx spinifer* and *Pseudemys scripta*. Comp. Biochem. Physiol. 32:363–365.

Berezanski DJ. 1999. Clutch sizes in Manitoba common snapping turtles. Blue Jay 57:50–55.

Bergounioux FM. 1932. Chéloniens fossiles conservés au Muséum d' Hist. Nat. de Munich. Bull. Soc. Hist. Nat. Toulouse 64: 525–537.

———. 1935. Contribution à l'étude paléontoloque des Chéloniens. Chéloniens fossiles du Bassin d'Aquitaine. Mém. Soc. Géol. Fr. N.S.. 25:1–216.

———. 1936. Monographie des Chéloniens fossiles conservés au laboratoire de Géologie de la Faculté des Sciences de Lyon. Trav. Lab. Géol. Fac. Sci. Lyon 31:1–40.

Bergounioux, FM and F Crouzel. 1965. Chéloniens de Sansan. Ann. Paléontol. (Vertébrés) 51:153–187.

Bernardo J. 1996a. Maternal effects in animal ecology. Am. Zool. 36:83–105.

———. 1996b. The particular maternal effect of propagule size, especially egg size: patterns, models, quality of evidence and interpretations. Am. Zool. 36:216–236.

Berthold AA 1827. Latreilles natürliche Familien des Thierreichs— aus dem Französichen mit. Ammerhungen und Zusäten, Weimer.

Bickham JW and RW Baker. 1976. Chromosome homology and evolution of emydid turtles. Chromosoma 54:201–219.

Bickham JW and JL Carr 1983. Taxonomy and phylogeny of the higher categories of cryptodiran turtles based on a cladistic analysis of chromosomal data. Copeia 1983:918–932.

Birchard GF. 2000. An ontogenetic shift in the response of heart rates to temperature in the developing snapping turtle (*Chelydra serpentina*). J. Therm. Biol. 25:287–291.

Birchard GF and D Marcellini. 1996. Incubation time in reptile eggs. J. Zool. Lond. 240:621–635.

Birchard GF and GC Packard. 1997. Cardiac activity in super-cooled hatchlings of the painted turtle (*Chrysemys picta*). J. Herpetol. 31:166–169.

Birchard GF and CL Reiber. 1993. A comparison of avian and reptilian chorioallantoic vascular density. J. Exp. Zool. 178:245–249.

———. 1995. Growth, metabolism, and chorioallantoic vascular density of developing snapping turtles (*Chelydra serpentina*): influence of temperature. Physiol. Zool. 68:799–811.

———. 1996. Heart rate during development in the turtle embryo: effect of temperature. J. Comp. Physiol. 166:461–466.

Bjersing L. 1967. On the ultrastructure of granulosa lutein cells in porcine corpus luteum: with special reference to endoplasmic reticulum and steroid hormone synthesis. Z. Zellforsch. 82:187–211.

Blaxter K. 1989. Energy Metabolism in Animals and Man. Cambridge University Press, Cambridge, U.K., 336 pp.

Bleakney JS. 1957. A snapping turtle, *Chelydra serpentina serpentina* containing eighty-three eggs. Copeia 1957:143.

Bligh J and KG Johnson. 1973. Glossary of terms for thermal physiology. J. Appl. Physiol. 35:941–961.

Bobyn M and RJ Brooks. 1994a. Interclutch and interpopulation variation in the effects of incubation conditions on sex, survival and growth of hatchling turtles (*Chelydra serpentina*). J. Zool. 233:233–257.

———. 1994b. Incubation conditions as potential factors limiting the northern distribution of snapping turtles, *Chelydra serpentina*. Can. J. Zool. 72:28–37.

Bocourt F. 1868. Description de quelques cheloniens nouveaux appartenant á la fauna Mexicaine. Ann. Sci. Nat. Zool. Paris (5)10:121–122.

Böhme W and M Lang. 1991. The reptilian fauna of the late Oligocene locality Rott near Born (Germany) with special reference to the taxonomic assignment of *"Lacerta" rottensis* von Meyer, 1856. Neues Jahrbuch für Geologie und Paläontologie, Manatschefte, no. 9:515–525.

Bojanus LH. 1819. Anatome *Testudinis europaeae*. [Reprinted 1970] Soc. Study of Amphibians and Reptiles, facsimile reprints in herpetology, no. 26.

Bonaparte CL. 1831. Saggai D'una distribuzione methodica degli animali vertebrati. Boulzaler, Rome.

Booth DT. 1991. A comparison of reptilian eggs with those of megapode birds. In Egg Incubation: Its Effect on Embryonic Development in Birds and Reptiles, ed. DC Deeming and MWJ Ferguson. Cambridge University Press, U.K., 325–344.

———. 1998a. Nest temperature and respiratory gases during natural incubation in the broad-shelled river turtle, *Chelodina expansa* (Testudinata: Chelidae). Aust. J. Zool. 46:183–189.

———. 1998b. Incubation of turtle eggs at different temperatures: do embryos compensate for temperature during development? Physiol. Zool. 71:23–26.

———. 1998c. Effects of incubation temperature on the energetics of embryonic development and hatchling morphology in the Brisbane river turtle *Emydura signata*. J. Comp. Physiol. B 168:399–404.

Booth DT and K Astill. 2001. Incubation temperature, energy expenditure and hatchling size in the green turtle (*Chelonia mydas*), a species with temperature-sensitive sex determination. Aust. J. Zool. 49:389–396.

Booth DT and MB Thompson. 1991. A comparison of reptilian eggs with those of megapode birds. In Egg Incubation: Its Effects on Embryonic Development in Birds and Reptiles, ed. DC Deeming and MWJ Ferguson. Cambridge University Press, Cambridge, U.K., 325–344.

Booth DT, MB Thompson, and S Herring. 2000. How incubation temperature influences the physiology and growth of embryonic lizards. J. Comp. Physiol. B 170:269–276.

Boulenger GA. 1889. Catalogue of the chelonians, rhychocephalians and crocodiles in the British Museum (Natural History). Brit. Mus. Nat. Hist., 309 pp.

Bour R. 1987. Type-specimen of the alligator snapping turtle, *Macroclemys temminckii* (Harlan, 1835). J. Herpetol. 21:340–343.

Bowden RM, MA Ewert, and CE Nelson. 2000. Environmental sex determination in a reptile varies seasonally and with yolk hormones. Proc. R. Soc. Lond. B 267:1745–1749.

Boyd CE. 1970. Amino acid, protein, and caloric content of vascular aquatic macrophytes. Ecology 51:902–906.

Boyer DR 1963. Hypoxia: effects on heart rate and respiration in the snapping turtle. Science 140:813–814.

———. 1965. Ecology of the basking habit in turtles. Ecology. 46:99–118.

———. 1966. Comparative effects of hypoxia on respiratory and cardiac function in reptiles. Physiol. Zool. 39:307–316.

Brady N. 1990. The Nature and Properties of Soils, 10th ed. Maxwell MacMillan International Editions, New York, NY.

Bragg WK, JD Fawcett, TB Bragg, and BE Viets. 2000. Nest-site selection in two eublepharid gecko species with temperature-dependent sex determination and one with genotypic sex determination. Biol. J. Linn. Soc. 69:319–332.

Brattstrom BH. 1962. Thermal control of aggregation behavior in tadpoles. Herpetologica 18:38–46.

———. 1965. Body temperatures of reptiles. Am. Midl. Nat. 73:376–422.

Breckenbridge WJ. 1944. Reptiles and Amphibians of Minnesota. University Minnesota Press, Minneapolis, MN.

———. 1960. A spiny soft-shelled nest study. Herpetologica 16:284–285.

Breitenbach GL, JD Congdon, and RC van Loben Sels. 1984. Winter temperatures of Chrysemys picta nests in Michigan: effects on hatchling survival. Herpetologica 40:76–81.

Brinkman DB. 1990. Paleoecology of the Judith River Formation (Campanian) of Dinosaur Provincial Park, Alberta, Canada: evidence from vertebrate microfossil localities. Palaeogeogr. Palaeoclimatol. Palaeoecol. 78:37–54.

Brinkman DB and X-C Wu. 1999. The skull of Ordosemys, an Early Cretaceous turtle from Inner Mongolia People's Republic of China, and the interrelationships of Eucryptodira (Chelonia, Cryptodira). Paludicola 2:134–147.

Brockelman W. 1975. Competition, the fitness of offspring and optimal clutch size. Am. Nat. 109:677–699.

Brody S. 1945. Bioenergetics and growth. Reinhold Publishing Corporation, New York, 1023 pp.

Broin F de. 1977. Contribution a l'étude des cheloniens continentaux du Tertiaire de France. Mém. Mus. Natl. Hist. Nat. Sér. C 38:1–423.

Bromham L. 2002. Molecular clocks in reptiles: life history influences rate of molecular evolution. Mol. Biol. Evol. 19:302–309.

Brooks RJ, ML Bobyn, DA Galbraith, JA Layfield, and EG Nancekivell. 1991a. Maternal and environmental influences on growth and survival of embryonic and hatchling snapping turtles (Chelydra serpentina). Can. J. Zool. 69:2667–2676.

Brooks RJ, GP Brown, and DA Galbraith. 1991b. Effects of a sudden increase in natural mortality of adults on a population of the common snapping turtle (Chelydra serpentina). Can. J. Zool. 69:1314–1320.

Brooks RJ, DA Galbraith, and CA Bishop. 1986. Reproductive output of nesting females in two Ontario populations of Chelydra serpentina. Abstr. Annu. Meet. Soc. Study Amph. Rept./Herpetologists' League, Southwest Missouri State Univ., Springfield.

Brooks R, DA Galbraith, EG Nancekivell, and CA Bishop. 1988. Developing management guidelines for snapping turtles. Proc. U.S. Dept. Agric. Forest Service Symp. Gen. Tech. Report RM-166, 174–179.

Brooks RJ, MA Krawchuk, C Stevens and N Koper. 1997. Testing the precision and accuracy of age estimation using lines in scutes of Chelydra serpentina and Chrysemys picta. J Herpetol. 31:521–529.

Brooks RJ, CM Shilton, GP Brown, and NWS Quinn. 1992. Body size, age distribution, and reproduction in a northern population of wood turtles (Clemmys insculpta). Can. J. Zool. 70:462–469.

Brown EE. 1992. Notes on amphibians and reptiles of the western Piedmont of North Carolina. J. Elisha Mitchell Sci. Soc. 108:38–54.

Brown GP, CA Bishop, and RJ Brooks. 1994a. Growth rate, reproductive output, and temperature selection of snapping turtles in habitats of different productivities. J. Herpetol. 28:405–410.

Brown GP and RJ Brooks. 1991. Thermal and behavioral responses to feeding in free-ranging turtles, Chelydra serpentina. J. Herpetol. 25:273–278.

———. 1993. Sexual and seasonal differences in activity in a northern population of snapping turtles, Chelydra serpentina. Herpetologica. 49: 311–318.

———. 1994. Characteristics of a fidelity to hibernacula in a northern population of snapping turtles, Chelydra serpentina. Copeia 1994:222–226.

Brown GP, RJ Brooks, and JA Layfield. 1990. Radiotelemetry of body temperatures of free-ranging snapping turtles (Chelydra serpentina) during summer. Can. J. Zool. 68:1659–1663.

Brown GP, RJ Brooks, ME Siddall, and SS Desser. 1994b. Parasites and reproductive output in the snapping turtle, Chelydra serpentina. Copeia 1994:228–231.

Brown PS, R Giuliano, and G Hough. 1974. Pituitary regulation of appetite and growth in the turtles Pseudomys scripta elegans and Chelydra serpentina. J. Exp. Zool. 187:205–216.

Bruce R and R. Luxmoore. 1986. Water retention: field methods. In Methods of Soil Analysis, ed. A Klute. Soil Science Society of America, Madison, WI, 663–686.

Budhabhatti J and EO Moll. 1990. Chelydra serpentina (common snapping turtle) feeding behavior. Herpetol. Rev. 21:19.

Buddington RK, JW Chen, and D Diamond. 1991. Dietary regulation of intestinal brush-border sugar and amino acid transport in carnivores. Am. J. Physiol. 261:R793–R801.

Buddington RK and JW Hilton. 1987. Intestinal adaptations of rainbow trout to changes in dietary carbohydrate. Am. J. Physiol. 253:G489–G496.

Buhlmann KA and G Coffmann. 2001. Fire ant predation of turtle nests and implications for the strategy of delayed emergence. J. Elisha Mitchell Sci. Soc. 117:94–100.

Bull JJ. 1980. Sex determination in reptiles. Q. Rev Biol. 55:3–21.

———. 1983. Evolution of Sex Determining Mechanisms. Benjamin/Cummings, Menlo Park, CA.

———. 1985. Sex ratio and nest temperature in turtles: comparing field and laboratory data. Ecology 66:1115–1122.

———. 1987. Temperature-dependent sex determination in reptiles: validity of sex diagnosis in hatchling lizards. Can. J Zool. 65:1421–1424.

Bull JJ and MG Bulmer. 1989. Longevity enhances selection of environmental sex determination. Heredity 63:315–320.

Bull, JJ, WHN Gutzke, and MG Bulmer. 1988a. Nest choice in a captive lizard with temperature-dependent sex determination. J. Evol. Biol. 2:177–184.

———. 1988b. Sex reversal by estradiol in three reptilian orders. Gen. Comp. Endocrinol. 70:425–428.

Bull JJ, DM Hillis and S O'Steen. 1988c. Mammalian ZFY

sequences exist in reptiles regardless of sex-determining mechanism. Science 242:567–569.

Bull JJ. and R. Shine. 1979. Iteroparous animals that skip opportunities for reproduction. Am. Nat. 114:296–303.

Bull JJ and RC Vogt. 1979. Temperature-dependent sex determination in turtles. Science 206:1186–1188.

Bull JJ, RC Vogt, and MG Bulmer. 1982a. Heritability of sex ratio in turtles with environmental sex determination. Evolution 36:333–341.

Bull JJ, RC Vogt, and CJ McCoy. 1982b. Sex determining temperatures in turtles: a geographic comparison. Evolution 36:326–332.

Bulmer MG and JJ Bull. 1982. Models of polygenic sex determination and sex ratio control. Evolution 36:13–26.

Burger J. 1976. Behavior of hatchling diamondback terrapins (Malaclemys terrapin) in the field. Copeia 1976:742–748.

Burghardt GM. 1967. The primacy effect of the first feeding experience in the snapping turtle. Psychon. Sci. 7:383–384.

Burghardt GM and EH Hess. 1966. Food imprinting in the snapping turtle, Chelydra serpentina. Science 151:108–109.

Burke AC. 1989a. Development of the turtle carapace: implications for the evolution of a novel bauplan. J. Morphol. 199:363–378.

———. 1989b. Epithelial-mesenchymal interactions in the development of the chelonian bauplan. Fortschr. Zool. 35:206–209.

———. 1991. The development and evolution of the turtle body plan: inferring intrinsic aspects of the evolutionary process from experimental embryology. Am. Zool. 31:616–627.

Burke AC and P Alberch. 1985. The development and homology of the chelonian carpus and tarsus. J. Morphol. 186:119–131.

Burke VJ, JL Greene, and JW Gibbons. 1995. The effect of sample size and study duration on metapopulation estimates for slider turtles (Trachemys scripta). Herpetologica 51:451–456.

Burke VJ, RD Nagle, MO Osentoski, and JD Congdon. 1993. Common snapping turtles associated with ant mounds. J. Herpetol. 27:114–115.

Burke VJ, SL Rathbun, JR Bodie, and JW Gibbons. 1998. Effect of density and predation rate for turtle nests in a complex landscape. Oikos 83:3–11.

Bury RB, AV Nebeker, and MJ Adams. 2000. Response of hatchling and yearling turtles to thermal gradients: comparison of Chelydra serpentina and Trachemys scripta. J. Therm. Biol.25:221–225.

Bustard HR. 1967. Mechanism of nocturnal emergence from the nest in green turtle hatchlings. Nature 214:317.

Butler BO and TE Graham. 1995. Early post-emergent behavior and habitat selection in hatchling Blanding's turtles, Emydoidea blandingii, in Massachusetts. Chelonian Conserv. Biol. 3:187–196.

Cagle FR 1944. Home range, homing behavior, and migration in turtles. Misc. Pub. Mus. Zool. Univ. Mich. 61:1–34.

———. 1944. Activity and winter changes of hatchling Pseudemys. Copeia 1944:105–109.

———. 1946. The growth of the slider turtle, Pseudemys scripta elegans. Am. Midl. Nat. 36:685–729.

———. 1950. The life history of the slider turtle, Pseudemys scripta troostii (Holbrook). Ecol. Monogr. 20:31–54.

Cagle KD, GC Packard, K Miller, and MJ Packard. 1993. Effects of the microclimate in natural nests on development of embryonic painted turtles, Chrysemys picta. Funct. Ecol. 7:653–660.

Cahn AR. 1937. The turtles of Illinois. Univ. Ill. Bull. 35:1–218.

Callard IP and M Hirsch. 1976. The influence of estradiol-17 and progesterone on the contractility of the oviduct of the turtle, Chrysemys picta, in vitro. J. Endocrinol. 68:147–152.

Callard IP, V Lance, AR Salhanuick, and D Barad. 1978. The annual ovarian cycle of Chrysemys picta: Correlated changes in plasma steroids and parameters of vitellogenesis. Gen. Comp. Endocrinol. 35:245–257.

Campbell G. 1985. Soil Physics with Basic. Elsevier Scientific Publishing, New York, NY.

Cannon ME, RE Carpenter, and RA Ackerman. 1986. Synchronous hatching and oxygen consumption of Darwin's rhea eggs (Pternocnemia pennata). Physiol. Zool. 59:95–108.

Carr A. 1952. Handbook of Turtles. Cornell University Press, Ithaca, NY.

Carr AF and HF Hirth. 1961. Social facilitation in green turtle siblings. Anim. Behav. 9:68–70.

Carroll DM. 1993. Trout Reflections—A Natural History of the Trout and Its World. St. Martin's Griffin, New York.

Carroll DM and GR Ultsch. 2007. Emergence season and survival in the nest of hatchling turtles in soutcentral New Hampshire. Northeast. Nat. 14:307–310.

Cassell D and A Klute. 1986. Water potential: tensiometry. In Methods of Soil Analysis, ed. A Klute. Soil Science Society of America, Madison, WI, 563–597.

Charnier M. 1966. Action de la température sur la sex-ratio chez l'embryon d'Agama agama (Agamidae, Lacertilien). C. R. Séances Soc. Biol. l'Ouest Africain 160:620–622.

Charnov EL. 1990. On evolution of age at maturity and adult lifespan. J. Evol. Biol. 3:139–144.

Charnov EL, D Berrigan, and R Shine. 1993. The M/k ratio is the same for fish and reptiles. Am. Nat. 142:707–711.

Charnov EL and J Bull. 1977. When is sex environmentally determined? Nature 266:828–830.

Christens E and JR Bider. 1987. Nesting activity and hatchling success of the painted turtle (Chrysemys picta marginata) in southwestern Quebec. Herpetologica 43:55–65.

Christiansen JL and RR Burken. 1979. Growth and maturity of the snapping turtle (Chelydra serpentina) in Iowa. Herpetologica 35:261–266.

Christiansen JL and EO Moll. 1973. Latitudinal reproductive variation within a single subspecies of painted turtle, Chrysemys picta bellii. Herpetologica 29:152–163.

Chrousos GP and PW Gold. 1992. The concepts of stress and stress disorders. JAMA 267:1244–1252.

Churchill TA and KB Storey. 1992. Natural freezing survival by painted turtles Chrysemys picta marginata and C. p. bellii. Am. J. Physiol. 262:R530–R537.

Ckhikvadze VM. 1971. [On the history of the tortoise family Chelydridae]. Akad. Nauk Gruz. SSR Soobshch. 61:137–240.

———. 1973. [Tertiary turtles from the Zaisan Depression]. Akad. Nauk Gruz. SSR Izdatel'stvo Metsniereba, Tbilisi, 101 pp.

———. 1982. [A large caiman turtle from Pliocene deposits of the northern Black Sea Area]. Vestn. Zool. 16:15–20.

———. 1985. Sur la classification et les charactères de certaines tortues fossiles d'Asia, rares et peu etudiees. Stud. Geol. Salmanticensia. Vol. Espec. 2:55–86.

———. 1999. [The history of the development of the Paleogene herpetofauna of the former Soviet Union territory.] Probl. Paleobiol. Tbilisi, Metsniereba 1:270–279.

Clark DB and JW Gibbons. 1969. Dietary shift in the turtle *Pseudemys scripta* (Schoepff) from youth to maturity. Copeia 1969:704–706.

Clark K, G Bender, BP Murray, K Panfilio, S Cook, R Davis, K Murnen, RS Tuan, and SF Gilbert. 2001. Evidence for the neural crest origin of turtle plastron bones. Genesis 31:111–117.

Clark N. 1967. Influence of estrogens upon the serum calcium, phosphate and protein concentrations of fresh water turtles. Comp. Biochem. Physiol. 20:823–824.

Clark PJ. 1986. Physiological aspects of temperature dependent sex determination in *Sternotherus odoratus,* the Common Musk Turtle. Unpublished M.S. thesis, Indiana University, Bloomington, IN.

———. 2000. Reproductive strategies of turtles: population, latitudinal, and phylogenetic comparisons. Unpublished Ph.D. thesis, Indiana University, Bloomington, IN.

Clarke A and KPP Fraser. 2004. Why does metabolism scale with temperature? Funct. Ecol. 18:243–251.

Cole LC. 1957. Sketches of general and comparative demography. Cold Spring Harbor Symp. Quant. Biol. 22:1–15.

Conant R and JT Collins. 1991. A Field Guide to Reptiles and Amphibians: Eastern and Central North America, 3rd ed. Houghton Mifflin, Boston, MA.

Congdon JD. 1989. Proximate and evolutionary constraints on energy relations of reptiles. Physiol. Zool. 62:356–373.

Congdon JD, GL Breitenbach, RC van Loben Sels, and DW Tinkle. 1987. Reproduction and nesting ecology of snapping turtle (*Chelydra serpentina*) in southeastern Michigan. Herpetologica 43:39–54.

Congdon JD, AE Dunham, and RC van Loben Sels. 1993. Delayed sexual maturity and demographics of Blanding's turtles (*Emydoidea blandingii*): Implications for conservation and management of long-lived organisms. Conserv. Biol. 7:826–833.

———. 1994. Demographics of common snapping turtles (*Chelydra serpentina*): implications for conservation and management of long-lived organisms. Am. Zool. 34:397–408.

Congdon JD and JW Gibbons. 1985. Egg components and reproductive characteristics of turtles: relationships to body size. Herpetologica 41:194–205.

———. 1987. Morphological constraint on egg size: A challenge to optimal egg size theory? Proc. Natl. Acad. Sci. USA 84:4145–4147.

———. 1990. Turtle eggs: their ecology and evolution. In Life History and Ecology of the Slider Turtle, ed. JW Gibbons. Smithsonian Institution Press, Washington, DC, 109–123.

Congdon JD, SW Gotte, and RW McDiarmid. 1992. Ontogenetic changes in habitat use by juvenile turtles, *Chelydra serpentina* and *Chrysemys picta*. Can. Field-Nat. 106:241–248.

Congdon JD, RD Nagle, AE Dunham, CW Beck, OM Kinney, and SR Yeomans. 1999. The relationship of body size to survivorship of hatchling snapping turtles (*Chelydra serpentina*): an evaluation of the "bigger is better" hypothesis. Oecologica 121:224–235.

Congdon JD, RD Nagle, OM Kinney, M Osentoski, HA Avery, RC van Loben Sels, and DW Tinkle. 2000. Nesting ecology and embryo mortality: implications for hatchling success and demography of Blanding's turtles (*Emydoidea blandingii*). Chelonian Conserv. Biol. 3:569–579.

Congdon, JD, RD Nagle, OM Kinney, and RC van Loben Sels. 2001. Hypotheses of aging in a long-lived vertebrate, Blanding's turtle (*Emydoidea blandingii*). Exp. Gerontol. 36:813–827.

Congdon JD, RD Nagle, OM Kinney, RC van Loben Sels, T Quinter, and DW Tinkle. 2003. Testing hypotheses of aging in long-lived painted turtles (*Chrysemys picta*). Exp. Gerontol. 38:765–772.

Congdon JD, DW Tinkle, GL Breitenbach, and RC van Loben Sels. 1983a. Nesting ecology and hatching success in the turtle *Emydoidea blandingii*. Herpetologica 39:417–429.

Congdon JD, DW Tinkle, and PC Rosen. 1983b. Egg components and utilization during development in aquatic turtles. Copeia 1983:264–268.

Conners JS. 1998. Testudines: *Chelydra serpentina* (common snapping turtle). Predation. Herpetol. Rev. 29:235

Conover DO. 1984. Adaptive significance of temperature-dependent sex determination in a fish. Am. Nat. 123:297–313.

Cope ED. 1872. Synopsis of the species of Chelydrinae. Proc. Acad. Nat. Sci. Philadelphia 1872:22–29.

Corgan JX. 1976. Vertebrate fossils of Tennessee. Tennessee Div. Geol. Bull. (77):1–100.

Costanzo JP, JB Iverson, MF Wright, and RE Lee. 1995. Cold hardiness and overwintering strategies of hatchlings in an assemblage of northern turtles. Ecology 76:1772–1785.

Costanzo JP, EE Jones, and RE Lee. 2001a. Physiological responses to supercooling and hypoxia in the hatchling painted turtle, *Chrysemys picta*. J. Comp. Physiol. B 171:335–340.

Costanzo JP, JD Litzgus, JB Iverson, and RE Lee. 2000a. Seasonal changes in physiology and development of cold hardiness in the hatchling painted turtle *Chrysemys picta*. J. Exp. Biol. 203:3459–3470.

———. 2000b. Ice nuclei in soil compromise cold hardiness in hatchling painted turtles, *Chrysemys picta*. Ecology 81:346–360.

———. 2001b. Cold-hardiness and evaporative water loss in hatchling turtles. *Physiol. Biochem. Zool.* 74:510–519.

Costanzo JP, JD. Litzgus, and RF. Lee. 1999. Behavioral responses of hatchling painted turtles (*Chrysemys picta*) and snapping turtles (*Chelydra serpentina*) at subzero temperatures. J. Therm. Biol. 24:161–166.

Coulter MW. 1958. Distribution, food, and weight of the snapping turtle in Maine. Maine Field Nat. 14:53–62.

Cowles RB and CM Bogert. 1944. A preliminary study of the thermal requirements of desert reptiles. Bull. Am. Mus. Nat. Hist. 83:261–296.

Coyer PE and CP Mangum 1973. Effect of temperature on active and resting metabolism of polychaetes. In Effects of Temperature on Ectothermic Organisms, ed. W Wieser. Springer-Verlag, New York, 173–180.

Crawford KM. 1991a. The effect of temperature and seasonal acclimatization on renal function of painted turtles, *Chrysemys picta*. Comp. Biochem. Physiol. 99A:375–380.

Crawford KM. 1991b. The winter environment of painted turtles, *Chrysemys picta;* temperature, dissolved oxygen, and potential cues for emergence. Can. J. Zool. 69:2493–2498.

Crawford KM, JR Spotila, and EA Standora. 1983. Operative environmental temperatures and basking behavior of the turtle *Pseudemys scripta*. Ecology 64:989–999.

Cree A, LJ Guillette Jr, JF Cockrem, MA Brown, and GK Chambers. 1990b. Absence of daily cycles in plasma sex steroids in male and female tuatara (*Sphenodon punctatus*) and the effect of acute capture stress on females. Gen. Comp. Endocrinol. 79:103–113.

Cree A, LJ Guillette Jr, JR Cockrem, and JMP Joss. 1990a. Effects of capture and temperature stresses on plasma steroid concentrations in male tuatara (*Sphenodon punctatus*). J. Exp. Zool. 253:38–46.

Cree A, MB Thompson, and CH Daugherty. 1995. Tuatara sex determination. Nature 375:543.

Crews D, T Wibbels, and WHN Gutzke. 1989. Action of sex steroid hormones on temperature-induced sex determination in the snapping turtle (*Chelydra serpentina*). Gen. Comp. Endocrinol. 76:159–166.

Crick, HQP and TH Sparks. 1999. Climate change related to egg-laying trends. Nature 399:423–424.

Crocker CE, RA Feldman, GR Ultsch, and DC Jackson. 2000a. Overwintering behavior and physiology of eastern painted turtles (*Chrysemys picta picta*) in Rhode Island. Can. J. Zool. 78:936–942.

Crocker CE, TE Graham, GR Ultsch, and DC Jackson. 2000b. Physiology of common map turtles (*Graptemys geographica*) hibernating in the Lamoille River, Vermont. J. Exp. Zool. 286:143–148.

Cuellar H. 1979. Disruption of gestation and egg shelling in deluteinized oviparous whiptail lizards *Cnemidophorus uniparens*. Gen. Comp. Endocrinol. 39:150–157.

Cunningham B. 1922. Some phases in the development of *Chrysemys cinerea*. J. Elisha Mitchell Sci. Soc. 38:51–73.

Cunnington DC and RJ Brooks. 1996. Bet-hedging theory and eigenelasticity: A comparison of the life histories of loggerhead sea turtles (*Caretta caretta*) and snapping turtles (*Chelydra serpentina*). Can. J. Zool. 74:291–296.

Cyrus RV, IY Mahmoud, and J Klicka. 1978. Fine structure of the corpus luteum of the snapping turtle, *Chelydra serpentina*. Copeia 4:622–627.

Cyrus RV, IY Mahmoud, DM Montag, and MJ Woller. 1982. Localization of progesterone receptors in the female reproductive tract of the snapping turtle, *Chelydra serpentina*. Am. Zool. 22:949.

Daly E. 1992. A list, bibliography and index of the fossil vertebrates of Mississippi. Mississippi Off. Geol. Bull. 128: 1–47.

Daniels CB, S Orgeig, and AW Smits. 1995. The composition and function of reptilian pulmonary surfactant. Respir. Physiol. 102:121–135.

Daniels CB, S Orgeig, AW Smits, and JD Miller. 1996. The influence of temperature, phylogeny, and lung structure on the lipid composition of reptilian pulmonary surfactant. Exp. Lung Res. 22:267–281.

Davies DG and JA Sexton. 1987. Brain ECF pH and central chemical control of ventilation during anoxia in turtles. Am. J. Physiol. 252:R848–R852.

DeBeer GR. 1926. Studies on the vertebrate head. II. The orbito-temporal region of the skull. Q. J. Microsc. Sci. 70:263–370.

———. 1937. The Development of the Vertebrate Head. Clarendon Press, Oxford [reprinted 1971].

deBraga M and O Rieppel. 1997. Reptile phylogeny and the interrelationships of turtles. Zool. J. Linn. Soc. 120:281–354.

Decker JD 1967. Motility of the turtle embryo, *Chelydra serpentina* (Linné). Science 157:952–954.

Deeming CD and MB Thompson. 1991. Gas exchange across reptilian eggshells. In Egg Incubation Its Effects on Embryonic Development in Birds and Reptiles, ed. CD Deeming and MWJ Ferguson. Cambridge University Press, New York, 277–284.

Deevey ES. 1947. Life tables for natural poplations of animals. Quart. Rev. Biol. 22:283–314.

Denver RJ and P Licht. 1991. Dependence of body growth on thyroid activity in turtles. J. Exp. Zool. 258:48–59.

DePari JA. 1996. Overwintering in the nest chamber by hatchling painted turtles, *Chrysemys picta*, in northern New Jersey. Chelonian Conserv. Biol. 2:5–12.

de Solla, SR and KJ Fernie. 2004. Is cost of locomotion the reason for prolonged nesting forays of Snapping Turtles, *Chelydra serpentina*? Can. Field Nat. 118:610–612.

Dick DAT. 1966. Cell Water. Butterworths, Washington, DC.

Dillon CD. 1998. The common snapping turtle, *Chelydra serpentina*. Tortuga Gaz. 34(3):1–4.

Dimond MT. 1954. The reactions of developing snapping turtles, *Chelydra serpentina serpentina* (Linné), to thiourea. J. Exp. Zool. 127:93–115.

———. 1983. Sex of turtle hatchlings as related to incubation temperature. In Proceedings of the 6th Reptile Symposium on Captive Propagation and Husbandry. Zoological Consortium, Thurmont, MD, 88–101.

Dobie JL. 1968. A new turtle species of the genus *Macroclemys* (Chelydridae) from the Florida Pliocene. Tulane Stud. Zool. Bot. 15:59–63.

———. 1971. Reproduction and growth in the alligator snapping turtle, *Macroclemys temminckii* (Troost). Copeia 1971:645–658.

Dodge CH and GE Folk Jr. 1963. Notes on comparative tolerance of some Iowa turtles to oxygen deficiency (hypoxia). Proc. Iowa Acad. Sci. 70:438–441.

Dorfman RI and F Ungar. 1965. Metabolism of Steroid Hormones. Academic Press, New York.

Doroff AM and LB Keith. 1990. Demography and ecology of an ornate box turtle (*Terrapene ornata*) population in south-central Wisconsin. Copeia 1990:387–399.

Du W-G and X Ji. 2003. The effects of incubation thermal environments on size, locomotor performance and early growth of hatchling soft-shelled turtles, *Pelodiscus sinensis*. J. Therm. Biol. 28:279–286.

Dubios W, J Pudney, and IP Callard. 1988. The annual testicular cycle in the turtle *Chrysemys picta*: a histochemical and electron microscopic study. Gen. Comp. Endocrinol. 71:191–204.

Duméril AMC and G Bibron 1835. Erpétologie génèrale ou histoire naturelle complete des reptiles, Vol. 2. Librairie Encyclopedique de Roret, Paris.

Duméril AMC and AHA Duméril 1851. Catalogue méthodique de la collection des reptiles du Muséum d'Histoire Naturelle de Paris. Gide and Boudry, Paris.

Dunn ER 1945. Anfibios y reptiles de Colombia, IV. Caldasia 3:316–317.

Dunson WA. 1960. Aquatic respiration in *Trionyx spinifer asper*. Herpetologica 16:277–283.

Duvall D, LJ Guillette Jr, and RE Jones. 1982. Environmental control of reptilian reproductive cycles. In Biology of the Reptilia, ed. C. Gans and FH Pugh, Vol. 13D. Academic Press, New York, 201–231

Dzialowski EM and MP O'Connor. 1999. Utility of blood flow to the appendages in physiological control of heat exchange in reptiles. J. Therm. Biol. 24:1–32.

———. 2001a. Physiological control of warming and cooling during simulated shuttling and basking in lizards. Physiol. Biochem. Zool. 74:679–693.

———. 2001b. Thermal time constant estimation in warming and cooling ectotherms. J. Therm. Biol. 26:231–245.

Eaton J, RL Cifelli, JH Hutchison, JI Kirtland, and JM Parrish. 1999a. Cretaceous vertebrate faunas from the Kapairowits Plateau, south-central Utah. Utah Geol. Surv. Misc. Publ. 99–1:345–353.

Eaton JG, JH Hutchison, PA Holroyd, WW Korth, and PM Goldstrand. 1999b. Vertebrates of the Turtle Basin local fauna, middle Eocene, Sevier Plateau, south-central Utah. Utah Geol. Surv. Misc. Publ. 99–1:463–468.

Eckert R, D Randall, and G Augustine. 1988. Animal Physiology. W. H. Freeman and Company, New York, 683 pp.

Edgren RA. 1949. Variation in the size of eggs of the turtles *Chelydra s. serpentina* (Linne) and *Sternotherus odoratus* (Latreille). Chic. Acad. Sci. 53:1.

Edmund AG. 1969. Dentition. In Biology of the Reptilia, ed. C Gans, Ad'A Bellairs, and TS Parsons, Vol. 1. Academic Press, New York, 117–200.

Ehrenfeld DW. 1979. Behavior Associated with Nesting, ed. M. Harless and H. Morlock. John Wiley and Sons, New York, 417–434.

Elf PK, JW Lang, and AJ Fivizzani. 2002a. Yolk hormone levels in the eggs of snapping turtles and painted turtles. Gen. Comp. Endocrinol. 127:26–33.

———. 2002b. Dynamics of yolk steroid hormones during development in a reptile with temperature-dependent sex determination. Gen. Comp. Endocrinol. 127:34–39.

Elgar MA and LJ Heaphy. 1989. Covariation between clutch size, egg weight and shape: Comparative evidence for chelonians. J. Zool. 219:137–152.

Erickson BR. 1973. A new chelydrid turtle *Protochelydra zangerli* from the late Paleocene of North Dakota. Sci. Publ. Sci. Mus. Minn. N.S. 2(2):1–16.

———. 1982. The Wannagan Creek Quarry and its reptilian fauna [Bullion Creek Formation, Paleocene] in Billings County, North Dakota. N.D. Geol. Surv. Rep. Invest. 72:1–17.

Ernst CH. 1966. Overwintering of hatchling *Chelydra serpentina* in southeastern Pennsylvania. Phila. Herpetol. Soc. Bull. 14:8–9.

———. 1968. Evaporative water-loss relationships of turtles. J. Herpetol. 2:159–161.

———. 1972. Temperature-activity relationship in the painted turtle, *Chrysemys picta*. Copeia 1972:217–222.

———. 1986. Environmental temperatures and activities in the wood turtle, *Clemmys insculpta*. J. Herpetol. 20:341–352.

Ernst CH and RW Barbour. 1972. Turtles of the United States. The University Press of Kentucky, Lexington.

———. 1989. Turtles of the World. Smithsonian Inst. Press, Washington, DC.

Ernst CH, SC Belfit, SW Sekscienski, and AF Laemmerzahl. 1997. The amphibians and reptiles of Ft. Belvoir and northern Virginia. Bull. Md. Herpetol. Soc. 33:1–62.

Ernst CH, JW Gibbons, and S Novak. 1988. *Chelydra*. Catalog. Am. Amphib. Rept. (419):1–4.

Ernst CH and HF Hamilton. 1969. Color preferences of some North American turtles. J. Herpetol 3:176–180.

Ernst CH, JE Lovich, and RW Barbour. 1994. Turtles of the United States and Canada. Smithsonian Inst. Press, Washington, DC.

Etchberger CR, MA Ewert, JB Phillips, CE Nelson, and HD Prange. 1992. Physiological responses to carbon dioxide in embryonic red-eared slider turtles, *Trachemys scripta*. J. Exp. Zool. 264:1–10.

Etchberger CR, JB Phillips, MA Ewert, CE Nelson, and HD Prange. 1991. Effects of oxygen concentration and clutch on sex determination and physiology in red-eared slider turtles (*Trachemys scripta*). J. Exp. Zool. 258:394–403.

Evermann BW and HW Clark. 1916. The turtles and batrachians of the Lake Maxinkuckee region. Proc. Indiana Acad. Sci. 1916:472–518.

Ewert MA. 1976. Nests, nesting and aerial basking of *Macroclemys* under natural conditions and comparisons with *Chelydra* (Testudines: Chelydridae). Herpetologica 32:150–156.

———. 1979. The embryo and its egg: development and natural history. In Turtles: Perspectives and Research, ed. M Harless and H Morlock. John Wiley & Sons, New York, 333–413.

———. 1985. Embryology of turtles. In Biology of the Reptilia, ed. C Gans, F Billett, and PA Maderson, Vol. 14. John Wiley & Sons, Chichester, 74–267.

———. 1991. Cold torpor, diapause, delayed hatching and aestivation in reptiles and birds. In Egg Incubation: Its Effects on Embryonic Development in Birds and Reptiles, ed. DC Deeming and MWJ Ferguson. Cambridge University Press, Cambridge, U.K., 173–191.

———. 2000. *Chelydra serpentina osceola* (Florida snapping turtle) reproduction. Herpetol. Rev. 31:172.

Ewert MA and DR Jackson. 1994. Nesting ecology of the alligator snapping turtle (*Macroclemys temminckii*) along the lower Apalachicola River, Florida. Florida Game and Freshwater Fish Commission, Non-Game Wildl. Prog., Final Report, 1–45.

Ewert MA, DR Jackson, and CE Nelson. 1994. Patterns of temperature-dependent sex determination in turtles. J. Exp. Zool. 270:3–15.

Ewert MA, JW Lang, and CE Nelson. 2005. Geographic variation in the pattern of temperature-dependent sex determination in the American snapping turtle (*Chelydra serpentina*). J. Zool. 265:81–95.

Ewert MA and CE Nelson. 1991. Sex determination in turtles: diverse patterns and some possible adaptive values. Copeia 1991:50–69.

Ewert MA and DS Wilson. 1996. Seasonal variation of embryonic diapause in the striped mud turtle (*Kinosternon baurii*) and general considertation for conservation planning. Chelonian Conserv. Biol. 2:43–54.

Falconer DS and TFC Mackay. 1996. Introduction to Quantitative Genetics. Longman Group, Essex.

Farrell RF and TE Graham. 1991. Ecological notes on the turtle *Clemmys insculpta* in northwestern New Jersey. J. Herpetol. 25:1–9.

Fay LP 1988. Late Wisconsinan Appalachian herpetofaunas: relative stability in the midst of change. Ann. Carnegie Mus. 57:189–220.

Feder ME, SL Satel, and AG Gibbs. 1982. Resistance of the shell membrane and mineral layer to diffusion of oxygen and water in flexible-shelled eggs of the snapping turtle (*Chelydra serpentina*). Respir. Physiol. 49:179–191.

Ferguson GW, KL Brown, and VG DeMarco. 1982. Selective basis for the evolution of variable egg and hatchling size in some Iguanid lizards. Herpetologica 38:178–188.

Feuer RC. 1971. Intergradation of the snapping turtles *Chelydra serpentina serpentina* (Linnaeus, 1758) and *Chelydra serpentina osceola* Stejneger, 1918. Herpetologica 27:379–384.

Figler RA, D Owens, DS MacKenzie, and P Licht. 1986. Changes in the plasma concentration of arginine vasotocin during oviposition in sea turtles. Am. Zool. 26:9 (abstract).

Filoramo NI and FJ Janzen. 2002. An experimental study of the influence of embryonic water availability, body size, and clutch on survivorship of neonatal red-eared sliders, *Trachemys scripta elegans*. Herpetologica 58:67–74.

Finkler MS. 1997. Impact of egg content on post-hatching size, body composition, and performance in the common snapping turtle (*Chelydra serpentina*). Linnaeus Fund research report. Chelonian Conserv. Biol. 2:452–455.

———. 1998. The influence of water availability during incubation on body size, body composition, locomotor performance, and desiccation tolerance in neonatal snapping turtles (*Chelydra serpentina*) from North-Central Nebraska. Unpublished Ph.D. dissertation, Miami University, Oxford, OH.

———. 1999. Influence of water availability during incubation on hatchling size, body composition, desiccation tolerance, and terrestrial locomotor performance in the snapping turtle *Chelydra serpentina*. Physiol. Biochem. Zool. 72:714–722.

———. 2001. Rates of water loss and estimates of survival time under varying humidity in juvenile snapping turtles (*Chelydra serpentina*). Copeia 2001:521–525.

———. 2006. Does variation in soil water content induce variation in the size of hatchling snapping turtles (*Chelydra serpentina*)? Copeia 2006:769–777.

Finkler MS, JT Bowen, TM Christman, and AD Renshaw. 2002. Effects of hydric conditions during incubation on body size and triglyceride reserves of overwintering hatchling snapping turtles (*Chelydra serpentina*). Copeia 2002:504–510.

Finkler MS and DL Claussen. 1997. Within and among clutch variation in the composition of *Chelydra serpentina* eggs with initial egg mass. J. Herpetol. 31:620–624.

Finkler MS, DL Knickerbocker, and DL Claussen. 2000. Influence of hydric conditions during incubation and population on overland movement of neonatal snapping turtles. J. Herpetol. 34:452–455.

Finkler MS and BT Kressley. 2002. Post-hatching yolk consumption and stored energy reserves in hatchling snapping turtles, *Chelydra serpentina*. Physiologist 45:345.

Finkler MS, AC Steyermark, and KE Jenks 2004. Geographic variation in snapping turtle (*Chelydra serpentina serpentina*) egg components across a longitudinal transect. Can. J. Zool. 82:102–109.

Finneran LC. 1947. A large clutch of eggs of *Chelydra serpentina serpentina* (Linnaeus). Herpetologica 3:182.

Fischer RU, FJ Mazzotti, JD Congdon, and RE Gatten Jr. 1991. Post-hatching yolk reserves: parental investment in American alligators from Louisiana. J. Herpetol. 25:253–256.

Fischer RU, EA Standora, and JR Spotila. 1987. Predator-induced changes in thermoregulation of bluegill, *Lepomis macrochirus*, from a thermally stressed reservoir. Can. J. Fish. Aquat. Sci. 44:1629–1634.

Fisher RA. 1930. The Genetical Theory of Natural Selection. Oxford University Press, Oxford.

Fitzinger LJ. 1835. Entwurgeiner systematischen Anordung der Schildkröten nach den Grundsatzen der natürlichen Methode. Ann. Mus. Wien 1:103–128.

———. 1843. Systema Reptilium. Braumuller et Seidel, Vienna.

Flausín LP, R Acuña-Mesén, and E Araya. 1997. Natalidad de *Chelydra serpentina* (Testudines: Chelydridae) in Costa Rica. Rev. Biol. Trop. 44/45:663–666.

Fleming A and D Crews. 2001. Estradiol and incubation temperature modulate regulation of steroidogenic factor 1 in the developing gonad of the red-eared slider turtle. Endocrinology 142:1403–1411.

Fleming J. 1822. The Philosophy of Zoology; or a General View of the Structure, Functions, and Classification of Animals, 2 vols. Constable, London.

Ford KM III. 1992. Herpetofauna of the Albert Ahrens local fauna (Pleistocene: Irvingtonian), Nebraska. Masters thesis, Michigan St. Univ., East Lansing.

Frair W. 1972. Taxonomic relations among chelydrid and kinosternid turtles elucidated by serological tests. Copeia 1972:97–108.

Frazer NB, JW Gibbons, and JL Greene. 1990. Life tables of a slider turtle population. In Life History and Ecology of the Slider Turtle, ed. JW Gibbons. Smithsonian Inst. Press, Washington, DC, 183–200.

Freedberg S, MA Ewert, and CE Nelson. 2001. Environmental effects on fitness and consequences for sex allocation in a reptile with environmental sex determination. Evol. Ecol. Res. 3:953–967.

Freedberg S and MJ Wade. 2001. Cultural inheritance as a mechanism for population sex-ratio bias in reptiles. Evolution 55:1049–1055.

Frische S, A Fago, and J Altimiras. 2000. Respiratory responses to short term hypoxia in the snapping turtle, *Chelydra serpentina*. Comp. Biochem. Physiol. A Mol. Integr. Physiol. 126:223–231.

Froese AD. 1978. Habitat preferences of the common snapping turtle, *Chelydra s. serpentina* (Reptilia, Testudines, Chelydridae). J. Herpetol. 12:53–58.

Froese AD and GM Burghardt. 1974. Food competition in captive juvenile snapping turtles, *Chelydra serpentina*. Anim. Behav. 22:735–740.

———. 1975. A dense natural population of the common snapping turtle (*Chelydra serpentina*). Herpetologica 31:204–208.

Fuchs E. 1938. Die Schidkrötenreste aus dem oberpfälzer Braunkolen tertiär. Palaeontographica 89, Abt. A:57–104.

Gaffney ES. 1972. An illustrated glossary of turtle skull nomenclature. Am. Mus. Novit. No. 2486:33 pp.

———. 1975. Phylogeny of the chelydrid turtles: a study of shared derived characters in the skull. Fieldiana Geol. 33:157–178.

———. 1979. Comparative cranial morphology of recent and fossil turtles. Bull. Am. Mus. Nat. Hist. 164(2):65–375.

———. 1983. The cranial morphology of the extinct horned turtle, *Meiolania platyceps,* from the Pleistocene of Lord Howe Island, Australia. Bull. Am. Mus. Nat. Hist. 175(4):326–479.

———. 1990. The Comparative osteology of the Triassic turtle *Proganochelys*. Bull. Am. Mus. Nat. Hist. 194.

———. 1996. The postcranial morphology of *Meiolania platyceps* and a review of the Meiolaniidae. Bull. Am. Mus. Nat. Hist. 229.

Gaffney ES, JH Hutchison, FA Jenkins, and LJ Meeker. 1987. Modern turtle origins: the oldest known cryptodire. Science 237:289–291.

Gaffney ES and PA Meylan. 1988. A phylogeny of turtles. In The Phylogeny and Classification of Tetrapods, ed. M. J. Benton. Clarendon Press, Oxford, 157–219.

Gaffney ES and HH Schleich. 1994. New reptile material from the German Tertiary. 16. On *Chelydropsis murchisoni*(Bell, 1892) from the middle Miocene locality of Unterwohlbach/South Germany. Courier Forsch.-Inst. Senckenberg 173:197–213.

Gaiduchenko LL and VM Ckhikvadze. 1985. [A new species of

cayman turtle from Pavlodar Priirtischia] (in Russian). Geol. Geof. (Novosibirsk) 1:116–118.

Galbraith DA. 1986. Age estimates, survival, growth and maturity of female *Chelydra serpentina* (Linnaeus) in Algonquin Provincial Park, Ontario. M.Sc. thesis, University of Guelph, Guelph, Ontario.

Galbraith DA, CA Bishop, RJ Brooks, WL Simse, and KP Lampman. 1988. Factors affecting the density of populations of common snapping turtles (*Chelydra serpentina serpentina*). Can. J. Zool. 66:1233–1240.

Galbraith DA and RJ Brooks. 1987. Addition of annual growth lines in adult snapping turtles *Chelydra serpentina*. J. Herpetol. 21:359–363.

———. 1989. Age estimates for snapping turtles. J. Wildl. Manage. 53:502–508.

Galbraith DA, RJ Brooks, and GP Brown. 1997. Can management intervention achieve sustainable exploitation of turtles? In Proceedings: Conservation, Restoration, and Management of Tortoises and Turtles—An International Conference, ed. J Van Abbema, PCH Pritchard, S Dohm, and MW Klemens. New York Turtle and Tortoise Society, New York, 186–194

Galbraith DA, RJ Brooks, and ME Obbard 1989. The influence of growth rate and age on body size at maturity in female snapping turtles (*Chelydra serpentina*). Copeia 1989:896–904.

Galbraith DA, CJ Graesser, and RJ Brooks. 1988. Egg retention by a snapping turtle, *Chelydra serpentina*, in Central Ontario. Can. Field Nat. 102:734.

Galbraith DA, BN White, RJ Brooks, and PT Boag. 1993. Multiple paternity in clutches of snapping turtles (*Chelydra serpentina*) detected using DNA fingerprints. Can. J. Zool. 71:318–324.

Galbreath EC 1948. Pliocene and Pleistocene records of fossil turtles from western Kansas and Oklahoma. Univ. Kans. Publ. Mus. Nat. Hist. 1:281–284.

Galtier N, F Depaulis, and NH Barton. 2000. Detecting bottlenecks and selective sweeps from DNA sequence polymorphism. Genetics 155:981–987.

Gasc J-P. 1981. Axial musculature. In Biology of the Reptilia, ed. C Gans and TS Parsons, Vol. 11: Morphology. Academic Press, London, 355–435.

Gatten RE Jr. 1974. Effect of nutritional status on the preferred body temperature of the turtles *Pseudemys scripta* and *Terrapene ornata*. Copeia 1974:912–917.

———. 1978. Aerobic metabolism in snapping turtles, *Chelydra serpentina* after thermal acclimation. Comp. Biochem. Physiol. 61A:325–337.

———. 1980. Aerial and aquatic oxygen uptake by freely-diving snapping turtles (*Chelydra serpentina*). Oecologia 46:266–271.

———. 1981. Anaerobic metabolism in freely diving painted turtles (*Chrysemys picta*). J. Exp. Zool. 216:377–385.

———. 1984. Aerobic and anaerobic metabolism of freely-diving loggerhead musk turtles (*Sternotherus minor*). Herpetologica 40:1–7.

———. 1985. The uses of anaerobiosis by amphibians and reptiles. Am. Zool. 25:945–954.

Gaunt AS and C Gans. 1969. Mechanics of Respiration in the Snapping Turtle, *Chelydra serpentina* (Linné). J. Morphol. 128:195–228.

George JC, J Bada, J Zeh, L Scott, SE Brown, T O'Hara, and R Suydam 1999. Age and growth estimates of bowhead whales (*Balaena mysticetus*) via aspartic acid racemization. Can. J. Zool. 77:571–580.

Georges A, C Limpus, and R Stoutjesdijk. 1994. Hatchling sex in the marine turtle *Caretta caretta* is determined by proportion of development at a temperature, not daily duration of exposure. J. Exp. Zool. 270:432–444.

Gerhart J and M Kirschner. 1997. Cells, Embryos, and Evolution. Blackwell Science, Malden, MA.

Gerholdt JE and B Oldfield. 1987. *Chelydra serpentina serpentina* (common snapping turtle), size. Herpetol. Rev. 18:73.

Gettinger RD, GL Paukstis, and WH Gutzke. 1984. Influence of hydric environment on oxygen consumption by embryonic turtles *Chelydra serpentina* and *Trionyx spiniferus*. Physiol. Zool. 57:468–473.

Gibbons JW. 1967. Variation in growth rates in three populations of the painted turtle, *Chrysemys picta*. Herpetologica 23:296–303.

———. 1968. Growth rates of the common snapping turtle, *Chelydra serpentina*. Herpetologica 24:266–267.

———. 1970. Reproductive dynamics of a turtle (*Pseudemys scripta*) population in a reservoir receiving heated effluent from a nuclear reactor. Can. J. Zool. 48:881–885.

———. 1982. Reproductive patterns in freshwater turtles. Herpetologica 38:222–227.

———. 1990. The slider turtle. In Life History and Ecology of the Slider Turtle, ed. JW Gibbons. Smithsonian Inst. Press, Washington, DC, 3–18.

Gibbons JW and JL Greene. 1978. Selected aspects of the ecology of the chicken turtle *Deirochelys reticularia* (Latreille) (Reptilia, Testudines, Emydidae). J. Herpetol. 12:237–241.

Gibbons JW, JL Greene, and KK Patterson. 1982. Variation in reproductive characteristics of aquatic turtles. Copeia 1982:776–784.

Gibbons JW and DH Nelson. 1978. The evolutionary significance of delayed emergence from the nest by hatchling turtles. Evolution 32:297–303.

Gibbons JW, SS Novak, and CH Ernst 1988. *Chelydra serpentina*. Cat. Am. Amph. Rept. 420:1–4.

Gibbons JW, DE Scott, TJ Ryan, KA Buhlmann, TD Tuberville, BS Metts, JL Greene, T Mills, Y Leiden, S Poppy, and CT Winne. 2000. The global decline of reptiles, déja vu amphibians. BioScience 50:653–666.

Gibbs JP and WG Shriver. 2002. Estimating the effects of road mortality on turtle populations. Conserv. Biol. 16:1647–1652.

Gibbs JP and DA Steen. 2005. Trends in sex ratios of turtles in the United States: implications of road mortality. Conserv. Biol. 19:552–556.

Gibert P, RB Huey, and GW Gilchrist. 2001. Locomotor performance of *Drosophila melanogaster*: interactions among developmental and adult temperatures, age, and geography. Evolution 55:205–209.

Gilbert B. 1993. The reptile that stakes its survival on snap decisions. Smithsonian 24:93–99.

Gilbert SF, GA Loredo, A Brukman, and AC Burke. 2001. Morphogenesis of the turtle shell: the development of a novel structure in tetrapod evolution. Evol. Dev. 3:47–58.

Gillooly JF, JH Brown, GB West, VM Savage, and EL Charnov. 2001. Effects of size and temperature on metabolic rate. Science 293:2248–2251.

Ginsberg L, F de Broin, F Criuzel, F Duranthon, F Escuillié, F Juillard, and S Lassaube. 1991. Les vertebras du Miocène inféreur de Barbotan-les-Thermes (Gers). Ann. Paléontol. 77:161–216.

Girgis S. 1961. Aquatic respiration in the common Nile turtle,

Trionyx triunguis (Forskal). Comp. Biochem. Physiol. 3:206–217.

Girondot M. 1999. Statistical description of temperature-dependent sex determination using maximum likelihood. Evol. Ecol. Res. 1:479–486.

Girondot M and C Pieau. 1999. A fifth hypothesis for the evolution of temperature-dependent sex determination in reptiles. Trends Ecol. Evol. 14:359–360.

Glass ML and SC Wood 1983. Gas exchange and control of breathing in reptiles. Physiol. Rev. 63:232–260.

Gloyd HK. 1928. The amphibians and reptiles of Franklin County, Kansas. Trans. Kans. Acad. Sci. 31:115–141.

Gmelin JF. 1789. Caroli a Linné, Systema naturae per regna tria naturae, secondum classes, ordines, genera, species, cum characteribus differentiis, synonymis, locis. Tomus Primus, Edito decim, tercia, aucta, reformata. G. E. Beer, Lipsiae.

Goodrich ES. 1930. Studies on the Structure and Development of Vertebrates. Macmillan, London.

Gordon MS, GA Bartholomew, AD Grinnell, CB Jorgensen, and FN White. 1972. Animal Physiology: Principles and Adaptations. MacMillan, New York.

Gordos M and CE Franklin. 2002. Diving behaviour of two Australian bimodally respiring turtles, *Rheodytes leukops* and *Emydura macquarii*, in a natural setting. J. Zool. Lond. 258:335–342.

Gotte SW. 1988. Nest site selection in the snapping turtle, mud turtle, and painted turtle. Masters thesis, George Mason Univ., Fairfax, VA.

Graham RW, JA Holman, and PW Parmalee. 1983. Taphonomy and paleoecology of the Christensen Bog mastodon bone bed, Hancock County, Indiana. Ill. State Mus. Rep. Invest. 38:1–29.

Graham TE. 1995. Habitat use and population parameters of the spotted turtle, *Clemmys guttata*, a species of special concern in Massachusetts. Chelonian Conserv. Biol. 1:207–214.

Graham TE and JE Forsberg. 1991. Aquatic oxygen uptake by naturally wintering wood turtles *Clemmys insculpta*. Copeia 1991:836–838.

Graham TE and AA Graham. 1992. Metabolism and behavior of wintering common map turtles, *Graptemys geographica*, in Vermont. Can. Field Nat. 106:517–519.

Graham TE and VH Hutchison. 1979. Effect of temperature and photoperiod acclimatization on thermal preferences of selected freshwater turtles. Copeia 1979:165–169.

Gray JE. 1825. A synopsis of the genera of reptiles and Amphibia, with a description of some new species. Ann. Philos. (n.s.) 10:193–217.

———. 1831a. A synopsis of the species of the class Reptilia. In The Animal Kingdom, ed. E. Griffith, 9(appendix):1–110. Geo. B. Whittaker, Treacher, & Co., London.

———. 1831b. Synopsis Reptilium; or Short Descriptions of the Species of Reptiles. Part I. Cataphracta. Tortoises, Crocodiles, Enaliosaurians. Truettle, Wurtz, and Co., London.

———. 1856a. On some new species of freshwater tortoises from North America, Ceylon and Australia, in the collections of the British Museum. Proc. Zool. Soc. Lond. 1855:197–202.

———. 1856b. Catalogue of the shield reptiles in the collection of the British Museum. Part I. Testudinata (tortoises). [1855] The Trustees (Br. Mus. Nat. Hist.), London.

———. 1870. Supplement to the catalogue of shield reptiles in the collection of the British Museum. Part 1. Testudinata (tortoises). Br. Mus. Nat. Hist. Lond. 1855:1–120.

Greenberg N and J Wingfield. 1987. Stress and reproduction: reciprocal relationships. In Hormones and Reproduction in Fishes, Amphibians and Reptiles, ed. DO Norris and RE Jones. Plenum Press, New York, 461–503.

Gregory LF, TS Gross, AB Bolten, KA Bjorndal, and LJ Guillette Jr. 1996. Plasma corticosterone concentrations associated with acute captivity stress in wild loggerhead sea turtles (*Caretta caretta*). Gen. Comp. Endocrinol. 104:312–320.

Gregory PT. 1982. Reptilian hibernation. In Biology of the Reptilia, ed. C Gans and FH Pough, Vol. 13. Physiology D. Academic Press, London, 53–154.

Groessens van Dyck M-C and HH Schleich. 1985. Nouveaux matérials des tortues (*Ptychogaster / Ergilemys*) de la localité Oligocène moyen de Ronheim (Sud de l'Allegmagne). Münch. Geowiss. Abh. (A) 4:17–66.

Guilday JE. 1962. The Pleistocene local fauna of the Natural Chimneys, Augusta County, Virginia. Ann. Carnegie Mus. 36:87–122.

Guillette LJ and SL Fox. 1985. Effect of deluteinizatoin on plasma progesterone concentration and gestation in the lizard, *Anolis carolinensis*. Comp. Biochem. Physiol. A 80:303–306.

Guillette LJ, LA Lavia Jr, NJ Walker, and DK Roberts. 1984. Luteolysis induced by prostaglandin F2 in the lizard, *Anolis carolinensis*. Gen. Comp. Endocrinol. 56:271–277.

Gunn DL and CA Cosway. 1938. Temperature and humidity relations of the cockroach: v. humidity preference of *Blatta orientalis*. J. Exp. Biol. 15:555–563.

Gutzke WHN and JJ Bull. 1986. Steroid hormones reverse sex in turtles. Gen. Comp. Endocrinol. 64:368–372.

Gutzke WHN and DB Chymiy. 1988. Sensitive periods during embryogeny for hormonally induced sex determination in turtles. Gen. Comp. Endocrinol. 71:265–267.

Gutzke WHN and D Crews. 1988. Embryonic temperature determines adult sexuality in a reptile. Nature 332:832–834.

Gutzke WHN and GC Packard. 1987. The influence of temperature on eggs and hatchlings of Blanding's Turtles, *Emydoidea blandingii*. J. Herpetol 21:161–163.

Gutzke WHN, GL Paukstis, and GC Packard. 1984. Pipping versus hatching as indices of time of incubation in reptiles. J. Herpetol. 18:494–496.

Guyot G, C Pieau, and S Renous. 1994. Développment embryonaire d'un tortue terrestre, la tortue d'Hermann, *Testudo hermanni* Gmelin, 1789. Ann. Sci. Nat. Zool. Paris 15:115–137.

Haiduk MW and JW Bickham. 1982. Chromosomal homologies and evolution of testudinoid turtles with emphasis on the systematic placement of *Platysternon*. Copeia 1982:60–66.

Halliday TR and PA Verrell. 1988. Body size and age in amphibians and reptiles. J. Herpetol. 22:253–265.

Hamburger V and HL Hamilton. 1951. A series of normal stages in the development of the chick embryo. J. Morphol. 88:49–92.

Hamilton WJ Jr. 1940. Observations on the reproductive behavior of the snapping turtle. Copeia 1940:124–126.

Hammer DA. 1969. Parameters of a marsh snapping turtle population, Lacreek Refuge, South Dakota. J. Wildl. Manage. 33:995–1005.

———. 1971. The durable snapping turtle. Nat. Hist. 80:59–65.

Hammerson GA. 1999. Amphibians and Reptiles in Colorado. Univ. Press Colorado, Niwet, CO.

Hammond KA, JR Spotila, and EA Standora. 1988. Basking behavior of the turtle, *Pseudemys scripta*—effects of digestive state, acclimation temperature, sex and season. Physiol. Zool. 61:69–77.

Hansen MC. 1992. Indian Trail Caverns: a window on Ohio's Pleistocene bestiary. Ohio Geol. (Spring):1, 3.

Hanson CB. 1996. Stratigraphy and vertebrate faunas of the Bridgerian-Duchesnean Clarno Formation, north-central Oregon. In The Terrestrial Eocene-Oligocene Transition in North America, ed. D. R. Prothero and R. J. Emry. Cambridge Univ. Press, New York, 206–239.

Hanson FB. 1919. The anterior cranial nerves of Chelydra serpentina. Wash. Univ. Stud. Sci. Ser. 7:13–41.

Harlan R. 1835. Genera of North American reptiles, and a synopsis of the species. In Medical and Physical Researches. Lydia R. Bailey, Philadelphia, 84–163.

Hart DR. 1983. Dietary and habitat shift with size of red-eared turtles (Pseudemys scripta) in a southern Louisiana population. Herpetologica 39:285–290.

Hartley LM, MJ Packard, and GC Packard. 2000. Accumulation of lactate by supercooled hatchling of the painted turtle (Chrysemys picta): implications for overwinter survival. J. Comp. Physiol. B 170:45–50.

Haskell A, TE Graham, CR Griffin, and JB Hestbeck. 1996. Size related survival of headstarted redbelly turtles (Pseudemys rubriventris) in Massachusetts. J. Herpetol. 30:524–527.

Haxton T. 2000. Road mortality of snapping turtles, Chelydra serpentina, in central Ontario during their nesting period. Can. Field Nat. 114:106–110.

Hay OP. 1899. Descriptions of two new species of tortoises from the tertiary of the United States. Proc. U.S. Nat. Mus. 22:21–24.

———. 1905. The fossil turtles of the Bridger basin. Am. Geol. 35:327–342.

———. 1907. Descriptions of seven new species of turtles from the Tertiary of the United States. Bull. Am. Mus. Nat. Hist. 23:847–863.

———. 1908a. The fossil turtles of North America. Carnegie Inst. Washington Publ. 75:1–568.

———. 1908b. Descriptions of five species of North American fossil turtles, four of which are new. Proc. U.S. Nat. Mus. 35:161–169.

———. 1911. A fossil specimen of the alligator snapper (Macrochelys temminckii) from Texas. Proc. Am. Philos. Soc. 50:452–455.

———. 1916. Descriptions of some Floridian fossil vertebrates, belonging mostly to the Pleistocene. Fla. State Geol. Surv. 8th Ann. Rep., 39–76.

———. 1917. Vertebrata mostly from Stratum No. 3 at Vero, Florida, together with descriptions of new species. Fla. State Geol. Surv. 9th Ann. Rep., 43–68.

———. 1923. The Pleistocene of North America and its vertebrated animals from the states east of the Mississippi River and from Canadian provinces east of longitude 95°. Carnegie Inst. Washington Publ. 322:1–499.

Heath JE. 1964. Reptilian thermoregulation: evaluation of field studies. Science 146:784–785.

Hennemann WW III. 1979. The influence of environmental cues and nutritional status on frequency of basking in juvenile Suwannee terrapins (Chrysemys concinna). Herpetologica 35:129–131.

Herbert CV and DC Jackson. 1985a. Temperature effects on the responses to prolonged submergence in the turtle Chrysemys picta bellii: I. Blood acid-base and ionic changes during and following anoxic submergence. Physiol. Zool. 58:655–669.

———. 1985b. Temperature effects on the responses to prolonged submergence in the turtle Chrysemys picta bellii: II. Metabolic rate, blood acid-base and ionic changes, and cardiovascular function in aerated and anoxic water. Physiol. Zool. 58:670–681.

Hernandez T and RA Coulson. 1957. Inhibition of renal tubular function by cold. Am. J. Physiol. 188:485–489.

Herter K. 1926. Thermotaxis und hydrotaxis bei tieren. Bethes Handb. Normal Pathol. Physiol. 11:173–180.

Hibbard CW. 1934. Two new genera of Felidae from the middle Pliocene of Kansas. Trans. Kans. Acad. Sci. 37:239–255.

———. 1939. Notes on additional fauna of Edson Quarry of the middle Pliocene of Kansas. Trans. Kans. Acad. Sci. 42:457–462.

———. 1963. The presence of Macroclemys and Chelydra in the Rexroad fauna from the Upper Pliocene of Kansas. Copeia 1963:708–709.

Hibbard CW and DW Taylor. 1960. Two late Pleistocene faunas from southwestern Kansas. Contrib. Mus. Paleontol. Univ. Mich. (16):1–223.

Hicks JMT and AP Farrell. 2000. The cardiovascular responses of the red-eared slider (Trachemys scripta) acclimated to either 22 or 5°C. J. Exp. Biol. 203:3765–3774.

Hildebrand M, DM Bramble, KF Liem, and DB Wake. 1985. Functional Vertebrate Morphology. The Belknap Press of Harvard University Press, Cambridge, MA.

Hillel D. 1971. Soil and Water: Physical Principles and Processes. Academic Press, New York.

———. 1980. Applications of Soil Physics. Academic Press, New York.

Hitzig BM and DC Jackson. 1978. Central chemical control of ventilation in the unanesthetized turtle. Am. J. Physiol. 235:R257–R264.

Ho S and IP Callard. 1984. High affinity binding of [3H]-R5020 and [3H]-progesterone by putative progesterone receptors in cytosol and nuclear extract of turtle oviduct. Endocrinology 114:70–79.

Hoffstetter R and J-P Gasc. 1969. Vertebrae and ribs of modern reptiles. In Biology of the Reptilia, ed. C Gans, Ad'A Bellairs, and T Parsons, Vol. 1. Academic Press, New York, 201–310.

Holman JA. 1964. Pleistocene amphibians and reptiles of Texas. Herpetologica 20:73–83.

———. 1966. Some Pleistocene turtles from Illinois. Trans Ill. State Acad. Sci. 59:214–216.

———. 1972. Herpetofauna of the Kanapolis local fauna (Pleistocene: Yarmouth) of Kansas. Mich. Acad. 5:87–98.

———. 1978. The late Pleistocene herpetofauna of Devil's Den Sinkhole, Levy County, Florida. Herpetologica 34:228–237.

———. 1986a. Butler Spring herpetofauna of Kansas (Pleistocene: Illinoian) and its climatic significance. J. Herpetol. 20:568–570.

———. 1986b. Turtles from the late Wisconsinan of westcentral Ohio. Am. Midl. Nat. 116:213–214.

———. 1988. The status of Michigan's Pleistocene herpetofauna. Mich. Acad. 20:125–132.

———. 1992. Late Quaternary herpetofauna of the central Great Lakes region, U.S.A.: zoogeographical and paleoecological implications. Quat. Sci. Rev. 11:345–351.

———. 1995. Pleistocene amphibians and reptiles in North America. Oxford Monogr. Geol. Geophys. No. 32. Oxford Univ. Press, New York.

———. 1997. Amphibians and reptiles from the Pleistocene (Late Wisconsinan) of Sheriden Pit Cave, northwestern Ohio. Mich. Acad. 29:1–20.

Holman JA and KD Andrews. 1994. Northern American Quaternary cold-tolerant turtles: distributional adaptations and constraints. Boreas 23:44–52.

Holman JA and DC Fisher. 1993. Late Pleistocene turtle remains (Reptilia: Testudines) from southern Michigan. Mich. Acad. 25:491–499.

Holman JA and JN McDonald. 1986. A late Quaternary herpetofauna from Saltville, Virginia. Brimleyana 12:85–100.

Holman JA and RL Richards. 1993. Herpetofauna of the Prairie Creek site, Daviess County, Indiana. Proc. Indiana Acad. Sci. 102:115–131.

Holman JA and ME Schloeder. 1991. Fossil herpetofauna of the Lisco C Quarries (Pliocene: early Blancan) of Nebraska. Trans. Nebraska Acad. Sci. 18:19–29.

Holman JA and RM Sullivan. 1981. A small herpetofauna from the type section of the Valentine Formation (Miocene: Barstovian), Cherry County, Nebraska. J. Paleontol. 55:138–144.

Holroyd PA and JH Hutchison. 2002. Patterns of geographic variation in latest Cretaceous vertebrates: evidence from the turtle component. Geol. Soc. Am. Spec. Vol. 361:177–190.

Holroyd PA JH Hutchison, and SG Strait. 2001. Turtle diversity and abundance through the lower Eocene Willwood Formation of the southern Bighorn Basin. Univ. Mich. Pap. Paleontol. No. 33, 97–107.

Hotaling EC. 1990. Temperature-dependent sex determination: factors affecting sex ratio in nests of a New Jersey population of Chelydra serpentina. Ph.D. dissertation, Rutgers University, Newark, NJ.

Hotaling EC, DC Wilhoft, and SB McDowell. 1985. Egg position and weight of hatchling snapping turtles (Chelydra serpentina) in natural nests. J. Herpetol. 19:534–536.

Huey RB. 1982. Temperature, physiology, and the ecology of reptiles. In. Biology of the Reptilia, ed. C Gans and FH Pough, Vol. 12. Academic Press, London, 25–91.

Hulbert RC and GS Morgan. 1989. Stratigraphy, paleoecology, and vertebrate fauna of the Leisey Sell Pit local fauna, early Pleistocene (Irvingtonian) of southwestern Florida. Pap. Fla. Paleontol. 2:1–19.

Hulse AC, CJ McCoy, and EJ Censky. 2001. Amphibians and Reptiles of Pennsylvania and the Northeast. Cornell Univ. Press, Ithaca, NY.

Humphrey OD. 1894. On the brain of the snapping turtle. J. Comp. Neurol. 4:73–116.

Hutchison JH. 1994. Nevada Neogene turtles: climatic and distributional significance. Geol. Soc. Am. Abstr. Prog. 26(7).

———. 1998. Turtles across the Paleocene/Eocene Epoch boundary in west-central North America. In Late Paleocene-Early Eocene Climatic and Biotic Events in the Marine and Terrestrial Records, ed. M-P Aubry, S Lucas, and WA Berggren. Columbia Univ. Press, New York, 401–408.

———. 2000. Diversity of Cretaceous turtle faunas of Eastern Asia and their contribution to the turtle faunas of North America. Paleontol. Soc. Korea, Spec. Publ. No. 4: 27–38.

Hutchison JH and JD Archibald. 1986. Diversity of turtles across the Cretaceous/Tertiary boundary in northeastern Montana. Palaeogeogr., Palaeoclimatol. Palaeoecol. 55:1–22.

Hutchison JH and DM Bramble. 1981. Homology of the plastral scales of the Kinosternidae and related turtles. Herpetologica 37:73–85.

Hutchison JH, JG Eaton, PA Holroyd, and MB Goodwin. 1998. Larger vertebrates of the Kaiparowits Formation (Campanian) in the Grand Staircase-Escalante National Monument and adjacent areas. Sci. Symp. Proc. 391–398.

Hutchison JH and PA Holroyd. 2003. Late Cretaceous and early Paleocene turtles of the Denver Basin. Rocky Mount. Rocky Mountain Geol. 38:1–22.

Hutchison JH and AD Pasch. 2004. First record of a turtle (Protochelydra, Chelydridae, Testudines) from the Cenozoic of Alaska (Chickaloon Formation, Paleocene-Eocene). PaleoBios 24:1–5.

Hutchison VH. 1961. Critical thermal maxima in salamanders. Physiol. Zool. 34:92–125.

———. 1979. Thermoregulation. In Turtles, Perspectives and Research, ed. M Harless and H Morlock. John Wiley and Sons, New York, NY, 207–228.

Hutchison VH and RK Dupré. 1992. Thermoregulation. In Environmental Physiology of the Amphibians, ed. ME Feder and WW Burggren. University of Chicago Press, Chicago, IL, 206–249.

Hutchison VH. and LG Hill. 1976. Thermal selection in the hellbender, Cryptobranchus alleganiensis, and the mudpuppy, Necturus maculosus. Herpetologica 32:327–331.

Hutchison VH, A Vinegar, and RJ Kosh. 1966. Critical thermal maxima in turtles. Herpetologica 22:32–41.

Hutton JM. 1987. Incubation temperatures, sex ratios and sex determination in a population of Nile crocodiles (Crocodylus niloticus). J. Zool. (Lond.) 211:143–155.

International Commission on Zoological Nomenclature (ICZN). 1963. Opinion 660. Suppression under the plenary powers of seven specific names of turtles (Reptilia, Testudines). Bull. Zool. Nomencl. 20:187–190.

International Commission on Zoological Nomenclature (ICZN). 1986. Opinion 1377. Chelydra osceola Stejneger, 1918 given nomenclatural precedence over Chelydra laticarinata Hay, 1916 and Chelydra sculpta Hay, 1916 (Reptilia, Testudines). Bull. Zool. Nomencl. 43:33–34.

Ishii K, K Ishii, and T Kusakabe. 1985. Electrophysiological aspects of reflexogenic area in the chelonian, Geoclemmys reevesii. Respir. Physiol. 59:45–54.

Ishimatsu A, JW Hicks, and N. Heisler 1996. Analysis of cardiac shunting in the turtle Trachemys (Pseudemys) scripta: application of the three vessel outflow model. J. Exp. Biol. 199:2667–2677.

Iverson JB. 1977. Reproduction in freshwater and terrestrial turtles of North Florida. Herpetologica 33:205–212.

———. 1982. Ontogenetic changes in relative skeletal mass in the painted turtle, Chrysemys picta. J. Herpetol. 16:412–414.

———. 1984. Proportional skeletal mass in turtles. Fla. Sci. 47:1–11.

———. 1992a. A revised checklist with distribution maps of the turtles of the world. Privately published. Richmond, IN.

———. 1992b. Reproduction in female razorback musk turtles (Sternotherus carinatus: Kinosternidae). Southwest. Nat. 77:215–224.

Iverson JB, CP Balgooyen, KK Byrd, and KK Lyddan. 1993. Latitudinal variation in egg and clutch size in turtles. Can. J. Zool. 71:2448–2461.

Iverson JB, EL Barthelmess, GR Smith, and CE deRivera. 1991. Growth and reproduction in the mud turtle Kinosternon hirtipes in Chihuahua, Mexico. J. Herpetol. 25:64–72.

Iverson JB, D Hearne, J Watters, D Croshaw, and J Larson. 2000. *Chelydra serpentina* (Common snapping turtle). Density and biomass. Herpetol. Rev. 31:238.

Iverson JB, H Higgins, A Sirulnik, and C Griffiths. 1997. Local and geographic variation in the reproductive biology of the snapping turtle (*Chelydra serpentina*). Herpetologica 53:96–117.

Iverson JB and GR Smith. 1993. Reproductive ecology of the painted turtle (*Chrysemys picta*) in the Nebraska Sandhills and across its range. Copeia 1993:1–21.

Jackson DC. 1968. Metabolic depression and oxygen depletion in the diving turtle. J. Appl. Physiol. 24:503–509.

———. 1997. Lactate accumulation in the shell of the turtle, *Chrysemys picta bellii*, during anoxia at 3 and 10°C. J. Exp. Biol. 200:2295–2300.

———. 2000. Living without oxygen: lessons from the freshwater turtle. Comp. Biochem. Physiol. A 125:299–315.

Jackson DC, CE Crocker, and GR Ultsch. 2000a. Bone and shell contribution to lactic acid buffering of submerged turtles *Chrysemys picta bellii* at 3°C. Am. J. Physiol. 278:R1564–R1571.

———. 2001. Mechanisms of homeostasis during long-term diving and anoxia in turtles. Zoology 103:150–156.

Jackson DC, Z Goldberger, S Visuri, and RN Armstrong. 1999. Ionic exchanges of turtle shell *in vitro* and their relevance to shell function in the anoxic turtle. J. Exp. Biol. 220:513–520.

Jackson DC and N Heisler. 1982. Plasma ion balance of submerged anoxic turtles at 3°C: the role of calcium lactate formation. Respir. Physiol. 49:159–174.

———. 1983. Intracellular and extracellular acid-base and electrolyte status of submerged anoxic turtles at 3°C. Respir. Physiol. 53:187–201.

Jackson DC, AL Ramsey, JM Paulson, CE Crocker, and GR Ultsch. 2000b. Lactic acid buffering by bone and shell in anoxic softshell and painted turtles. Physiol. Biochem. Zool. 73:290–297.

Jackson DC and K Schmidt-Nielsen. 1966. Heat production during diving in the fresh water turtle, *Pseudemys scripta*. J. Cell. Physiol. 67:225–231.

Jackson DR. 1994. Overwintering of hatchling turtles in northern Florida. J. Herpetol. 28:401–402.

Jackson DR and MA Ewert. 1997. *Chelydra serpentina* (snapping turtle). Reproduction. Herpetol. Rev. 28:87.

Janzen FJ. 1992. Heritable variation for sex ratio under environmental sex determination in the common snapping turtle (*Chelydra serpentina*). Genetics 131:155–161.

———. 1993a. An experimental analysis of natural selection on body size of hatchling turtles. Ecology 74:332–341

———. 1993b. The influence of incubation temperature and family on eggs, embryos, and hatchlings of the smooth softshell turtle (*Apalone mutica*). Physiol. Zool. 66:349–373.

———. 1994a. Climate change and temperature-dependent sex determination in reptiles. Proc. Natl. Acad. Sci. USA 91:7487–7490.

———. 1994b. Vegetational cover predicts the sex ratio of hatchling turtles in natural nests. Ecology 75:1593–1599.

———. 1995. Experimental evidence for the evolutionary significance of temperature-dependent sex determination. Evolution 49:864–873.

Janzen FJ and JG Krenz. 2004. Phylogenetics: which was first, TSD or GSD? In Temperature-Dependent Sex Determination in Vertebrates, ed. N Valenzuela and VA Lance. Smithsonian Books, Washington, DC, 121–130.

Janzen FJ and CL Morjan. 2001. Repeatability of microenvironment-specific nesting behaviour in a turtle with environmental sex determination. Anim. Behav. 62:73–82.

———. 2002. Egg size, incubation temperature, and posthatching growth in painted turtles (*Chrysemys picta*). J. Herpetol. 36:308–311.

Janzen FJ and S O'Steen. 1990. An instance of male combat in the common snapping turtle (*Chelydra serpentina*). Bull. Chicago Herpetol. Soc. 25:11.

Janzen FJ, GC Packard, MJ Packard, TJ Boardman, and JR zumBrunnen. 1990. Mobilization of lipid and protein by embryonic snapping turtles in wet and dry environments. J. Exp. Zool. 255:155–162.

Janzen FJ and GL Paukstis. 1988. Environmental sex determination in reptiles. Nature 332:790.

———. 1991a. Environmental sex determination in reptiles: ecology, evolution, and experimental design. Q. Rev. Biol. 66:149–179.

———. 1991b. A preliminary test of the adaptive significance of environmental sex determination in reptiles. Evolution 45:435–440.

Janzen FJ, JF Tucker, and GL Paukstis. 2000a. Experimental analysis of an early life-history stage: selection on size of hatchling turtles. Ecology 81:2290–2304.

———. 2000b. Experimental analysis of an early life-history stage: avian predation selects for larger body size of hatchling turtles. J. Evol. Biol. 13:947–954.

Janzen FJ, ME Wilson, JK Tucker, and SP Ford. 1998. Endogenous yolk steroid hormones in turtles with different sex-determining mechanisms. Gen. Comp. Endocrinol. 111:306–317.

Jarocki FP 1822. Zoology or a General Description of Animals According to the Most Recent System, Vol. III. Reptiles and amphibians. Latkiewicz, Warsaw, 6, 184 (in Polish).

Jessop TS, M Hamann, MA Read, and CJ Limpus. 2000. Evidence for a hormonal tactic maximizing green turtle reproduction in response to a pervasive ecological stressor. Gen. Comp. Endocrinol. 118:407–417.

Johnson SM, RA Johnson, and GS Mitchell. 1998. Hypoxia, temperature and pH/CO$_2$ effects on respiratory discharge from a turtle brainstem preparation. J. Appl. Physiol. 84:649–660.

Johnson SM, JER Wilkerson, MR Wenninger, DR Henderson, and GS Mitchell. 2002. Role of synaptic inhibition in turtle respiratory rhythm generation. J. Physiol. 544:253–265.

Johnston SD, CB Daniels, D Cenzato, JA Whitsett, and S Orgeig. 2002. The pulmonary surfactant system matures upon pipping in the freshwater turtle *Chelydra serpentina*. J. Exp. Biol. 205:415–425.

Jones RE and LJ Guillette Jr. 1982. Hormonal control of oviposition and parturition in lizards. Herpetologica 38:80–93.

Jones RE, LJ Guillette Jr, M Norman, and JJ Roth. 1982. Corpus luteum-uterine relationships in the control of uterine contraction in the lizard *Anolis carolinensis*. Gen. Comp. Endocrinol. 48:104–112.

Joyal LA, M McCullough, and ML Hunter Jr. 2001. Landscape ecology approaches to wetland species conservation: a case study of two turtle species in southern Maine. Conserv. Biol. 15:1755–1762.

Junk WJ and VMF da Silva. 1997. Mammals, reptiles and amphibians. In The Central Amazon Floodplain, ed. WJ Junk. Springer, Heidelberg, 409–417.

Jury W, W Gardner, and W Gardner. 1991. Soil Physics, 5th ed. John Wiley & Sons, New York.

Kam Y-C. 1993. Physiological effects of hypoxia on metabolism and growth of turtle embryos. Respir. Physiol. 92:127–138.

Kam Y-C. 1994. Effects of simulated flooding on metabolism and water balance of turtle eggs and embryos. J. Herpetol. 28:173–178.

Kam Y-C and RA Ackerman. 1990. The effect of incubation media on the water exchange of snapping turtle (Chelydra serpentina) eggs and hatchlings. J. Comp. Physiol. B 160:317–324.

Kam Y-C and HB Lillywhite 1994. Effects of temperature and water on critical oxygen tension of turtle embryos. J. Exp. Zool. 268:1–8.

Karasov WH and JM Diamond. 1988. Interplay between physiology and ecology in digestion. Bioscience 38:602–611.

Karl H-V. 1990. Erstnachweis einer fossilen Schnappschildfrööten (Testudines, Chelydridae) in marinen Mitteloligozän der DDR. Mauritiana 12:477–481.

Kennett R, A Georges, and M Palmer-Allen. 1993. Early developmental arrest during immersion of eggs of a tropical freshwater turtle, Chelodina rugosa (Testudinata:Chelidae), from Northern Australia. Aust. J. Zool. 41:37–45.

Kepenis V and JJ McManus. 1974. Bioenergetics of young painted turtles, Chrysemys picta. Comp. Biochem. Physiol. 48A:309–317.

Khosatsky LI and OI Redkozubov. 1989. [Neogene turtles of Moldavia] Neogeneovye cherepaci Moldavii. "Shtiinca" ed. Kichinev. Acad. Nauk Mold. SSR 1–94.

Killebrew FC. 1977. Mitotic chromosomes of turtles. V. The Chelydridae. Southwest. Nat. 21:547–548.

King JM, G Kuchling, and SD Bradshaw. 1998. Thermal environment, behavior, and body condition of wild Pseudemydura umbrina (Testudines: Chelidae) during late winter and early spring. Herpetologica 54:103–112.

Kinneary JJ. 1992. The effect of water salinity on growth and oxygen consumption of snapping turtle (Chelydra serpentina) hatchlings from an estuarine habitat. J. Herpetol. 26:461–467.

Kiviat E. 1980. A Hudson River tidemarsh snapping turtle population. Trans. NE Sect. Wildl. Soc. 37:158–168.

Klemens MW. 1993. Amphibians and reptiles of Connecticut and adjacent areas. Bull. State Geol. Nat. Hist. Surv. Connecticut 112:1–318.

Klicka J and IY Mahmoud. 1972. Conversion of pregnenolone-[4–14C] to progesterone-[4–14C] by turtle corpus luteum. Gen. Comp. Endocrinol. 19:367–369.

Klicka J and IY Mahmoud. 1973. Conversion of cholesterol to progesterone by turtle corpus luteum. Steroids 21:483–495.

Klicka J and IY Mahmoud. 1977. The effects of hormones on the reproductive physiology of the painted turtle, Chrysemys picta. Gen. Comp. Endocrinol. 31:407–413.

Klimstra WD. 1951. Notes on late summer snapping turtle movements. Herpetologica 7:140.

Klute A. 1986. Methods of Soil Analysis. Soil Science Society of America, Madison, WI.

Knight TW, JA Layfield, and RJ Brooks. 1990. Nutritional status and mean selected temperature of hatchling snapping turtles (Chelydra serpentina): is there a thermophilic response to feeding? Copeia 1990:1067–1072.

Kolbe JJ and FJ Janzen. 2001. The influence of propagule size and maternal nest-site selection on survival and behaviour of neonate turtles. Func. Ecol. 15:772–781.

Kolbe JJ and FJ Janzen. 2002a. Impact of nest-site selection on nest success and nest temperature in natural and disturbed habitats. Ecology 83:269–281.

Kolbe JJ and FJ Janzen. 2002b. Experimental analysis of an early life-history stage: water loss and migrating hatchling turtles. Copeia 2002:220–226.

Koopman P, J Gubbay, J Collignon, and R Lovell-Badge. 1989. Zfy gene expression patterns are not compatible with a primary role in mouse sex determination. Nature 342:940–942.

Koorevaar P, G Menelik, and C Dirksen. 1983. Elements of Soil Physics. Elsevier, New York.

Korpelainen H. 1990. Sex ratios and conditions required for environmental sex determination in animals. Biol. Rev. 65:147–184.

Kraemer JE and SH Bennett. 1981. Utilization of post-hatching yolk in loggerhead sea turtles, Caretta caretta. Copeia 1981:406–411.

Krawchuk MA and RJ Brooks. 1998. Basking behavior as a measure of reproductive cost and energy allocation in the painted turtle, Chrysemys picta. Herpetologica 54:112–121.

Krebs CJ 1972. Ecology. Harper & Row, New York.

Krenz JG, GJP Naylor, HB Shaffer, and FJ Janzen. 2005. Molecular evolution and phylogenetics of turtles. Mol. Phylogenet. Evol. 37:178–191.

Kuchling G. 1998. The Reproductive Biology of the Chelonia. Springer-Verlag, Berlin.

Kuchling GR, R Skolek-Winnishch, and E Bamber. 1981. Histochemical and biochemical investigation on the annual cycle of testis, epididymis and plasma testosterone of the tortoise, Testudo hermanni hermanni. Gen. Comp. Endocrinol. 44:194–201.

Kumar S and SB Hedges. 1998. A molecular timescale for vertebrate evolution. Nature 392:917–920.

Kunkel BW. 1912. The development of the skull of Emys lutaria. J. Morphol. 23:693–780.

Kutchai H and J Steen. 1971. Permeability of the shell and shell membranes of hen's eggs during development. Respir. Physiol. 11:265–278.

Lacèpéde BGE. 1788. Histoire naturelle des quadrupèdes ovipares et des serpens, Vol. 1 (Ovipares). Paris.

Lagler KF. 1943a. Food habits and economic relations of the turtles of Michigan with special reference to fish management. Am. Midl. Nat. 29:257–312.

———. 1943b. Methods of collecting freshwater turtles. Copeia 1943:21–25.

Lagler KF and VC Applegate. 1943. Relationship between the length and weight of the snapping turtle, Chelydra serpentina Linnaeus. Am. Nat. 77:476–478.

Lamb T and JD Congdon. 1985. Ash content: relationships to leathery and brittle egg shell types of turtles. J. Herpetol. 19:527–530.

Lance V and IP Callard. 1978. Hormonal control of ovarian steroidogenesis in nonmammalian vertebrates. In The Vertebrate Ovary: Comparative Biology and Evolution, ed. RE Jones. Plenum, New York, 361–407.

Lance VA. 1994. Life in the slow lane: hormones, stress, and the immune system in reptiles. In Perspectives in Comparative Endocrinology, ed. KG Davey, RE Peter, and SS Tobe. National Research Council of Canada, Ottawa, 529–534.

———. 1997. Sex determination in reptiles: an update. Am. Zool. 37:504–513.

Lance VA and RM Elsey. 1986. Stress-induced suppression of testosterone secretion in male alligators. J. Exp. Zool. 239:241–246.

———. 1999. Plasma catecholamines and plasma corticosterone following restraint stress in juvenile alligators. J. Exp. Zool. 283:559–565.

Landers JL, JA Garner, and WA McRae. 1980. Reproduction of gopher tortoieses (Gopherus polyphemus) in southwestern Georgia. Herpetologica 36:353–361.

Lang JW and HV Andrews. 1994. Temperature-dependent sex determination in crocodilians. J. Exp. Zool. 270:28–44.

Lapparent de Broin F de. 2000. Les Chéloniens de Sansan. Mém. Mus. Nat. Hist. Nat. 183:219–261.

———. 2001. The European turtle fauna from the Triassic to the Present. Dumerilia 4:155–218.

Laube GC. 1900. Neue Schildkröten und Fische aus der böhmischen Braukohlenformation. Abh. Dtsch. Naturwiss.-med. Ver. Böhmen "Lotos" 2:37–56.

Lauder GV and T Prendergast. 1992. Kinematics of aquatic prey capture in the snapping turtle, Chelydra serpentina. J. Exp. Biol. 164:55–78.

Lefevre K and RJ Brooks. 1995. Effects of sex and body size on basking behavior in a northern population of the painted turtle, Chrysemys picta. Herpetologica 51:217–224.

Legler J. 1954. Nesting habits of the wester painted turtle, Chrysemys picta bellii (Gray). Herpetologica 10:137–144.

Leshem A, A Ar, and RA Ackerman. 1991. Growth, water, and energy metabolism of the soft-shelled turtle (Trionyx triunguis) embryo: effects of temperature. Physiol. Zool. 64:568–594.

Lewis J, IY Mahmoud, and J Klicka. 1979. Seasonal fluctuations of plasma concentrations of progesterone and estradiol-17 in the female snapping turtle, Chelydra serpentina. J. Endocrinol. 80:127–131.

Lewis TL and J Ritzenhaler. 1997. Characteristics of hibernacula use by spotted turtles, Clemmys guttata, in Ohio. Chelonian Conserv. Biol. 2:611–615.

Lewontin R. 1974. The analysis of variance and the analysis of causes. Am. J. Human Genet. 26:400–411.

Licht P. 1972. Actions of mammalian pituitary gonadotropins (FSH and LH) in reptiles. II. Turtles. Gen. Comp. Endocrinol. 19:282–228

———. 1975. Temperature dependence of the actions of mammalian and reptilian gonadotropins in a lizard. Comp. Biochem. Physiol. A 50:221–222.

———. 1982. Endocrine patterns in the reproductive cycles of turtles. Herpetologica 38:51–61.

———. 1984. Reptiles. In Marshall's Physiology of Reproduction, ed. GE Lamming. Churchill, London, 206–282.

Licht P, GL Breitenbach, and JD Congdon. 1985a. Seasonal cycles in testicular activity, gonadotropin and thyroxine in the painted turtle Chrysemys picta, under natural conditions. Gen. Comp. Endocrinol. 59:130–139.

Licht P, WR Dawson, VH Shoemaker, and AR Main. 1966. Observations on the thermal relations of western Australian lizards Copcia 1966 97 110.

Licht P, P Khorrami-Yaghoobi, and DA Porter. 1985b. Effects of gonadectomy and steroid treatment on plasma gonadotropins and the response of superfused pituitaries to gonadotropin releasing hormone (GnRH) in the turtle Sternotherus odoratus. Gen. Comp. Endocrinol. 60:441–449.

Licht P, BR McCreery, R Barnes, and R Pang. 1983. Seasonal and stress related changes in plasma gonadotropins, sex steroids, and corticosterone in the bullfrog, Rana catesbeiana. Gen. Comp. Endocrinol. 50:124–145.

Licht P and H Papkoff. 1985. Reevaluation of the relative activities of the pituitary glycoprotein hormones (FSH, LH, and TSH) from the green sea turtle, Chelonia mydas. Gen. Comp. Endocrinol. 58:443–451.

Licht P, W Rainey, and K Cliffton. 1980. Serum gonadotropin and steroids associated with breeding activities in the green sea turtle Chelonia mydas. II. Mating and nesting in natural populations. Gen. Comp. Endocrinol. 40:116–122.

Licht P, J Wood, DW Owens, and F Wood. 1979. Serum gonadotropins and steroids associated with breeding activities in the green sea turtle Chelonia mydas. I. Captive animals. Gen. Comp. Endocrinol. 39:274–289.

Lillywhite HB. 1971. Temperature selection by the bullfrog Rana catesbeiana. Comp. Biochem. Physiol. 40A:213–227.

Limpus CJ, P Reed, and JD Miller. 1985. Temperature dependent sex determination in Queensland sea turtles: intraspecific variation in Caretta caretta. In Biology of Australasian Frogs and Reptiles, ed. G Grigg, R Shine, and H Ehmann. Royal Zoological Society, New South Wales, 343–351.

Lindeman PV. 1991. Survivorship of overwintering hatchling painted turtles, Chrysemys picta, in northern Idaho. Can. Field-Nat. 105:263–266.

Linnaeus C. 1758. Systema Naturae per Regna Tria Naturae, Secondum Classes, Ordines, Genera, Species cum Characteribus, Differentis, Synonymis, Locis, 10th ed., Vol. 1. L. Salvius, Stockholm.

Litzgus JD, JP Costanzo, RJ Brooks, and RE Lee. 1999. Phenology and ecology of hibernation in spotted turtles (Clemmys guttata) near the northern limit of their range. Can. J. Zool. 77:1348–1357.

Liu S-S, G-M Zhang, and J Zhu. 1995. Influence of temperature variations on rate of development in insects: analysis of case studies from the entomological literature. Ann. Entomol. Soc. Am. 88:107–119.

Lofts B. 1972. The Sertoli cell. Gen. Comp. Endocrinol. Suppl. 3:636–648.

———. 1977. Patterns of spermiogenesis and steroidogenesis in male reptiles. In Reproduction and Evolution, ed. CH Tynedale-Biscoe. Proc. 4th Int. Symp. Comp. Biol. Reprod., Canberra, 1976.

Lofts B and HA Bern. 1972. The functional morphology of steroidogenic tissues. In Steroids in Nonmammalian Vertebrates, ed. DR Idler. Academic Press, New York.

Lofts B and HW Tsui. 1977. Histological and histochemical changes in the gonads and epididymides of the male soft-shelled turtle Trionyx sinensis. J. Zool. (Lond.) 181:57–68.

Loncke DJ and ME Obbard. 1977. Tag success, dimensions, clutch size and nesting site fidelity for the snapping turtle, Chelydra serpentina, (Reptilia, Testudines, Chelydridae) in Algonquin Park, Ontario, Canada. J. Herpetol. 11:243–244.

Lott DB. 1998. Egg water exchange and temperature dependent sex determination in the common snapping turtle Chelydra serpentina. Unpublished Ph.D. thesis, Iowa State University, Ames, IA.

Lovich JE. 1993. Macroclemys, M. temminckii. Cat. Am. Amph. Rept. (562):1–4.

Lynn WG and T von Brand. 1945. Studies on the oxygen consumption and water metabolism of turtle embryos. Biol. Bull. 88:112–125.

Maack GA. 1869. Die bis jetzt bekannten fossilen Schilköten und im oberen Jura bei Kelheim (Bayern) und Hannover neu aufgefundenen ältesten Arten derselben. Palaeontographica 18:193–338.

Macarovici N and S Vancea. 1960. Sur les restes de Tortues de la fauna de Malusteni de la Moldavie méridionale (R. P. Romanie). An. Stünt. Univ. Cuza, Sect. 2, Stünte Nat., Jassy, 2, n.s. 6:377–386.

Mahmoud IY, AE Colas, MJ Woller, and RV Cyrus. 1986. Cytoplasmic progesterone receptors in uterine tissue of the snaping turtle (Chelydra serpentina). J. Endocr. 109:385–392.

Mahmoud IY and RV Cyrus. 1992. The testicular cycle of the common snapping turtle, Chelydra serpentina in Wisconsin. Herpetologica 48:193–201.

Mahmoud IY, RV Cyrus, TM Bennett, MJ Woller, and DM Montag. 1985a. Ultrastructural changes in testes of the snapping turtle, Chelydra serpentina, in relation to plasma testosterone, 3β-hydroxysteroid dehydrogenase and choles-terol. Gen. Comp. Endocrinol. 57:454–464.

Mahmoud IY, RV Cyrus, M McAsey, C Cady, M Woller, and DL Wright. 1984. The effect of ovarian steroids and arginine vasotocin (AVT) on the contractility of the uterus in the snapping turtle (Chelydra serpentina). Am. Zool. 24:394 (abstract).

Mahmoud IY, RV Cyrus, ME McAsey, C Cady, and MJ Woller. 1988. The role of arginine vasotocin and prostaglandin F2α on oviposition and luteolysis in the common snapping turtle Chelydra serpentina. Gen. Comp. Endocrinol. 69:56–64.

Mahmoud IY, RV Cyrus, MJ Woller, and A Bieber. 1985b. Development of the ovarian follicles in relation to changes in plasma parameters and Δ⁵-3β-HSD in snapping turtle, Chelydra serpentina. Comp. Biochem. Physiol. A 83:131–136.

Mahmoud IY, RV Cyrus, and DL Wright. 1987. The effect of arginine, vasotocin and ovarian steroids on uterine contractil-ity in the snapping turtle, Chelydra serpentina. Comp. Biochem. Physiol. A 86:559–564.

Mahmoud IY, LJ Guillette Jr, ME McAsey, and C Cady. 1989. Stress-induced changes in serum testosterone, estradiol-17β and progesterone in the turtle, Chelydra serpentina. Comp. Biochem. Physiol. 93:433–427.

Mahmoud IY and J Klicka. 1971. Post-hatching changes in glycogen concentration in tissues of fed and unfed snapping turtles. Am. Midl. Nat. 86:248–252.

———. 1979. Feeding, drinking, and excretion. In Turtles, perspectives and research Harless, ed. M. and H. Morlock. John Wiley and Sons, New York, 229–243.

Mahmoud IY, J Klicka, RV Cyrus, and AE Colas. 1980. The rate of conversion of [4–¹⁴C] progesterone by corpora lutea of the snapping turtle, Chelydra serpentina. Gen. Comp. Endocrinol. 21:569–572.

Mahmoud IY and P Licht. 1997. Seasonal changes in gonadal activity and the effects of stress on reproductive hormones in the common snapping turtle, Chelydra serpentina. Gen. Comp. Endocrinol. 107:359–372.

Mahmoud IY, K Vliet, LJ Guilette, and JL Plude. 1996. Effect of stress and ACTH1–24 on hormonal levels in male alliga-tors, Alligator mississippiensis. Comp. Biochem. Physiol. A 115:57–62.

Major PD. 1975. Density of snapping turtles, Chelydra serpentina, in western West Virginia. Herpetologica 31:332–335.

Maloney JE, C Darian-Smith, Y Takahashi, and CJ. Limpus. 1990. The environment for development of the embryonic loggerhead turtles (Caretta caretta) in Queensland. Copeia 1990:378–387.

Manning B and GC Grigg. 1997. Basking is not of thermoregula-tory significance in the "basking" freshwater turtle Emydura signata. Copeia 1997:579–584.

Marchand MN and JA Litvaitis. 2004. Effects of landscape compostiion, habitat features, and nest distribution on predation rates of simulated turtle nests. Biol. Conserv. 117:243–251.

Matthews WD. 1924. Third contribution to the Snake Creek fauna. Bull. Amer. Mus. Nat. Hist. 50:59–210.

Mayeaux MH, DD Culley Jr, and RC Reigh. 1996. Effects of dietary energy: protein ration and stocking density on growth and survival of the common snapping turtle Chelydra serpentina. J. World Aquacult. Soc. 27:64–73.

McGehee MA. 1990. Effects of moisture on eggs and hatchlings of loggerhead sea turtles (Caretta caretta). Herpetologica 46:251–258.

McKenna MC. 1983. Cenozoic paleogeography of North America land bridges. In Structure and Development of the Greenland-Scotland Ridge, ed. MHP Bott, S Saxov, M Talwani, and J Thiede. Plenum Publishing Corporation, New York, 351–399.

McKnight CM and WHN Gutzke. 1993. Effects of the embryonic environment and of hatchling housing conditions on growth of young snapping turtles (Chelydra serpentina). Copeia 1993:475–482.

McPherson RJ and KR Marion. 1981. Seasonal testicular cycle of the stinkpot turtle (Sternotherus odoratus) in Alabama. Herpetologica 37:33–10.

McPherson RJ, LR Boots, R MacGreger, and KR Marion. 1982. Plasma steroids associated with seasonal reproductive changes in a multiclutched freshwater turtle, Sternotherus odoratus. Gen. Comp. Endocrinol. 48:440–451.

Medem F. 1962. La distribución geográfica y ecología de los Crocodylia y Testudinata en el Departamento del Choco. Rev. Acad. Columb. Cienc. Ex. Fis. Nat. 11(44):279–303.

———. 1977. Contribución al conocimiento sobre la taxonomía, distribución geográfica y ecológica de la tortuga "Bache" (Chelydra serpentina acutirostris). Caldasia 12:41–98.

Meeks RL and GR Ultsch. 1990. Overwintering behavior of snapping turtles. Copeia 1990:880–884.

Meier S and DS Packard Jr. 1984. Morphogenesis of the cranial segments and distribution of neural crest in the embryos of the snapping turtle, Chelydra serpentina. Dev. Biol. 102:309–323.

Mendonca MT and P Licht. 1986a. Seasonal cycles in gonadal activity and plasma gonadotropin in the musk turtle Sternotherus odoratus. Gen. Comp. Endocrinol. 62:459–469.

Mendonca MT and P Licht. 1986b. Photothermal effects on the testicular cycle of the musk turtle, Sternotherus odoratus. J. Exp. Zool. 239:117–130.

Mendonca MT, P Licht, MJ Ryan, and R. Barnes. 1985. Changes in hormaone levels in relation to breeding behavior in male bullfongs (Rana catesbaiana) at the individual and population levels. Gen. Comp. Endocrinol. 58:270–279.

Merrem B. 1820. Tentamen Systematis Amphiborum (Versuch eines Systems der Amphibien). Johann Christian Krieger, Marburg.

Mertens R, L Müller, and HT Rust. 1934. Systematische Liste der lebenden Schildkröten. Bl. Aquar.-v. Terrarienk. 45:42–45, 59–67.

Mesner PW, IY Mahmoud, and RC Cyrus. 1993. Seasonal testosterone levels in Leydig and Sertoli cells of the snapping turtle (*Chelydra serpentina*) in natural populations. J. Exp. Zool. 266:266–276.

Messina FJ. 1998. Maternal influences on larval competition in insects. In Maternal Effects as Adaptations, ed. TA Mousseau and CW Fox. Oxford University Press, New York, 227–243.

Meuschen FC. 1778. Museum Gronovianum, sive Index Rerum naturalium . . . Th. Haak, Lugduni Batavorum (Leiden).

Milam JC and SM Melvin. 2001. Density, habitat use, movements, and conservation of spotted turtles (*Clemmys guttata*) in Massachusetts. J. Herpetol. 35:418–427.

Miller JD. 1985. Embryology of marine turtles. In Biology of the Reptilia, ed. C Gans, F Billett, and PA Maderson, Vol. 14. John Wiley & Sons, Chichester, 269–328.

Miller K. 1993. The improved performance of snapping turtles (*Chelydra serpentina*) hatched from eggs incubated on a wet substrate persists through the neonatal period. J. Herpetol. 27:228–233.

Miller K, GF Birchard, and GC Packard. 1989. *Chelydra serpentina* (common snapping turtle). Fecundity. Herpetol. Rev. 20:69.

Miller K and GC Packard. 1992. The influence of substrate water potential during incubation on the metabolism of embryonic snapping turtles (*Chelydra serpentina*). Physiol. Zool. 65:172–187.

Miller K, GC Packard, and MJ Packard. 1987. Hydric conditions during incubation influence locomotor performance of hatchling snapping turtles. J. Exp. Biol. 127:401–412.

Minton SA. 2001. Amphibians and Reptiles of Indiana. Indiana Acad. Sci., Indianapolis, IN.

Mitchell JC. 1985a. Variation in the male reproductive cycle in a population of painted turtles, *Chrysemys picta,* from Virginia. Herpetologica 41:45–51.

———. 1985b. Variation in the male reproductive cycle in a population of stinkpot turtles, *Sternotherus odoratus,* in Virginia. Copeia 1985:50–56.

———. 1994. The Reptiles of Virginia. Smithsonian Inst. Press, Washington, DC.

Mitchell JC and CA Pague. 1991. Ecology of freshwater turtles in Back Bay, Virginia. *In* Proc. Back Bay Ecol. Symp., ed. H. G. Marshall and M. D. Norman. Old Dominion University, Norfolk, VA, 183–187.

Mitchell SL. 1990. The mating system genetically affects offspring performance in Woodhouses's toad (*Bufo woodhousei*). Evolution 44:502–519.

Mitro MG. 2003. Demography and viability analyses of a diamondback terrapin population. Can. J. Zool. 81:716–726.

Mlynarski M. 1963. Die plio-pleistoczänen Wierbeltierfauna von Hajnacka und Ivanorce (Slowakei) CSSR. IV. Schildkröten. Testudine. Neues Jahrb. Geol. Paläontol. Abh. 118:231–244.

———. 1966. Die fossilen Schlkröten in den ungarischen Sammlungen. Acta Zool. Cracov 11:223–288.

———. 1968. Die plio-pleistozänen Schildkröten mitteeuropas. Ber. Dtsch. Ges. Geol. Wiss. A 13:351–356

———. 1976. Handbuch der Paläoherpetologie. Part 7. Testudines. Gustav Fischer Verlag, Stuttgart.

———. 1980a. Die Schildkröten des Steinheimer Beckens. B. Chelydridae mit einem Nachtrag zu den Testudinoidea. Palaeontographica 8(2) Suppl. B:1–35.

———. 1980b. Die Pleistocänen Schildkröten mittel-und osteruropas (Bestimmungschlüssel). Folia Quat. 52:2–41.

———. 1981a. Chelydropsinae, the Eurasiatic fossil snapping turtles (Chelydridae). Chelonologica 2:57–63.

———. 1981b. *Chelydropsis murchisoni* (Bell, 1832) (Testudines, Chelydridae) from the Miocene of Przeworno in Silesia (Poland). Acta Zool. Cracov 25:219–226.

Moberg GP. 1985. Influence of stress on reproduction, measure of well-being. In Animal Stress, ed. GP Moberg. American Physiology Society, Bethesda, MD, 246–267.

Moll D. 1976. Environmental influence on growth rate in the Ouachita map turtle, *Graptemys pseudogeographica ouachitensis.* Herpetologica 32:439–443.

———. 1997. Ecological characteristics of the snapping turtle, *Chelydra serpentina acutirostris,* in Costa Rica, In Abstracts of the Third World Congress of Herpetology. Prague, Czech Republic, 145.

Moll D and E Moll. 1990. The slider turtle in the Neotropics: adaptations of a temperate species to a tropical environment. In Life History and Ecology of the Slider Turtle, ed. JW Gibbons. Smithsonian Inst Press, Washington, DC, 152–161.

Moll D and EO Moll. 2004. The ecology, exploitation and conservation of river turtles. Oxford University Press, New York, 393 pp.

Moll EO. 1973. Latitundinal and intersubspecific variation in repdocution of the painged turtle, *Chrysemys picta.* Herpetologica 29:307–318.

———. 1979. Reproductive cycles and adaptations. In Turtles: Perspectives and Research, ed. M. Harless and H. Morlock. John Wiley and Sons, Toronto, Canada, 305–331.

Moll EO and JM Legler. 1971. The life history of a neotropical pond slider turtle, *Pseudemys scripta* (Schoepff), in Panama. Bull. Los Angeles County Mus. Nat. Hist. Sci. 11:1–102.

Moran KL, KA Bjorndal, and AB Bolten. 1999. Effects of the thermal environment on the temporal pattern of emergence of hatchling loggerhead turtles, *Caretta caretta.* Mar. Ecol. Progr. Ser. 189:251–261.

Morici LA, RM Elsey, and VA Lance. 1997. Effects of long-term corticosterone implants on growth and immune function in juvenile alligators, *Alligator mississippiensis.* J. Exp. Zool. 279:156–162.

Morjan CL. 2003. How rapidly can maternal behavior affecting primary sex ratio evolve in a reptile with environmental sex determination? Am. Nat. 162:205–219.

Morjan CL and JN Stuart. 2001. Nesting record of a Big Bend slider turtle (*Trachemys gaigeae*) in New Mexico, and overwintering of hatchlings in the nest. Southwest. Nat. 46:230–234.

Morris KA, GC Packard, TJ Boardman, GL Paukstis, and MJ Packard. 1983. Effect of the hydric environment on growth of embryonic snapping turtles (*Chelydra serpentina*). Herpetologica 39:272–285.

Mosimann JE and JR Bider. 1960. Variation, sexual dimorphism, and maturity in a Quebec population of the common snapping turtle, *Chelydra serpentina.* Can. J. Zool. 38:19–38.

Mousseau TA and CW Fox. 1998. Maternal Effects as Adaptations. Oxford University Press, New York.

Mrosovsky N. 1968. Nocturnal emergence of hatchling sea turtles: control by thermal inhibition of activity. Nature 220:1338–1339.

———. 1988. Pivotal temperatures for loggerhead turtles (*Caretta caretta*) from northern and southern nesting beaches. Can. J. Zool. 66:661–669.

———. 1994. Sex ratios of sea turtles. J. Exp. Zool 270:16–27.

Mrovosky N and J Provancha. 1992. Sex ratio of hatchling

loggerhead sea turtles: data and estimates from a 5-year study. Can. J. Zool. 70:530–538.

Müller L 1939. Ueberdie Verbreitung der Chelonier auf dem südamerikanischen Kontinent. Physics 16:89–102.

Murelaga X, F de Lapparent de Broin, XP Suberbiola, and H Astibia. 1999. Two new chelonian species from the Lower Miocene of the Ebro basin (Bardenas Reales of Navarre). Acad. Sci. (Paris) C. R. Sci. Terre Planét. 328:423–429.

Murelaga X, XP Suberbiola, F de Lapparent de Broin, J-C Rage, S Duffaud, H Astibia, and A Badiola. 2002. Amphibians and reptiles from the early Miocene of the Bardenas Reales of Navarre (Ebro Basin, Iberian Peninsula). Geobios 35:347–365.

Nagle RD, VJ Burke, and JD Congdon. 1998. Egg components and hatchling lipid reserves: parental investment in kinosternid turtles from the southeastern United States. Comp. Biochem. Physiol. 120B:145–152.

Nagle RD, OM Kinney, JD Congdon, and CW Beck. 2000. Winter survivorship of hatchling painted turtles (Chrysemys picta) in Michigan. Can. J. Zool. 78:226–233.

Nagle D., MV Plummer, JD Congdon, and RU Fischer. 2003. Parental investment, embryo growth, and hatchling lipid reserves in softshell turtles (Apalone mutica) from Arkansas. Herpetologica 59:145–154.

Near TJ, PA Meylan, and HB Shaffer. 2005. Assessing concordance of fossil calibration points in molecular clock studies: an example using turtles. Am. Nat. 165:137–146.

Newman HH. 1906. The significance of scute and plate "abnormalities" in Chelonia. Biol. Bull. 10:68–114.

Newman RA. 1994. Genetic variation for phenotypic plasticity in the larval life history of spadefoot toads (Scaphiopus couchii). Evolution 48:1773–1785.

Nick L. 1912. Das Kopfskelett von Dermochelys coriacea. L. Zool. Jahrb. Abt. Anat. 33:1–238.

Niewiarowski PH and W Roosenburg. 1993. Reciprocal transplant reveals sources of variation in growth rates of the lizard Sceloporus undulatus. Ecology 74:1992–2002.

Nobel PS. 1991. Physicochemical and Environmental Plant Physiology. Academic Press, San Diego, CA.

Noble GK and AM Breslau. 1938. The senses involved in the migration of young fresh-water turtles after hatching. J. Comp. Psychol. 25:175–193.

Noborio K, R Horton, and C Tan. 1999. Time domain reflectometry probe for simultaneous measurement of soil matric potential and water content. Soil Sci. Soc. Am. J. 63:1500–1505.

Norris-Elye LTS. 1949. The common snapping turtle (Chelydra serpentina) in Manitoba. Can. Field Nat. 63:145–147.

Obbard ME. 1983. Population ecology of the common snapping turtle, Chelydra serpentina, in north-central Ontario. Unpublished Ph.D. thesis, University of Guelph, Guelph, Ontario.

Obbard ME and RJ Brooks. 1979. Factors affecting basking in a northern population of the common snapping turtle, Chelydra serpentina. Can. J. Zool. 57:435–440.

———. 1980. Nesting migrations of the snapping turtle (Chelydra serpentina). Herpetologica 36:158–162.

———. 1981a. A radio telemetry and mark recapture study of activity in the common snapping turtle, Chelydra serpentina. Copeia 1981:630–637.

———. 1981b. Fate of overwintered clutches of the common snapping turtle (Chelydra serpentina) in Algonquin Park, Ontario. Can. Field-Nat. 95:350–352.

———. 1987. Prediction of the onset of the annual nesting season of the common snapping turtle, Chelydra serpentina. Herpetologica 43:324–328.

Obst FJ. 1986. Turtles, Tortoises and Terrapins. St. Martin's Press, New York.

O'Connor MP. 1999. Physiological and ecological implications of a simple model of heating and cooling in reptiles. J. Therm. Biol. 24:113–136.

Ogilby JD. 1905. Catalogue of emydosaurian and testudinian reptiles of New Guinea. Proc. R. Soc. Queensland 19:1–31.

Orgeig S, AW Smits, CB Daniels, and JK Herman. 1997. Surfactant regulates pulmonary fluid balance in reptiles. Am. J. Physiol. 273:R2013–R2021.

O'Steen S. 1998. Embryonic temperature influences juvenile temperature choice and growth rate in snapping turtles Chelydra serpentina. J. Exp. Biol. 201:439–449.

O'Steen S and FJ Janzen. 1999. Embryonic temperature affects metabolic compensation and thyroid hormones in hatchling snapping turtles. Physiol. Biochem. Zool. 72:520–533.

Packard GC. 1991. Physiological and ecological importance of water to embryos of oviparous reptiles. In Egg Incubation: Its Effects on Embryonic Development in Birds and Reptiles, ed. DC Deeming and MJW Ferguson. Cambridge Univ. Press, New York, 213–228.

———. 1997. Temperatures during winter in nests with hatchling painted turtles (Chrysemys picta). Herpetologica 53:89–95.

———. 1999. Water relations of chelonian eggs and embryos: is wetter better? Am. Zool. 39:289–303.

Packard GC, SL Fasano, MB Attaway, LD Lohmiller, and TL Lynch. 1997. Thermal environment for overwintering hatchlings of the painted turtle (Chrysemys picta). Can. J. Zool. 75:401–406.

Packard GC, K Miller, and MJ Packard. 1992. A protocol for measuring water potential in subterranean nests of reptiles. Herpetologica 48:202–209.

———. 1993. Environmentally induced variation in body size of turtles hatching in natural nests. Oecologia 93:445–448.

Packard GC, K Miller, MJ Packard, and GF Birchard. 1999. Environmentally induced variation in body size and condition in snapping turtles (Chelydra serpentina). Can. J. Zool. 77:278–289.

Packard GC and MJ Packard. 1984. Coupling of physiology of embryonic turtles to the hydric environment. In Respiration and Metabolism of Embryonic Vertebrates, ed. R Seymour. Kluwer Academic Press, Boston, 99–119.

———. 1988a. Physiological ecology of reptilian eggs and embryos. in Biology of the Reptilia, ed. C. Gans and R.B. Huey, Vol. 16. Alan R. Liss, New York, 523–605.

———. 1988b. Water relations of embryonic snapping turtles (Chelydra serpentina) exposed to wet or dry environments at different times in incubation. Physiol. Zool. 61:95–106.

———. 1989a. Control of metabolism and growth in embryonic turtles: a test of the urea hypothesis. J. Exp. Biol. 147:203–216.

———. 1993. Sources of variation in laboratory measurements of water relations of reptilian eggs and embryos. Physiol. Zool. 66:115–127.

———. 2001. The overwintering strategy of hatchling painted turtles, or how to survive in the cold without freezing. BioScience 51:199–207.

———. 2002. Wetness of the nest environment influences cardiac development in pre-and post-natal snapping turtles (Chelydra serpentina). Comp. Biochem. Physiol. A 132:905–912.

Packard GC, MJ Packard, and GF Birchard. 2000. Availability of water affects organ growth in prenatal and neonatal snapping turtles (*Chelydra serpentina*). J. Comp. Physiol. 170:69–74.

Packard GC, MJ Packard, and TJ Boardman. 1984. Effects of the hydric environment on metabolism of embryonic snapping turtles do not result from altered patterns of sexual differentiation. Copeia 1984:547–550.

———. 1984a. Influence of hydration of the environment on the pattern of nitrogen excretion by embryonic snapping turtles. J. Exp. Biol. 108:195–204.

Packard GC, MJ Packard, TJ Boardman, and MD Ashen. 1981. Possible adaptive value of water exchanges in flexible-shelled eggs of turtles. Science 213:471–473.

Packard GC, MJ Packard, TJ Boardman, KA Morris, and RD Shuman. 1983. Influence of water exchanges by flexible-shelled eggs of painted turtles *Chrysemys picta* on metabolism and growth of embryos. Physiol. Zool. 56:217–230.

Packard GC, MJ Packard, JW Lang, and JK Tucker. 1999. Tolerance for freezing in hatchling turtles. J. Herpetol. 33:536–543.

Packard GC, MJ Packard, and K Miller. 1990. *Chelydra serpentina* (Common Snapping Turtle). Fecundity. Herpetol. Rev. 21:92.

Packard GC, MJ Packard, K Miller, and TJ Boardman. 1987. Influence of moisture, temperature, and substrate on snapping turtle eggs and embryos. Ecology 68:983–993.

———. 1988. Effects of temperature and moisture during incubation on carcass composition of hatchling snapping turtles (*Chelydra serpentina*). J. Comp. Physiol. 158:117–125.

Packard GC, G Paukstis, T Boardman, and WHM Gutzke. 1985. Daily and seasonal variation in hydric conditions and temperatures inside nests of common snapping turtles (*Chelydra serpentina*). Can. J. Zool. 63:2422–2429.

Packard G, T Taigen, M Packard, and T Boardman. 1980. Water relations of pliable-shelled eggs of common snapping turtles (*Chelydra serpentina*). Can. J. Zool. 58:1404–1411.

Packard GC, T Taigen, MJ Packard, and R Shuman. 1979. Water vapor conductance of testudinian and crocodilian eggs (class Reptilia). Respir. Physiol. 38:1–10.

Packard GC, C Tracy, and J Roth. 1977. The physiological ecology of reptilian egg and embryos and the evolution of viviparity within the class Reptilia. Biol. Rev. 52:71–105.

Packard M and R Seymour. 1997. Evolution of the amniote egg. In Amniote Origins Completing the Transition to Land, ed. S Sumida and K Martin. Academic Press, New York, 265–290.

Packard MJ. 1980. Ultrastructural morphology of the shell and shell membrane of eggs of common snapping turtles (*Chelydra serpentina*). J. Morphol. 167:187–204.

Packard MJ and VG DeMarco. 1991. Eggshell structure and formation in eggs of oviparous reptiles. In Egg Incubation Its effects on Embryonic Development in Birds and Reptiles, ed. DC Deeming and MWJ Ferguson. Cambridge University Press, New York, 53–70.

Packard MJ and GC Packard. 1986. Effect of water balance and calcium mobilization of embryonic painted turtles (*Chrysemys picta*). Physiol. Zool. 59:398–405.

———. 1989b. Environmental modulation of calcium and phosphorus metabolism in embryonic snapping turtles (*Chelydra serpentina*). J. Comp. Physiol. 159:501–508.

Packard MJ, TM Short, GC Packard, and TA Gorell. 1984b. Sources of calcium for embryonic development in eggs of the snapping turtle *Chelydra serpentina*. J. Exp. Zool. 230:81–87.

———. 1984c. Sources of calcium for embryonic development in eggs of the snapping turtle *Chelydra serpentina*. J. Exp. Zool. 230:81–87.

Paganelli CV. 1991. The avian eggshell as a mediating barrier: respiratory gas fluxes and pressures during development. In Egg Incubation: Its Effects on Embryonic Development in Birds and Reptiles, ed. CD Deeming and MWJ Ferguson. Cambridge University Press, New York, 261–276.

Palmer WM and AL Braswell. 1994. Reptiles of North Carolina. Univ. North Carolina Press, Chapel Hill, NC.

Pappas MJ, BJ Brecke, and JD Congdon. 2000. The Blanding's turtles (*Emydoidea blandingii*) of Weaver Dunes, Minnesota. Chelonian Conserv. Biol. 3:557–568.

Parmalee PW and RD Oesch. 1972. Pleistocene and Recent faunas from the Brynjulfson Caves, Missouri. Ill. State Mus. Rep. Invest. 25:1–52.

Parmenter RR. 1980. Effects of food availability and water temperature on the feeding ecology of pond sliders (*Chrysemys s. scripta*). Copeia 1980:503–514.

———. 1981. Digestive turnover rates in freshwater turtles: the influence of temperature and body size. Comp. Biochem. Physiol. 70A:235–238.

Parmenter RR and HW Avery. 1990. The feeding ecology of the slider turtle. In Life History and Ecology of the Slider Turtle, ed. JW Gibbons. Smithsonian Inst. Press, Washington, DC, 257–266.

Parmley D 1992. Turtles from the late Hemphillian (latest Miocene) of Knox County, Nebraska. Texas J. Sci. 44:339–348.

Parren SG and MA Rice. 2004. Terrestrial overwintering of hatchling turtles in Vermont nests. Northeast. Nat. 11:229–233.

Parris KM and MA McCarthy 2001. Identifying effects of toe clipping on anuran return rates: the importance of statistical power. Amphibia-Reptilia 22:275–289.

Parsons TS and JE Cameron. 1977. Internal relief of the digestive tract. In Biology of the Reptilia, ed. C Gans and TR Parsons, Vol 6. Academic Press, London, 159–223.

Pasteels J. 1937. Etudies sur la gastrulation des vertébrés méroblastiques. II. Reptiles. Arch. Biol. 48:105–184.

Patnaik BK. 1994. Ageing in reptiles. Gerontology 40:200–220.

Paukstis GL and FJ Janzen. 1990. Sex determination in reptiles: summary of effects of constant temperatures of incubation on sex ratios of offspring. Smithsonian Herpetol. Info. Serv. 83:1–28.

Pawley R. 1987. Measurements of a large alligator snapping turtle, *Macroclemys temmincki*, in the Brookfield Zoo. Bull. Chic. Herpetol. Soc. 22:134.

———. 1987. Measurements of a large alligator snapping turtle, *Macroclemys temmincki*, in the Brookfield Zoo. Bull. Chic. Herpetol. Soc. 22:134.

Pell SM. 1941. Notes on the habits of the common snapping turtle, *Chelydra serpentina* (Linn.) in central New York. M.S. thesis, Cornell University, Ithaca, NY.

Peng J-H and DB Brinkman. 1993. New material of *Xinjiangchelys* (Reptilia: Testudines) from the Late Jurassic Qigu Formation (Shishugou Group) of the Pingfengshan Locality, Junggar basin, Xinjiang. Can. J. Earth Sci. 30:2013–2026.

Peng J, AP Russell, and DB Brinkman. 2001. Vertebrate microsite assemblages (exclusive of mammals) from the Foremost and Oldman Formations of the Judith River Group (Campanian) of southeastern Alberta: an illustrated guide. Prov. Mus. Alberta, Nat. Hist. Occ. Pap. no. 25:54 p.

Pennant T. 1787. Arctic Zoology. Supplement. London.

Peters KF. 1855. Schildkrötenreste aus den österreichischen Tertiär-Ablagerungen. Denkschrift. Akad. Wiss. Wien. Math.-Naturwiss. Kl. 9:1–22.

———. 1868. Zur Kenntniss der Wirbelthiere aus den Miocen-schichten von Eibiswald in Steiermark. I. Die Schildkröten-reste (Auszug). Sitzunsgber. Akad Wiss. Wien, Math.-Naturwiss. Kl. 27:72–74.

———. 1869. Zur Kenntniss der Wirbelthiere aus den Miocän-schichten von Eibiswald in Steiermark. I. Die Schilkrotenreste. Denkschrift. Akad. Wiss. Wien. Math.-Naturwiss. Kl. 29:111–124.

Peters WKH. 1862. [No title.] Monatsber. Acad. Wiss. Berlin 1862:626–627.

Petokas PJ and MM Alexander. 1980. The nesting of *Chelydra serpentina* in northern New York. J. Herpetol. 14:239–244.

Phillips CA, WW Dimmick and JL Carr. 1996. Conservation genetics of the common snapping turtle (*Chelydra serpentina*). Conserv. Biol. 10:397–405.

Pidoplichko IG and BI Tarashchuk. 1960. [New genus of large-headed (macrocephalous) turtle from the Pontian beds in the environs of Odessa.] Zbirn. Prats. Zool. Akad. Nauk URSR 29:105–110.

Pieau C. 1971. Sur la proportion sexuelle chez les embryons de deux chéloniens (*Testudo graeca* L. et *Emys orbicularis* L.) issus d'oeufs incubés artificiellement. C. R. Acad. Sci. Paris 272:3071–3074.

———. 1974. Sur la différenciation sexuelle chez des embryons d'*Emys orbicularis* L. (Chélonien) issus d'oeufs incubés dans le sol au cours de l'été 1973. Bull. Soc. Zool. Fr. 99:363–376.

Pieau C, M Dorizzi, and N Richard-Mercier. 1999. Temperature-dependent sex determination and gonadal differentiation in reptiles. Cell. Mol. Life. Sci. 55:887–900.

Pinsof JD. 1998. The American Falls local fauna: late Pleistocene (Sangamonian) vertebrates from southeastern Idaho. Idaho Mus. Nat. Hist. Occ. Pap. 36:121–145

Place AR and VA Lance. 2004. The temperature-dependent sex determination drama: same cast, different stars. In Tempera-ture-Dependent Sex Determination in Vertebrates, ed. N Valenzuela and VA Lance. Smithsonian Books, Washington, DC, 99–110.

Place AR, J Lang, S Gavasso, and P Jeyasuria. 2001. Expression of P450arom in *Malaclemys terrapin* and *Chelydra serpentina*: a tale of two sites. J. Exp. Zool. 290:673–690.

Plummer M and Snell H. 1988. Nest site selection and water relations of eggs in the snake, *Opheodrys aestivus*. Copeia 1988:58–64.

Plummer MV. 1976. Some aspects of nesting success in the turtle, *Trionyx muticus*. Herpetologica 32:353–359.

Plummer MV and JC Burnley. 1997. Behavior, hibernacula, and thermal relations of softshell turtles (*Apalone spinifera*) overwintering in a small stream. Chelonian Conserv. Biol. 2:489–493.

Pluto TG and ED Bellis. 1988. Seasonal and annual movements of riverine map turtles, *Graptemys geographica*. J. Herpetol. 22:152–158.

Pomel A. 1846. Mémoire pour servir à la Géologie paléon-tologique des terrains tertiaires du Département de l'Allier. Bull. Soc. Géol. Fr. (sér. 2) 3:353–373.

Popper KR. 1999. All Life is Problem Solving. Routledge, New York.

Pough FH. 1980. The advantages of ectothermy for tetrapods. Am. Nat. 115:92–112.

Pough FH and C Gans. 1982. The vocabulary of reptilian thermoregulation. In Biology of the Reptilia, Physiology C, Physiological Ecology, ed. C Gans and FH Pough, Vol. 12, Academic Press, New York, NY, 17–23.

Prange HD and RA Ackerman. 1974. Oxygen consumption and mechanisms of gas exchange of green turtle (*Chelonia mydas*) eggs and hatchlings. Copeia 1974:758–763.

Preston RE. 1979. Late Pleistocene cold-blooded vertebrate faunas from the mid-continental United States. I. Reptilia, Testudines, Crocodilia. Univ. Mich. Mus. Paleontol. Pap. Paleontol. 19:1–53.

Pritchard PCH. 1979. Encyclopedia of Turtles. T.F.H. Publica-tions, Neptune, NJ.

———H. 1980. Record size turtles from Florida and South America. Chelonologica 1:113–123.

———. 1989. The alligator snapping turtle: biology and conservation. Milwaukee Public Museum, Milwaukee, WI.

Prosser CL. 1973. Comparative Animal Physiology. W.B. Saunders Co., Philadelphia, PA.

Przeworski M. 2002. The signature of positive selection at randomly chosen loci. Genetics 160:1179–1189.

Punzo F. 1975. Studies on the feeding behavior, diet, nesting habits and temperature relationships of *Chelydra serpentina osceola* (Chelonia: Chelydridae). J. Herpetol. 9:207–210.

Rabl C. 1910. Bausteine zu einer Theorie der Extremitäten der Wirbeltiere. I Teil. Verlag von Wilhelm Engelmann, Leipzig.

Rafinesque CS. 1832. Description of two new genera of softshell turtles of North America. Atlantic J. Friend Knowledge Philadelphia 1:64–65.

Rahn H. 1938. The corpus luteum of reptiles. Anat. Rec. 72:55.

Rahn H and Ar A. 1974. The avian egg: Incubation time and water loss. Condor 76:147–152.

Rahn H, A Ar, and C Paganelli. 1979. How bird eggs breathe. Sci. Am. 1979:38–47.

Ramkissoon Y and P Goodfellow. 1996. Early steps in mam-malian sex determination. Curr. Opin. Genet. Dev. 6:316–321.

Ratterman R and R Ackerman. 1989. The water exchange and hydric microclimate of painted turtle (*Chrysemys picta*) eggs incubating in field nests. Physiol. Zool. 62: 1059–1079.

Rawlins S and G Campbell. 1986. Water potential: Thermocou-ple psychrometry. In *Methods of Soil Analysis, ed.* A Klute. Soil Science Society of America, Madison, WI, 597–618.

Reese SA, CE Crocker, ME Carwile, DC Jackson, and GR Ultsch. 2001. The physiology of hibernation in common map turtles (*Graptemys geographica*). Comp. Biochem. Physiol. A 130:331–340.

Reese SA, CE Crocker, DC Jackson, and GR Ultsch. 2000. The physiology of hibernation among painted turtles: the midland painted turtle (*Chrysemys picta marginata*). Respir. Physiol. 124:43–50.

Reese SA, DC Jackson, and GR Ultsch. 2002. The physiology of overwintering in a turtle that occupies multiple habitats, the common snapping turtle (*Chelydra serpentina*). Physiol. Biochem. Zool. 75:432–438,

———. 2003. Hibernation in freshwater turtles: softshell turtles (*Apalone spinifera*) are the least intolerant of anoxia among North American species. J. Comp. Physiol. B 173:263–268.

Reese SA, GR Ultsch, and DC Jackson. 2004. Lactate accumula-tion, glycogen depletion, and shell composition of hatchling turtles during simulated aquatic hibernation. J. Exp. Biol. 207:2889–2895.

Reinhold K. 1998. Nest-site philopatry and selection for environmental sex determination. Evol. Ecol. 12:245–250.

Ren T, K Noborio, and R Horton. 1999. Measuring soil water content, electrical conductivity, and thermal properties with a thermo-time domain reflectometry probe. Soil Sci. Soc. Am. J. 63:450–457.

Resetarits WJ Jr. 1996. Oviposition site choice and life history evolution. Am. Zool. 36:205–215.

Rhen T, PK Elf, AJ Fivizzani, and JW Lang. 1996. Sex-reversed and normal turtles display similar sex steroid profiles. J. Exp. Zool. 274:221–226.

Rhen T and JW Lang. 1994. Temperature-dependent sex determination in the snapping turtle: manipulation of the embryonic sex steroid environment. Gen. Comp. Endocrinol. 96:243–254.

———. 1995. Phenotypic plasticity for growth in the common snapping turtle: effects of incubation temperature, clutch, and their interaction. Am. Nat. 146:726–747.

———. 1998. Among-family variation for environmental sex determination in reptiles. Evolution 52:1514–1520.

———. 1999a. Temperature during embryonic and juvenile development influences growth in hatchling snapping turtles, Chelydra serpentina. J. Therm. Biol. 24:33–41.

———. 1999b. Incubation temperature and sex affect mass and energy reserves of hatchling snapping turtles, Chelydra serpentina. Oikos 86:311–319.

———. 2004. Phenotypic effects of incubation temperature in reptiles. In Temperature-Dependent Sex Determination in Vertebrates, ed. N. Valenzuela and V. Lance. Smithsonian Books, Washington, DC, 90–98.

Richards RL, DR Whitehead, and DR Cochran. 1987. The Dollans mastodon (Mammutha americanum) locality, Madison County, East Central Indiana. Proc. Indiana Acad. Sci. 97:571–581.

Richmond ND. 1945. Nesting habits of the mud turtle. Copeia 1945:217–219.

———. 1958. The status of the Florida snapping turtle, Chelydra osceola Stejneger. Copeia 1958:41–43.

———. 1964. The mechanical functions of the testudinate plastron. Am. Midl. Nat. 72:50–56.

Ricklefs RE. 1979. Ecology, 2nd ed. Chiron, New York, NY, 966 pp.

Rieppel O. 1976. Die orbitotemporale Region im Schädel von Chelydra serpentina Linnaeus (Chelonia) und Lacerta sicula Rafinesque (Lacertilia). Acta Anat. 96:309–320.

———. 1977. Über die Entwicklung des Basicranium bei Chelydra serpentina Linnaeus (Chelonia) und Lacerta sicula Rafinesque (Lacertilia). Verhandl. Naturf. Gesell. 86:153–170.

———. 1990. The structure and development of the jaw adductor musculature in the turtle Chelydra serpentina. Zool. J. Linn. Soc. 98:27–62.

———. 1993. Studies on skeleton formation in reptiles: Patterns of ossification in the skeleton of Chelydra serpentina (Reptilia, Testudines). J. Zool. Lond. 231:487–509.

Rieppel O and RR Reisz. 1999. The origin and early evolution of turtles. Annu. Rev. Ecol. Syst. 30:1–22.

Rimkus TA. 1996. Water exchange in reptile eggs: mechanisms for transportation, driving forces behind movement, and the effects on hatchling size. Unpublished Ph.D. dissertation, Iowa State University, Ames.

Rimkus TA, N Hruska, and RA Ackerman. 2002. Separating the effects of vapor pressure and heat exchange on water exchange by snapping turtle (Chelydra serpentina) eggs. Copeia 2002:857–864

Risley PL. 1933. Observations on the natural history of the common musk turtle, Sternotherus odoratus (Latreille). Papers Mich. Acad. Sci. Arts Lett. 52:685–711.

———. 1938. Seasonal changes in the testes of the musk turtle Sternotherus odoratus. J. Morphol. 63:301–317.

Ritland K. 1996. A marker-based method for inferences about quantitative inheritance in natural populations. Evolution 50:1062–1073.

Robertson SL and EN Smith. 1982. Evaporative water loss in the spiny soft-shelled turtle Trionyx spiniferus. Physiol. Zool. 55:124–129.

Robinson C. 1989. Orientation and survival of hatchlings and reproductive ecology of the common snapping turtles (Chelydra serpentina) in Southern Quebec. Unpublished M.Sc. thesis, McGill University, Montreal.

Robinson C and JR Bider. 1988. Nesting synchrony—A strategy to decrease predation of snapping turtle (Chelydra serpentina) nests. J. Herpetol. 22:470–473.

Rodriquez-de la Rosa RA and SRS Cevallos-Ferriz. 1998. Vertebrates of the El Pelillal locality (Campanian, Cerro del Pueblo Formation), southeastern Coahuila, Mexico. J. Vert. Paleontol. 18:751–764

Roman J, SD Santhuff, PE Moler, and BW Bowen. 1999. Population structure and cryptic evolutionary units in the alligator snapping turtle. Conserv. Biol. 13:135–142.

Rome LC, ED Stevens, and HB John-Alder. 1992. The influence of temperature and thermal acclimation on physiological function. In Environmental Physiology of the Amphibians, ed. ME Feder and WW Burggren. University of Chicago Press, Chicago, IL, 183–205.

Romer AS. 1957. Origin of the amniotic egg. Sci. Mon. 85:57–63.

Roosenburg WM. 1996. Maternal condition and nest site choice: an alternative for the maintenance of environmental sex determination? Am. Zool. 36:157–168.

Roosenburg WM and KC Kelley. 1996. The effect of egg size and incubation temperature on growth in the turtle, Malaclemys terrapin. J. Herpetol. 30:198–204.

Rosenberg N, B Blad, and S Verma. 1983. Microclimate: The Biological Environment, 2nd ed. John Wiley & Sons, New York.

Roth JJ, RE Jones, and A Gerrard. 1973. Corpora lutea and oviposition in the lizard Sceloporus undulates. Gen. Comp. Endocrinol. 21:569–572.

Rowe JW. 1994. Egg size and shape variation within and among Nebraskan painted turtle (Chrysemys picta bellii) populations: Relationships to clutch and maternal body size. Copeia 1994:1034–1040.

Rowe JW, L Holy, RE Ballinger, and D Stanley-Samuelson. 1995. Lipid provisioning of turtle eggs and hatchlings: total lipid, phospholipid, triacylglycerol, and triacylglycerol fatty acids. Comp. Biochem. Physiol. 112B:323–330.

Ruckes H. 1929. Studies in chelonian osteology. Part I. Truss and arch analogies in chelonian pelves. Ann. NY Acad. Sci. 31:31–80.

Saba VS. 2001. The survival and behavior of freshwater turtles after rehabilitation from an oil spill. Masters thesis, Drexel University, Philadelphia, PA.

St. Clair RC and PT Gregory. 1990. Factors affecting the northern range limit of painted turtles (Chrysemys picta): winter acidosis or freezing? Copeia 1990:1083–1089.

St. Juliana JR, RM Bowden, and FJ Janzen. 2004. The impact of behavioral and physiological maternal effects on offspring sex ratio in the common snapping turtle, *Chelydra serpentina*. Behav. Ecol. Sociobiol. 56:270–278.

SAS Institute. 2000. JMP User's Guide, Version 4. SAS Institute, Cary, NC.

Saumure RA. 2001. Limb mutilations in snapping turtles, *Chelydra serpentina*. Can. Field Nat. 115:182–184.

Schleich HH. 1981. Jungtertiäre Schildkröten Süddeutschlands unter besonderer Berücksichtigung der Fundstelle Sandelzhausen. Courier Forsch.-Inst. Senckenberg, 48:1–372.

———. 1985 Zur Verbeitung tertiärer and quartärer Reptilien und Amphibien. I. Süddeutschl. Münchner Geowiss. Abh. (A) 4:67–149.

———. 1986. Vorläufige Mittleilungen zur Bearbeitung der fossilen Schildkröten der Fundstelle Höwenegg. Carolinea 44:47–50.

———. 1994. Fossile Schldkröten-und Krockdilreste aus dem Tertiär Thrakiens (W-Türkei). Courier Forsch.-Inst. Senckenberg 1973:137–151.

Schleich HH and MC Groessens van Dyck. 1988. Nouveaux materiels de reptiles du Tertiaire d'Allemagne. 9. Des tortues de l'Oligocène de Allemagne du Sud (Testudines: Testudinidae, Emydidae, Chelydridae). Stud. Salmanticensia Spec. Vol. 3:7–84.

Schmidt H. 1966. Eine Entwicklungreihe bei Schildkröten Gattung Chelydra. N. Jahrb. Geol. Paläontol. Abh. 125:19–28.

Schmidt-Nielsen K. 1979. Animal Physiology: Adaptation and Environment. Cambridge University Press, Cambridge, U.K.

Schoolfield RM, PJH Sharpe, and CE Magnuson. 1981. Nonlinear regression of biological temperature-dependent rate models based on absolute reaction-rate theory. J. Theor. Biol. 88:719–731.

Schubauer J and RR Parmenter. 1981. Winter feeding by aquatic turtles in a southeastern reservoir. J. Herpetol. 15:444–447.

Schuett GW and RE Gatten Jr. 1980. Thermal preference in snapping turtles (*Chelydra serpentina*). Copeia 1980: 149–152.

Schultz GE. 1965. Pleistocene vertebrates from the Butler Spring local fauna, Meade County, Kansas. Pap. Mich. Acad. Sci. Arts Lett. 50:235–265.

Schumacher G-H. 1973. The head muscles and hyolaryngeal skeleton of turtles and crocodilians. In Biology of the Reptilia, ed. C Gans and TS Parsons, Vol. 4. Academic Press, London, 101–199.

Schwarzkopf L and RJ Brooks. 1985. Sex determination in northern painted turtles: effect of incubation at constant and fluctuating temperatures. Can. J. Zool. 63:2543–2547.

Schweigger F. 1812. Prodromus monographie cheloniorum. Kongisberger Arch. Naturwiss. Math. 1:271–368, 406–468.

Scribner KT, JW Arntzen, N Cruddace, RS Oldham, and T. Burke. 2001. Environmental correlates of toad abundance and population genetic diversity. Biol. Conserv. 98:201–210.

Scribner KT and RK Chesser. 2001. Group-structured genetic models in analyses of the population and behavioural ecology of poikelothermic vertebrates. J. Hered. 92:180–189.

Scribner KT, JD Congdon, RK Chesser, and MH Smith 1993. Annual differences in female reproductive success affect spatial and cohort-specific genotypic heterogeneity in painted turtles. Evolution 47:1360–1373.

Scribner KT, JE Evans, SJ Morreale, MH Smith, and JW Gibbons. 1986. Genetic divergence among populations of the yellow-bellied slider turtle (*Pseudemys scripta*) separated by aquatic and terrestrial habitats. Copeia 1986:691–700.

Scribner KT, SJ Morreale, MH Smith, and JW Gibbons. 1995. Factors contributing to temporal and age-specific genetic variation in the freshwater turtle *Trachemys scripta*. Copeia 1995:970–977.

Scribner KT, MH Smith, and JW Gibbons. 1984. Genetic differentiation among local populations of the yellow-bellied slider turtle (*Pseudemys scripta*). Herpetologica 40:382–387.

Secor SM and J Diamond. 1999. Maintenance of digestive performance in the turtles *Chelydra serpentina*, *Sternotherus odoratus,* and *Trachemys scripta*. Copeia 1999:75–84.

Secor SM and JM Diamond. 1995. Adaptive response to feeding in Burmese pythons: pay before pumping. J. Exp. Biol. 198:1313–1325.

———. 1997. Determinants of the post-feeding metabolic response of Burmese pythons (*Python molurus*). Am. J. Physiol. (Regul. Integr. Comp. Physiol.) 272:R902–R912.

Secor SM and J Phillips. 1997. Specific dynamic action in a large carnivorous lizard, *Varanus albigularis*. Comp. Biochem. Physiol. A 117:515–522.

Secor SM, ED Stein, and J Diamond. 1994. Rapid upregulation of snake intestine in response to feeding: a new model of intestinal adaptation. Am. J. Physiol. 266:G695–G705.

Semple RE, D Sigsworth, and JT Stitt. 1970. Seasonal observations on the plasma, red cell, and blood volumes of two turtle species native to Ontario. Can. J. Physiol. Pharmacol. 48:282–290.

Sexton OJ. 1959. Spatial and temporal movements of a population of the painted turtle, *Chrysemys picta marginata* (Agassiz). Ecol. Monogr. 29:113–140.

Seymour R and J Piiper. 1988. Aeration of the shell membranes of avian eggs. Respir. Physiol. 71:101–116.

Seymour RS. 1982. Physiological adaptations to aquatic life. In Biology of the Reptilia, ed. C Gans and FH Pough, Vol. 13. Physiology D. Academic Press, London, 1–51.

Shaffer HB, P Meylan, and ML McKnight. 1997. Tests of turtle phylogeny: molecular, morphological, and paleontological approaches. Syst. Biol. 46:235–268.

Shaw G. 1802. General Zoology, or Systematic Natural History, Vol. 3, Part 1. G. Kearsley, London.

Shine R. 1995. A new hypothesis for the evolution of viviparity in reptiles. Am. Nat. 145:809–823.

———. 1999. Why is sex determined by nest temperature in many reptiles? Trends Ecol. Evol. 14:186–189.

Shine R and PS Harlow. 1996. Maternal manipulation of offspring phenotypes via nest-site selection in an oviparous lizard. Ecology 77:1808–1817.

Siebenrock F. 1897. Das Kopfskelett der Schildkröten. Sitzber. K. Akad. Wiss. (Wien) Math.-Naturwiss. 106:245–28.

Sieglbauer F. 1909. Zur Anatomie der Schildkrötenextremität. Arch. Anat. Physiol. 1909:183–280.

Silva AMR, GS Moraes, and GF Wasserman. 1984. Seasonal variations of testicular morphology and plasma levels of testosterone in the turtle *Chrysemys dorbigni*. Comp. Biochem. Physiol. 78:153–157.

Simandle ET, RE Espinoza, KE Nussear, and CR Tracy. 2001. Lizards, lipids, and dietary links to animal function. Physiol. Biochem. Zool. 74:625–640.

Sims PA, GC Packard, and PL Chapman. 2001. The adaptive strategy for overwintering by hatchling snapping turtles (*Chelydra serpentina*). J. Herpetol. 35:514–517.

Sinclair AH, P Berta, MS Palmer, JR Hawkins, BL Griffiths, MJ Smith, JW Foster, A-M Frischauf, R Lovell-Badge, and PN

Goodfellow. 1990. A gene from the human sex-determining region encodes a protein with homology to a conserved DNA-binding motif. Nature 346:240–244.

Sinervo B. 1990a. The evolution of maternal investment in lizards: an experimental and comparative analysis of egg size and its effects on offspring performance. Evolution 44:279–294.

———. 1990b. Evolution of thermal physiology and growth rate between populations of the western fence lizard (*Sceloporus occidentalis*). Oecologia 83:228–237.

Sinervo B and SC Adolph. 1989. Thermal sensitivity of growth rate in hatchling *Sceloporus* lizards: environmental, behavioral and genetic aspects. Oecologia 78:411–419.

Sinervo B, P Doughty, RB Huey, and K Zamudio. 1992. Allometric engineering: a causal analysis of natural selection on offspring size. Science 258:1927–1930.

Sinervo B and KD Dunlap. 1995. Thyroxine affects behavioral thermoregulation but not growth rate among populations of the western fence lizard (*Sceloporus occidentalis*). J. Comp. Physiol. 164:509–517.

Sites JW Jr and KA Crandall 1997. Testing species boundaries in biodiversity studies. Conserv. Biol. 11:1289–1297.

Skulan J. 2000. Has the importance of the amniote egg been overstated? Zoological J. Linnean Soc 130:235–261.

Smith CA, PJ McClive, PS Western, KJ Reed, and AH Sinclair. 1999. Conservation of a sex-determining gene. Nature 402:601–602.

Smith CC and SD Fretwell. 1974. The optimal balance between size and number of offspring. Am. Nat. 108:499–506.

Smith DD, TR Johnson, R Powell, and HL Gregory. 1983. Life history observations of Missouri amphibians and reptiles with recommendations for standardized data collection. Trans. Mo. Acad. Sci. 17:37–58.

Smith HM and RB Smith. 1979[1980]. Synopsis of the herpetofauna of Mexico, Vol. 6, Guide to Mexican turtles. Bibliographic addendum III. John Johnson, North Bennington, VT.

———. 1983. *Chelydra osceola* Stejneger, 1918 (Reptilia, Testudines): proposed conservation by use of the plenary powers. Z.N.(S.)2282. Bull. Zool. Nomencl. 40:225–228.

Sobolik KD and D G Steele. 1996. A Turtle Atlas to Facilitate Archaeological Identifications. Hot Springs, SD: Mammoth Site of Hot Springs, SD, in conjunction with the Office of Research and Public Services, University of Maine.

Soliman MA. 1964. Die Kopfnerven der Schildkröten. Z. Wiss. Zool. 169:216–312.

Spangler M and R Handy. 1982. Soil Engineering, 4th ed. Harper & Row, New York.

Spencer RJ. 2001. The Murray River turtle, *Emydura macquarii:* population dynamics, nesting ecology and impact of the introduced red fox, *Vulpes vulpes.* Ph.D. thesis, University of Sydney, Sydney.

Spencer RJ and MB Thompson. 2003. The significance of predation in nest site selection of turtles: an experimental consideration of macro-and microhabitat preferences. Oikos 102:592–600.

Spencer R J, MB Thompson, and PB Banks. 2001. Hatch or wait: a dilemma in reptilian incubation. Oikos 93:401–406.

Spindel EL, JL Dobie, and DF Buxton. 1987. Functional Mechanisms and Histologic Composition of the Lingual Appendage in the Alligator Snapping Turtle, *Macroclemys temmincki* (Troost) (Testudines: Chelydridae). J. Morphol. 194:287–301.

Spotila JR. 1995. Metabolism, physiology, and thermoregulation. In Biology and Conservation of Sea Turtles, ed. KA Bjorndal, Rev. ed. Smithsonian Inst. Press, Washington, DC, 591–592.

Spotila JR, RE Foley, JP Schubauer, RD Semlitsch, KM Crawford, EA Standora, and JW Gibbons. 1984. Opportunistic behavioral thermoregulation of turtles, *Pseudemys scripta,* in response to microclimatology of a nuclear reactor cooling reservoir. Herpetologica 40:299–308.

Spotila JR, RE Foley, and EA Standora. 1990. Thermoregulation and climate space of the slider turtle. In Life History and Ecology of the Slider Turtle, ed. JW Gibbons. Smithsonian Inst. Press, Washington, DC, 288–298.

Spotila JR, MP O'Connor, and FV Paladino. 1997. Thermal biology. In The Biology of Sea Turtles, ed. PL Lutz and JA Musick. CRC Press, Boca Raton, FL, 299–316.

Spotila JR and EA Standora. 1985. Environmental constraints on the thermal energetics of sea turtles. Copeia 1985:694–702.

Spotila LD, NF Kaufer, E Theriot, KM Ryan, D Penick, and JR Spotila. 1994. Sequence analysis of the ZFY and Sox genes in the turtle, *Chelydra serpentina.* Mol. Phylogenet. Evol. 3:1–9.

Spray DC and ML May. 1972. Heating and cooling rates in four species of turtles. Comp. Biochem. Physiol. 41A:491–494.

Stamper DL, RJ Denver, and P Licht. 1990. Effects of thyroidal status on metabolism and growth of juvenile turtles, *Pseudemys scripta elegans.* Comp. Biochem. Physiol. 96A:67–73.

Standing KL, TB Herman, DD Hurlburt, and IP Morrison. 1997. Postemergence behavior of neonates in a northern peripheral population of Blanding's turtle, *Emydoidea blandingii,* in Nova Scotia. Can. J. Zool. 75:1387–1395.

Standing KL, TB Herman, and IP Morrison. 1999. Nesting ecology of Blanding's turtle (*Emydoidea blandingii*) in Nova Scotia, the northeastern limit of the species' range. Can. J. Zool. 77:1609–1614.

Standora EA. 1982. A telemetric study of the thermoregulation and climate space of free-ranging yellow-bellied turtles, *Pseudemys scripta.* Ph.D. dissertation, University of Georgia, Athens, GA.

Starkey DE. 1997. Molecular systematics and biogeography of the New World turtle genera *Trachemys* and *Kinosternon.* Unpublished Ph.D. dissertation, Texas A&M University, College Station, TX.

Steen DA and JP Gibbs. 2004. Effects of roads on the structure of freshwater turtle populations. Conserv. Biol. 18:1143–1148.

Stejneger L. 1918. Description of a new lizard and a new snapping turtle from Florida. Proc. Biol. Soc. Washington 31:89–92.

Stewart J. 1997. Morphology and evolution of the egg of oviparous amniotes. In Amniote Origins Completing the Transition to Land, ed. S Sumida and K Martin. Academic Press, New York, 291–326.

Stewart PA. 1981. How to Understand Acid-Base. Elsevier, New York.

Steyermark AC. 2002. A high standard metabolic rate constrains juvenile growth. Zoology 105:147–151.

Steyermark AC and JR Spotila. 2000. Effects of maternal identity and incubation temperature on snapping turtle (*Chelydra serpentina*) metabolism. Physiol. Biochem. Zool. 73:298–306.

———. 2001a. Effects of maternal identity and incubation temperature on hatching and hatchling morphology in snapping turtles, *Chelydra serpentina.* Copeia 2001:129–135.

———. 2001b. Body temperature and maternal identity affect snapping turtle (*Chelydra serpentina*) righting response. Copeia 2001:1050–1057.

———. 2001c. Effects of maternal identity and incubation

temperature on snapping turtle (*Chelydra serpentina*) growth. Funct. Ecol. 15:624–632.

Stitt JT and RE Semple. 1971. Sites of plasma sequestration induced by body cooling in turtles. Am. J. Physiol. 221:1189–1191.

Stitt JT, RE Semple, and DW Sigsworth. 1970. Effect of changes in body temperature on circulating plasma volume of turtles. Am. J. Physiol. 219:683–686.

———. 1971. Plasma sequestration produced by acute changes in body temperature in turtles. Am. J. Physiol. 221:1185–1188.

Stock AD. 1972. Karyological relationships in turtles (Reptilia: Chelonia). Can. J. Genet. Cytol. 14:859–868.

Stone PA and JB Iverson. 1999. Cutaneous surface area in freshwater turtles. Chelonian Conserv. Biol. 3:512–515.

Stoner D. 1925. The toll of the automobile. Science 61:56–57.

Storey KB, JM Storey, SPJ Brooks, TA Churchill, and RJ Brooks. 1988. Hatchling turtles survive freezing during winter hibernation. Proc. Natl. Acad. Sci. USA 85:8350–8454.

Suhashini H and CJ Parmenter. 2002. Egg components and utilization of lipids during development of the flatback turtle *Natator depressus*. J. Herpetol. 36:43–50.

Sukhanov VB. 2000. Mesozoic turtles of Middle and Central Asia. In. Age of Dinosaurs in Russian and Mongolia, ed. JM Benton, MA Shishkin, DM Unwin, and EN Kurochkin. Cambridge University Press, New York, 309–367.

Sullivan LC, S Orgeig, PG Wood, and CB Daniels. 2001. The ontogeny of pulmonary surfactant secretion in the embryonic green sea turtle (*Chelonia mydas*). Physiol, Biochem. Zool. 74:493–501.

Swainson W. 1839. On the natural history and classification of fishes, amphibians, and reptiles. The Cabinet Cyclopedia 2:1–452 + 135 figs.

Swingland IR and M Coe. 1979. The natural regulation of giant tortoise populations on Aldabra Atoll, Indian Ocean: recruitment. Phil. Trans. R.l Soc. Lond. B 286:177–188.

Szalai T. 1934. Die fossilen Schldkröten Ungarns. Folia Zool. Hydrobiol. 6:97–192.

Tarashchuk VI. 1971. [Turtles of neogene and anthropogene deposits in the Ukraine. Communication I. Platysternidae family]. Vestn. Zool. 2:56–62.

Taylor GM and E Nol. 1989. Movements and hibernation sites of overwintering painted turtles in southern Ontario. Can. J. Zool. 67:1877–1881.

Tazawa H. 1980. Adverse effect of failure to turn the avian egg on the embryo oxygen exchange. Respir. Physiol. 41:137–142.

Thompson M. 1985. Functional significance of the opaque white patch in eggs of *Emydura macquarii*. In The Biology of Australasian Frogs and Reptiles, ed. G Grigg, R Shine, and H Ehmann. Surrey Beatty and Sons, Sydney, NSW, 387–395.

———. 1987. Water exchange in reptilian eggs. Physiol. Zool. 60:1–8.

Thompson MB. 1989. Patterns of metabolism in embryonic reptiles. Respir. Physiol. 76:243–256.

———. 1993. Oxygen consumption and energetics of development in eggs of the leatherback turtle, *Dermochelys coriacea*. Comp. Biochem. Physiol. 104A:449–453.

Thomson JS. 1932. The anatomy of the tortoise. Sci. Proc. R. Soc. Dublin 20:359–461.

Tinkle DW, JD Congdon, and PC Rosen. 1981. Nesting frequency and success: implications for the demography of painted turtles. Ecology 62:1426–1432.

Toerien MJ. 1965a. An experimental approach to the development of the ear capsule in the turtle, *Chelydra serpentina*. J. Embryol. Exp. Morphol. 13:141–149.

———. 1965b. Experimental studies on the columella-capsular interrelationship in the turtle, *Chelydra serpentina*. J. Embryol. Exp. Morphol. 14:265–272.

Toner GC. 1940. Delayed hatching in the snapping turtle. Copeia 1940: 265.

Tracy C, GC Packard, and MJ Packard. 1978. Water relations of chelonian eggs. Physiol, Zool, 51:378–387.

Travis J. 1980. Genetic variation for larval specific growth rate in the frog *Hyla gratiosa*. Growth 44:167–181.

Trewartha GT and LH Horn. 1980. An Introduction to Climate, Vol. 2. McGraw Hill, New York.

Troyer K. 1983. Posthatching yolk energy in a lizard: utilization pattern and interclutch variation. Oecologia 58:340–344.

Tsui HW and P Licht. 1977. Gonadotropin regulation of in vitro androgen production by reptilian testes. Gen. Comp. Endocrinol. 31:422–434.

Tucker JK. 1997. Natural history notes on nesting, nests, and hatchling emergence in the red-eared slider turtle, *Trachemys scripta elegans*, in west-central Illinois. Ill. Nat. Hist. Surv. Notes 140:1–13.

———. 1999. Environmental correlates of hatchling emergence in the red-eared turtle, *Trachemys scripta elegans*, in Illinois. Chelonian Conserv. Biol. 3:401–406.

———. 2000. Body size and migration of hatchling turtles: Inter- and intraspecific comparisons. J. Herpetol. 34:541–546.

Tucker JK, NI Filoramo, GL Paukstis, and FJ Janzen. 1998. Residual yolk in captive and wild-caught hatchlings of the red-eared slider turtle (*Trachemys scripta elegans*). Copeia 1998:488–492.

Tucker JK and GC Packard. 1998. Overwinter survival by hatchling sliders (*Trachemys scripta*) in west-central Illinois. J. Herpetol. 32:431–434.

Tucker JK, GL Paukstis, and FJ Janzen. 1998. Annual and local variation in reproduction in the red-eared slider, *Trachemys scripta* elegans. J. Herpetol. 32:515–526.

Ultsch GR. 1988. Blood gases, hematocrit, plasma ion concentrations, and acid-base status of musk turtles (*Sternotherus odoratus*) during simulated hibernation. Physiol. Zool. 61:78–94.

———. 1989. Ecology and physiology of hibernation and overwintering among freshwater fishes, turtles, and snakes. Biol. Rev. 64:435–516.

———. 2006. The ecology of overwintering among turtles: Where turtles overwinter and its consequences. Biol. Rev. 81:339–367.

Ultsch GR and JF Anderson. 1986. The respiratory microenvironment within the burrows of Gopher Tortoises (*Gopherus polyphemus*). Copeia 1986:787–795.

Ultsch GR, ME Carwile, CE Crocker, and DC Jackson. 1999. The physiology of hibernation among painted turtles: the eastern painted turtle (*Chrysemys picta picta*). Physiol. Biochem. Zool.: 72:493–501.

Ultsch GR and BM Cochran. 1994. Physiology of northern and southern musk turtles (*Sternotherus odoratus*) during simulated hibernation. Physiol. Zool. 67:263–281.

Ultsch GR, TE Graham, and CE Crocker. 2000. An aggregation of overwintering leopard frogs, *Rana pipiens*, and common map turtles, *Graptemys geographica*, in northern Vermont. Can. Field-Nat. 114:314–315.

Ultsch GR, CV Herbert, and DC Jackson 1984. The comparative

physiology of diving in North American freshwater turtles. I. Submergence tolerance, gas exchange, and acid-base balance. Physiol. Zool. 57:620–631.

Ultsch GR and DC Jackson. 1982a. Long-term submergence at 3°C of the turtle *Chrysemys picta bellii* in normoxic and severely hypoxic water. I. Survival, gas exchange, and acid-base status. J. Exp. Biol. 96:11–28.

———. 1982b. Long-term submergence at 3°C of the turtle *Chrysemys picta bellii* in normoxic and severely hypoxic water. III. Effects of changes in ambient Po₂ and subsequent air breathing. J. Exp. Biol. 97:87–99.

———. 1995. Acid-base status and ion balance during simulated hibernation in freshwater turtles from the northern portions of their ranges. J. Exp. Zool. 273:482–493.

Ultsch GR and D Lee. 1983. Radiotelemetric observations of wintering snapping turtles (*Chelydra serpentina*) in Rhode Island. J. Ala. Acad Sci. 54: 200–206.

Ultsch GR and JS Wasser. 1990. Plasma ion balance of North American freshwater turtles during prolonged submergence in normoxic water. Comp. Biochem. Physiol. A 97:505–512.

Valenzuela N. 2001a. Constant, shift, and natural temperature effects on sex determination in *Podocnemis expansa* turtles. Ecology 82:3010–3024.

———. 2001b. Maternal effects on life-history traits in the Amazonian giant river turtle *Podocnemis expansa*. J. Herpetol. 35:368–378.

Valenzuela N and FJ Janzen. 2001. Nest-site philopatry and the evolution of temperature-dependent sex determination. Evol. Ecol. Res. 3:779–794.

Vallén E. 1942. Beiträge zur Kenntnis der Ontogenie und der vergleichenden Anatomie des Schildkrötenpanzers. Acta Zool. Stockholm 23:1–127.

Van Devender TR and NT Tessman. 1975. Late Pleistocene snapping turtles (*Chelydra serpentina*) from southern Nevada. Copeia 1975:249–253.

Van Loben Sels RC, JD Congdon, and JT Austin. 1997. Life history and ecology of the Sonoran mud turtle (*Kinosternon sonoriense*) in southeastern Arizona: a preliminary report. Chelonian Conserv. Biol. 2:338–344.

Van Sommers P. 1963. Air motivated behaviour in the turtle. J. Comp. Physiol. Psychol. 56:590–596.

Vermersch TG. 1992. Lizards and Turtles of South-Central Texas. Eakin Press, Austin, TX.

Viets BE, MA Ewert, LG Talent, and CE Nelson. 1994. Sex-determining mechanisms in squamate reptiles. J. Exp. Zool. 270:45–56.

Vleck CM and DF Hoyt. 1991. Metabolism and energetics of reptilian and avian embryos. In Egg incubation: Its Effects on Embryonic Development in Birds and Reptiles, ed. DC Deeming and MWJ Ferguson. Cambridge University Press, Cambridge, U.K., 285–306.

Vogt RC. 1980. Natural history of the map turtles *Graptemys pseudogeographica* and *G. ouachitensis* in Wisconsin. Tulane Stud. Zool. Bot. 22:17–48.

———. 1981. Turtle egg (*Graptemys*: Emydidae) infestation by fly larvae. Copeia 1981:457–459.

Vogt RC and JJ Bull. 1982. Temperature controlled sex-determination in turtles: ecological and behavioral aspects. Herpetologica 38:156–164.

———. 1984. Ecology of hatchling sex ratio in map turtles. Ecology 65:582–587.

Vogt RC and O Flores-Villela. 1992. Effects of incubation temperature on sex determination in a community of neotropical freshwater turtles in southern Mexico. Herpetologica 48:265–270.

von Brand T and WG Lynn. 1947. Chemical changes in the developing turtle embryo. Proc. Soc. Exp. Biol. Med. 64:61–62.

von Meyer H. 1845. Fossile Säugethiere, Vigel und Reptilien aus dem Molasse-Mergel von Oeningen. "Zur Fauna der Vorwelt", Frankfurt am Main, 52.

———. 1852. Ueber *Chelydra Murchisoni* und *Chelydra Decheni*. Palaeontographica 2:237–247.

———. 1856. Ueber den Jugendzustand der *Chelydra Decheni* aus der Braunkohl des Siebengebirges. Palaeontographica 4:56–60.

———. 1868 (1856–68). Zu *Chelydra Decheni* aus der Braunkohl des Siebengebirges. Palaeontographica 15:41–47.

Vu Hai MT, F Logeat, M Warembourg, and E Milgrom. 1977. Control of progesterone receptors. Ann. NY Acad. Sci. 286:199–209.

Wagner TL, H-I Wu, PJH Sharpe, RM Schoolfield, and RN Coulson. 1984. Modeling insect development rates: a literature review and application of a biophysical model. Ann. Entomol. Soc. Am. 77:208–225.

Waldschmidt SR, SM Jones, and WP Porter. 1987. Reptilia. In Animal Energetics, ed. TJ Pandian and FJ Vernberg, Vol. 2. Academic Press, San Diego, 553–619.

Walker D and JC Avise. 1998. Principles of phylogeography as illustrated by freshwater and terrestrial turtles in the southeastern United States. Annu. Rev. Ecol. Syst. 29:23–58.

Walker D, PE Moler, KA Buhlmann, and JC Avise. 1998. Phylogenetic uniformity in mitochondrial DNA of the snapping turtle (*Chelydra serpentina*). Anim. Conserv. 1:55–60.

Walker WF Jr. 1973. The locomotor apparatus of Testudines. In Biology of the Reptilia, ed. C Gans and TS Parsons, Vol. 4. Academic Press, New York, 1–100.

Wang T, CJ Brauner, and WK Milsom. 1999. The effect of isovolemic anemia on blood O₂ affinity and red-cell triphosphate concentrations in the painted turtle (*Chrysemys picta*). Comp. Biochem. Physiol. A 122:341–345.

Wang T and JW Hicks. 1996. The interaction of pulmonary ventilation and right-left shunt on arterial oxygen levels. J. Exp. Biol. 199:2121–2129.

Warburton SJ and DC Jackson. 1995. Turtle (*Chrysemys picta bellii*) shell mineral content is altered by exposure to prolonged anoxia. Physiol. Zool. 68:783–798.

Weathers WW and FN White. 1971. Physiological thermoregulation in turtles. Am. J. Physiol. 221:704–710.

Webb GJW, D Choquenot, and PJ Whitehead. 1986. Nests, eggs, and embryonic development of *Carettochelys insculpta* (Chelonia: Carettochelidae) from Northern Australia. J. Zool. Lond. B 1:521–550.

Webb GJW and CR Johnson. 1972. Head-body temperature differences in turtles. Comp. Biochem. Physiol. 43A:593–611.

Webb GJW, SC Manolis, PJ Whitehead, and K Dempsey. 1987. The possible relationship between embryo orientation, opaque banding and the dehydration of albumen in crocodile eggs. Copeia 1987:252–257.

Webb GJW and GJ Witten. 1973. Critical thermal maxima of turtles, validity of body temperature. Comp. Biochem. Physiol. 45A:829–832.

Webb RG. 1995. The date of publication of Gray's Catalogue of Shield Reptiles. Chelonian Conserv. Biol. 1:322–323.

Weber JA. 1928. Herpetological observations in the Adirondack Mountains, New York. Copeia 1928:106–112.

Weigel RD. 1962. Fossil vertebrates of Vero, Florida. Fla. Geol. Surv. Spec. Publ. (10):1–59.

Weisrock DW and FJ Janzen. 1999. Thermal and fitness-related consequences of nest location in painted turtles (Chrysemys picta). Funct. Ecol. 13:94–101.

Wermuth H and R Mertens. 1961. Schildkröten, Krokodile, Brükenechsen. Gustav Fischer Verlag, Jena.

———. 1977. Liste de rezenten Amphibien und Reptilien. Testudines, Crocodylia, Rhynchocephalia. Das Tierreich, Berlin 28:1–174.

West NH, PJ Butler, and RM Bevan. 1992. Pulmonary blood flow at rest and during swimming in the green turtle, Chelonia mydas. Physiol. Zool. 65:287–310.

West NH, AW Smits, and WW Burggren. 1989. Factors terminating nonventilatory periods in the turtle, Chelydra serpentina. Respir. Physiol. 77:337–349.

West NH, ZL Topor, and BN Van Vliet. 1987 Hypoxemic threshold for lung ventilation in the toad. Respir. Physiol. 70:377–390.

Whetstone KN. 1978a. Additional record of the fossil snapping turtle Macroclemys schmidti from the Marshland Formation (Miocene) of Nebraska with notes on interspecific skull variation within the genus Macroclemys. Copeia 1978:159–162.

———. 1978b. A new genus of cryptodiran turtles (Testudinoidea, Chelydridae) from the Upper Cretaceous Hell Creek Formation of Montana. Univ. Kans. Sci. Bull. 51:539–563.

White HBI. 1991. Maternal diet, maternal proteins and egg quality. In Egg Incubation: Its Effect on Embryonic Development in Birds and Reptiles, ed. DC Deeming and MWJ Ferguson. Cambridge University Press, Cambridge, U.K., 1–15.

White JB and GG Murphy. 1973. The reproductive cycle and sexual dimorphism of the common snapping turtle Chelydra serpentina serpentina. Herpetologica 29:240–246.

Whitehead PJ and RS Seymour. 1990. Patterns of metabolic rate in embryonic crocodilians Crocodylus johnstoni and Crocodylus porosus. Physiol. Zool. 63:334–352.

Wibbels T, JJ Bull, and D Crews. 1991. Chronology and morphology of temperature-dependent sex determination. J. Exp. Zool. 260:371–381.

Wibbels T and D Crews. 1995. Steroid-induced sex determination at temperatures producing mixed sex ratios in a turtle with TSD. Gen. Comp. Endocrinol. 100:53–60.

Wibbels T, DW Owens, and M Amoss. 1987. Seasonal changes in serum testosterone of loggerhead sea turtles captured along the Atlantic coast of the United States. In Ecology of East Florida Sea Turtles, ed. WN Witzell. U.S. Department of Commerce, NOAA Tech. Rep. NMFS 53, 59–64.

Wibbels T, DW Owens, P Licht, C Limpus, PC Reed, and MS Amoss. 1992. Serum gonadotropin and gonadal steroids associated with ovulation and egg production in sea turtles. Gen. Comp. Endocrinol. 87:71–78.

Wibbels T, DW Owens, CJ Limpus, PC Reed, and MS Amoss Jr. 1990. Seasonal Changes in serum gonadal steroids associated with migration, mating and nesting n the loggerhead sea turtle (Caretta caretta). Gen. Comp. Endocrinol. 79:154–164.

Wieser W. 1994. Cost of growth in cells and organisms: general rules and comparative aspects. Biol. Rev. 68:1–33.

Wikelski M and C Thom. 2000. Marine iguanas shrink to survive El Nino. Nature 403:37.

Wilhoft DC. 1986. Eggs and hatchling components of the snapping turtle (Chelydra serpentina). Comp. Biochem. Physiol. A 84:483–486.

Wilhoft DC, MG Del Baglivo, and MD Del Baglivo. 1979. Observations on mammalian prediation of snapping turtle nests (Reptilia, Testudines, Chelydridae). J. Herpetol. 13:435–438.

Wilhoft DC, E Hotaling, and P Franks. 1983. Effects of temperature on sex determination in embryos of the snapping turtle, Chelydra serpentina. J. Herpetol. 17:38–42.

Willard R, GC Packard, MJ Packard, and JK Tucker. 2000. The role of the integument as a barrier to penetration of ice into overwintering hatchlings of the painted turtle (Chrysemys picta). J. Morphol. 246:150–159.

Williams EE. 1950. Variation and selection in the cervical central articulations of living turtles. Bull. Am. Mus. Nat. Hist. 94:505–562.

———. 1952. A staurotypine skull from the Oligocene of South Dakota (Testudinata, Chelydridae). Breviora 2:1–16.

Williams EE and SB McDowell. 1952. The plastron of soft-shelled turtles (Testudinata, Trionychidae): A new interpretation. J. Morphol. 90:263–280.

Williams GC. 1966, Adaptation and Natural Selection: A Critique of Some Current Evolutionary Thought. Princeton University Press, Princeton, NJ.

Williamson LU, JR Spotila, and EA Standora. 1989. Growth, selected temperature and CTM of young snapping turtles, Chelydra serpentina. J. Therm. Biol. 14:33–39.

Wilson DS, CR Tracy, and CR Tracy 2003. Estimating age of turtles from growth rings: a critical evaluation of the technique. Herpetologica 59:178–194.

Wilson RL 1967. The Pleistocene vertebrates of Michigan. Pap. Mich. Acad. Sci. Arts Lett. 52:197–234.

Winter T. 1983. The interaction of lakes with variably saturated porous media. Water Resour. Res. 19: 1203–1218.

———. 1986. Effect of ground-water recharge on configuration of the water table beneath sand dunes and on seepage in lakes in the Sandhills of Nebraska, USA. J. Hydrol. 86: 221–237.

———. 1989. Hydrologic studies of wetlands in the northern prairie. In Northern Prairie Wetlands, ed. A Van der Valk. Iowa State University Press, Ames, IA, 16–54.

Witherington BE, KA Bjorndal, and CM McCabe. 1990. Temporal pattern of nocturnal emergence of loggerhead turtle hatchlings from natural nests. Copeia 1990:1165–1168.

Yeomans SR. 1995. Water-finding in adult turtles: random search or oriented behaviour? Anim. Behav. 49:977–987

Yntema CL. 1960. Effects of various temperatures on the embryonic development of Chelydra serpentina. Anat. Rec.136:305–306.

———. 1964. Procurement and use of turtle embryos for experimental procedures. Anat. Rec. 149:557–586.

———. 1966. Depletions in the cervical and thoracolumbar sympathetic system following removal of neural crest from the embryo of Chelydra serpentina. J. Morphol. 120:203–217.

———. 1968. A series of stages in the embryonic development of Chelydra serpentina. J. Morphol. 125:219–252.

———. 1970a. Extirpation experiments on the embryonic rudiments of the carapace of Chelydra serpentina. J. Morphol. 132:235–244.

———. 1970b. Survival of xenogenic grafts of embryonic pigment and carapace rudiments in embryos of Chelydra serpentina. J. Morphol. 132:353–360.

———. 1970c. Observations on females and eggs of the common snapping turtle, *Chelydra serpentina*. Am. Midl. Nat. 84:69–84.

———. 1974a. Incidence and progress of rejection of embryonic limb bud transplants in the turtle, *Chelydra serpentina*. J. Morphol. 144:453–462.

———. 1974b. Survival of ski allografts following embryonic limb bud transplants in the turtle, *Chelydra serpentina*. J. Morphol. 144:463–468.

———. 1976. Effects of incubation temperatures on sexual differentiation in the turtle, *Chelydra serpentina*. J. Morphol. 150:453–462.

———. 1978. Incubation times for eggs of the turtle *Chelydra serpentina* (Testudines: Chelydridae) at various temperatures. Herpetologica 34:274–277.

———. 1979. Temperature levels and periods of sex determination during incubation of eggs of *Chelydra serpentina*. J. Morphol. 159:17–27.

———. 1981. Characteristics of gonads and oviducts in hatchlings and young of *Chelydra serpentina* resulting from three incubation temperatures. J. Morphol. 167:297–304.

Yokosuka H, M Ishiyama, S Yoshie, and T Fujita. 2000a. Villiform processes in the pharynx of the soft-shelled turtle, *Trionyx sinensis japonicus*, functioning as a respiratory and presumably salt uptaking organ in the water. Arch. Histol. Cytol. 63:181–192.

Yokosuka H, T Murakami, M Ishiyama, S Yoshie, and T Fujita. 2000b. The vascular supply of the villiform processes in the pharynx of the soft-shelled turtle, *Trionyx sinensis japonicus*.

A scanning electron microscopic study of corrosion casts. Arch. Histol. Cytol. 63:193–198.

Zangerl R. 1945. Fossil specimens of *Macrochelys* from the Tertiary of the plains. Fieldiana: Geol. 10:5–12.

———. 1953. The vertebrate fauna of the Selma Formation of Alabama. Part 3. The turtles of the family Protostegidae. Part 4. The turtles of the family Toxochelyidae. Fieldiana Geol. Mem. 3:61–277.

———. 1969. The turtle shell. In Biology of the Reptilia, ed. IC Gans, A Bellairs, and TS Parsons, Vol. I. Academic Press, New York, 311–339.

Zhao-Xian W, S Ning-Zhen, and S Wen-Feng. 1989. Aquatic respiration in soft-shelled turtles, *Trionyx sinensis*. Comp. Biochem. Physiol. A 92:593–598.

Zug GR. 1966. The Penial Morphology and the Relationships of Cryptodiran Turtles. Occ. Pap. Mus. Zool., Univ. Mich. 647:1–24.

———. 1971. Buoyancy, locomotion, morphology of the pelvic girdle and hindlimb, and systematics of cryptodiran turtles. Misc. Pub. Mus. Zool., Univ. Mich. 142:1–98.

———. 1993. Herpetology. Academic Press, New York.

Zuurmand F and H Netten 1983. Huisvesting en verzorging van waterschildpadden (6); *Chelydra serpentina serpentina*, de bijtschildpad. Lacerta 42:52–58.

Zweifel RG. 1968. Reproductive biology of anurans of the arid southwest, with emphasis on adaptation of embryos to temperature. Bull. Am. Mus. Nat. Hist. 140:1–64.

INDEX

Page numbers in **bold type** indicate location of a figure, those in *italic type,* a table.